An Introduction to Ore Geology

GEOSCIENCE TEXTS

SERIES EDITOR

A. HALLAM

Lapworth Professor of Geology
University of Birmingham

GEOSCIENCE TEXTS

An Introduction to Ore Geology

ANTHONY M. EVANS
BSc, PhD, CEng, MIMM, MIGeol, FGS
Senior Lecturer in Mining Geology
University of Leicester

SECOND EDITION

BLACKWELL SCIENTIFIC PUBLICATIONS
OXFORD LONDON EDINBURGH
BOSTON PALO ALTO MELBOURNE

To Jo, Nick, Caroline and Jason

Editorial offices:
Osney Mead, Oxford, OX2 0EL
8 John Street, London, WC1N 2ES
23 Ainslie Place, Edinburgh, EH3 6AJ
52 Beacon Street, Boston
Massachusetts 02108, USA
667 Lytton Avenue, Palo Alto
California 94301, USA
107 Barry Street, Carlton
Victoria 3053, Australia

First published 1980
Reprinted 1982, 1983, 1984
Second edition 1987

DISTRIBUTORS

USA and Canada
Blackwell Scientific Publications Inc
PO Box 50009, Palo Alto
California 94303

Australia
Blackwell Scientific Publications
(Australia) Pty Ltd
107 Barry Street,
Carlton, Victoria 3053

Set by Setrite Typesetters Ltd, Hong Kong
Printed and bound in Great Britain

British Library
Cataloguing in Publication Data

Evans, Anthony M.
An introduction to ore geology. — 2nd ed.
— (Geoscience texts; v.2)
1. Ore deposits
I. Title II. Series
553 TN263

ISBN 0-632-01654-X
ISBN 0-632-01655-8 Pbk

Library of Congress
Cataloging-in-Publication Data
Evans, Anthony M.
An introduction to ore geology.

(Geoscience texts)
Bibliography: p.
Includes index.
I. Ore-deposits. I. Title. II. Series.
QE390.E92 1987 553'.1 86-23302

ISBN 0-632-01654-X
ISBN 0-632-01655-8 (pbk.)

Contents

Preface to the second edition

This revision appears in response to what the media are pleased to call popular demand. The publishers and I were quite astonished by the impressive sales figures for the first edition, the flattering reviews, the 'fan mail' from places as far apart as France, California, Japan, New Zealand and Spain and the offers to translate it into both French and Japanese.

I would like to express my thanks to the many readers who have been kind enough to comment on the first edition, instead of making the usual exclamation marks in the margins of their copies when they objected to my prose, or caught me out in some fact, or disagreed with my interpretation of the evidence. Many of what I hope will be seen as improvements to the text owe their presence to the kindness of readers and reviewers, and I hope that none of them will feel that any of their constructive criticism has been ignored.

I have attempted a thorough revision and many sections have been rewritten. A chapter on diamonds has been added to meet many requests. Chapters on greisen and pegmatite deposits have also been added, the former in response to the changing situation in tin mining following the recent tin crisis and the latter in response to suggestions from geologists in a number of overseas countries. Some chapters have been considerably expanded and new sections added; in particular on disseminated gold deposits and unconformity-associated uranium deposits. The chapter on ore genesis has been enlarged and I am grateful to Dr A.D. Saunders for his comments on it.

To emphasize still further the importance of viewing mineral deposits from an economic standpoint, I have expanded Chapter 1 considerably and I am grateful to Mr M.K.G. Whateley for reviewing it. I have continued my policy of the first edition of peppering the text with grade and tonnage figures and other allusions to mineral economics in a further attempt to create commercial awareness in the tyro.

As in the first edition bibliographic references generally direct attention to works in English. The student should note that this, in itself, is misleading; for much significant work in the field is written in French, German, Russian and other languages. But works in English are much more widely accessible and the main aim has been to help the reader find works that will amplify the discussions this book has begun.

Much of the success of the first edition was due to Sue Aldridge's fine artwork and I am deeply grateful to her for the pains she has taken. Once again I lovingly acknowledge the encouragement, editorial and typing skills which my wife has contributed and without which this edition would still be awaiting attention.

Anthony Evans
Burton on the Wolds
July 1986

Preface to the first edition

This book is an attempt to provide a textbook in ore geology for second and third year undergraduates which, in these days of inflation, could be retailed at a reasonable price. The outline of the book follows fairly closely the undergraduate course in this subject at Leicester University which has evolved over the last twenty years. It assumes that the student will have adequate practical periods in which to handle and examine hand specimens, and thin and polished sections of the common ore types and their typical host rocks. Without such practical work students often develop erroneous ideas of what an orebody looks like, ideas often based on a study of mineralogical and museum specimens. In my opinion, it is essential that the student handles as much run-of-the-mill ore as possible during his course and makes a start on developing such skills as visual assaying, the ability to recognize wall rock alteration, using textural evidence to decide on the mode of genesis, and so on.

In an attempt to keep the reader aware of financial realities I have introduced some mineral economics into Chapter 1 and sprinkled grade and tonnage figures here and there throughout the book. It is hoped that this will go some way towards meeting that perennial complaint of industrial employers that the new graduate has little or no commercial awareness, such as a realization that companies in the West operate on the profit motive. This little essay into mineral economics only scratches the surface of the subject, and the intending practitioner of mining geology would be well advised to accompany his study of ore geology by dipping into such journals as *World Mining**, the *Mining Journal*, the *Engineering and Mining Journal* and the *Mining Magazine*, to watch the latest trends in metal and mineral prices and to gain knowledge of mining methods and recent orebody discoveries.

In order to produce a reasonably priced book, a strict word limit had to be imposed. As a result, the contents are necessarily selective and no doubt some teachers of this subject will feel that important topics have either received rather scanty treatment or have been omitted altogether. To these folk I offer my apologies, and hope that they will send me their ideas for improving the text, always remembering that if the price is to be kept down additions must be balanced by subtractions!

I would like to thank Mr Robert Campbell of Blackwell Scientific Publications for his help and encouragement, and not least for his tact in leaving me

*No longer available.

to get on with the job. My colleagues Dr J.G. Angus and Dr J. O'Leary read some of the chapters and made helpful suggestions for their improvement, and I thank them for their kindness. To my wife I owe an inestimable debt for the care with which she checked my manuscript and then produced the typescript.

Part I
Principles

'Here is such a vast variety of phenomena and these many of them so delusive, that 'tis very hard to escape imposition and mistake'.

These words, written about ore deposits by John Woodward in 1695, are every bit as true today as when he wrote them.

1

Introductory Definitions and Discussions

Ore, gangue and protore

'What is ore geology?' Unfortunately, it is not possible to give an unequivocal answer to this question if one wishes to go beyond saying that it is a branch of economic geology. The difficulty is that there are a number of distinctly different definitions of ore. A definition which has been current in capitalist economies for nearly a century runs as follows: 'Ore is a metalliferous mineral, or an aggregate of metalliferous minerals, more or less mixed with gangue, which from the standpoint of the miner can be won at a profit, or from the standpoint of the metallurgist can be treated at a profit. The test of yielding a metal or metals *at a profit* seems to be the only feasible one to employ.' Thus wrote J.F. Kemp in 1909. There are many similar definitions of ore which all emphasize (a) that it is material from which we extract a metal, and (b) that this operation must be a profit-making one. Thus there are geologists who divide minerals of economic importance into two main groups: the ore minerals from which a metal (or metals) is extracted, and the industrial minerals in which the mineral itself is used for one or more industrial purposes. Examples of ore minerals are chalcopyrite and galena from which we extract copper and lead respectively, and important among industrial minerals are baryte and asbestos. Over the last two decades there has, however, been a tendency among those who search for and mine industrial minerals to call them 'ore' and to measure 'ore reserves'. Some professional institutions have widened their definitions of ore to take note of this trend. This book will, however, follow the more common usage and consequently will be almost entirely confined to a study of metallic deposits.

A further complication, which we may note in passing, is that in socialist economies ore is often defined as mineral material that can be mined for the benefit of mankind. Such an altruistic definition is necessary to cover those examples in both capitalist and socialist countries where minerals are being worked at a loss. Such operations are carried on for various good or bad reasons depending on one's viewpoint! These include a government's reluctance to allow large isolated mining communities to be plunged into unemployment because a mine or mines have become unprofitable, a need to earn foreign currency and other reasons.

A definition about which there is little argument is that of gangue. This is simply the unwanted material, minerals or rock, with which ore minerals are usually intergrown. Mines commonly possess mineral dressing plants in which the raw ore is milled before the separation of the ore minerals from the gangue minerals by various processes, which provide ore concentrates, and tailings which are made up of the gangue.

Another word that must be introduced at this stage is protore. This is mineral material in which an initial but uneconomic concentration of metals has occurred that may by further natural processes be upgraded to the level of ore. Economically mineable aggregates of ore minerals are termed orebodies, ore shoots, ore deposits or ore reserves.

Economic considerations

PRINCIPAL STEPS IN THE ESTABLISHMENT AND OPERATION OF A MINE

These may be briefly summarized as follows:
(a) *mineral exploration*: to discover an orebody;
(b) *feasibility study*: to prove its commercial viability;
(c) *mine development*: establishment of the entire infrastructure;
(d) *mining*: extraction of ore from the ground;
(e) *ore dressing* (mineral processing): milling of the ore, separation of ore minerals from gangue, separation of the ore minerals into concentrates, e.g. copper concentrate;
(f) *smelting*: recovering metals from the mineral concentrates;
(g) *refining*: purifying the metal;
(h) *marketing*: shipping the metal (or metal concentrate if not smelted and refined at the mine) to the buyer, e.g. custom smelter.

SOME IMPORTANT FACTORS IN THE EVALUATION OF A POTENTIAL OREBODY

(a) *Ore grade*. The concentration of a metal in an orebody is called its grade, usually expressed as a percentage or in parts per million (ppm). The process of determining these concentrations is called assaying. Various economic and sometimes political considerations will determine the lowest grade of ore which can be produced from an orebody; this is termed the cut-off grade. In order to delineate the boundaries of an orebody in which the level of mineralization gradually decreases to a background value many samples will have to be collected and assayed. The boundaries thus established are called assay limits. Being entirely economically determined, they may not be marked by any particular geological feature. If the price received for the product increases, then it may be possible to lower the value of the cut-off grade and thus increase the tonnage of the ore reserves; this will have the effect of lowering the overall grade of the orebody, but for the same daily production, it will increase the life of the mine.

Grades vary from orebody to orebody and, clearly, the lower the grade, the greater the tonnage of ore required to provide an economic deposit. The tendency in metalliferous mining during this century has been to mine lower and lower grade ores. This has led to the development of more large scale operations with outputs of 40 kt of ore per day being not unusual. Technological advances may also transform waste into ore. For example the introduction of solvent extraction has enabled Nchanga Consolidated Copper Mines in Zambia to treat 9 Mt of tailings to produce 80 kt of copper. It will also be necessary to estimate, if possible by comparison with similar orebodies, what the head grade will be. This is the grade of the ore as delivered to the mill (mineral dressing plant). Often the head grade is lower than the measured ore grade because of mining dilution — the inadvertent or unavoidable incorporation of barren wall rock into the ore during mining.

(b) *By-products*. In some ores several metals are present and the sale of one may help finance the mining of another. For example, silver and cadmium can be by-products of the mining of lead-zinc ores and uranium is an important by-product of many South African gold ores.

(c) *Commodity prices*. The price of metals is a vital factor. Prices vary from metal to metal and for most of them daily fluctuations may occur. The prices of many metals are governed by supply and demand and the prices on the London Metal Exchange are quoted daily by many newspapers, whilst more comprehensive guides to current prices can be found in the *Mining Journal*, *Engineering and Mining Journal* and other technical journals. The prices of most of the common metals have not kept pace with inflation as can be seen from the examples in Fig. 1.1. This has had drastic effects on the level of recent mineral exploration activity, the profitability of many mines and the economy of whole nations, such as Zambia and Chile, which are heavily dependent on their mineral industries.

Short and long term contracts between seller and buyer may be based on these fluctuating prices. On the other hand, the parties concerned may agree on a contract price in advance of production, with clauses to allow for price changes because of such factors as inflation or currency exchange rate fluctuations. Contracts of this nature are very common in the case of iron and uranium ore production. Whatever the form of sale is to be, the mineral economists of a mining company must try to forecast the future demand for, and hence the price of, the mine product, well in advance of mine development.

(d) *Mineralogical form*. The properties of a mineral govern the ease with which existing technology can extract and refine certain metals and this may affect the cut-off grade. Thus nickel is far more readily recovered from sulphide than from silicate ores and sulphide ores can be worked down to about 0.5%, whereas silicate ores must assay about 1.5% to be economic.

Tin may occur in a variety of silicate minerals such as andradite and axinite, from which it is not recoverable, as well as in its main ore mineral form, cassiterite. Aluminium is of course abundant in many silicate rocks, but it must be in the form of hydrated aluminium oxides, the rock called bauxite, for economic recovery. The mineralogical nature of the ore will also place limits

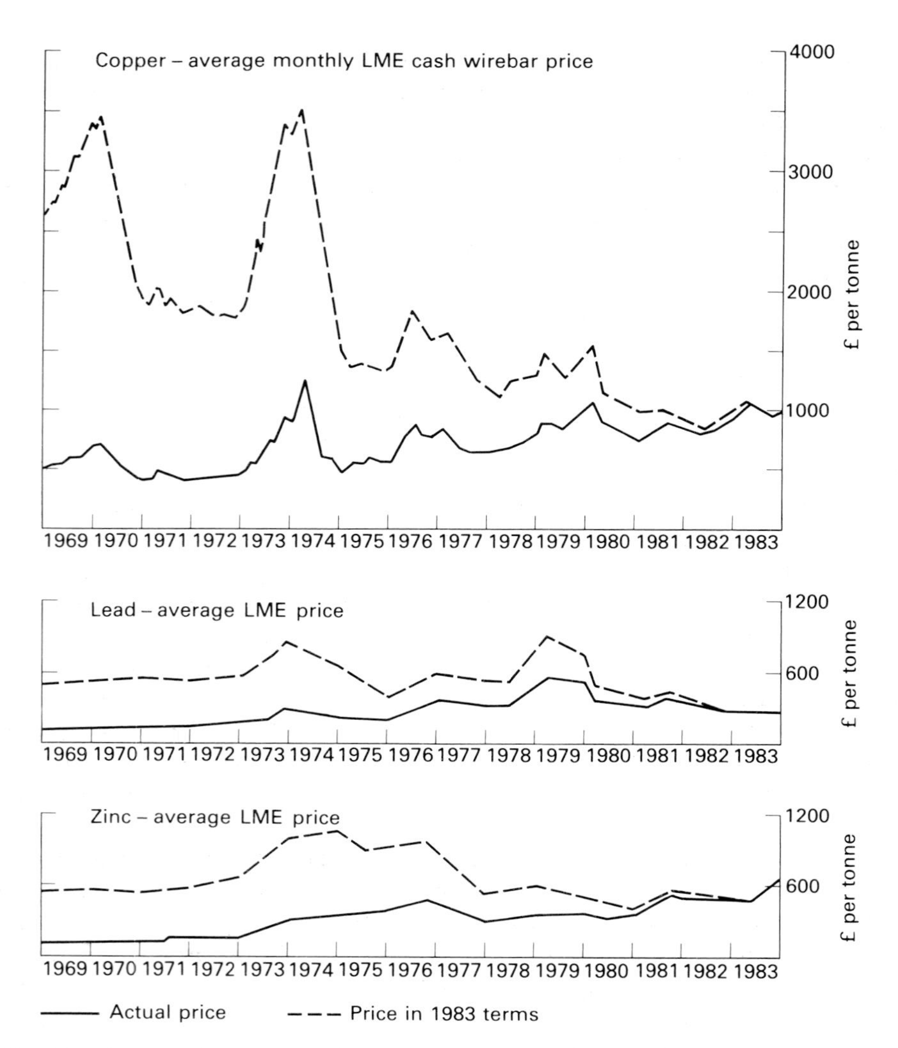

Fig. 1.1. Average prices of copper, lead and zinc during recent years showing actual prices and prices in 1983 terms. LME = London Metal Exchange. (Modified from RTZ Annual Reports).

on the maximum possible grade of the concentrate. For example, in an ore containing native copper it is theoretically possible to produce a concentrate containing 100% Cu but, if the ore mineral was chalcopyrite ($CuFeS_2$), the best concentrate would only contain 34.5% Cu.

(e) *Grain size and shape.* The *recovery* is the percentage of the *total* metal contained in the ore that is recovered in the concentrate; a recovery of 90% means that 90% of the metal in the ore is recovered in the concentrate and 10% is lost in the tailings. It might be thought that if one were to grind ores to a sufficiently fine grain size then complete separation of mineral phases might occur to make 100% recovery possible. In the present state of technology this is not the case, as most mineral processing techniques fail in the ultra-fine size range. Small mineral grains and grains finely intergrown with other minerals are difficult or impossible to recover in the processing plant, and recovery may be poor. Recoveries from primary (bedrock) tin deposits are traditionally poor, ranging over 40–80% with an average around 65%, whereas recoveries from copper ores usually lie in the range 80–90%. Sometimes fine grain size and/or complex intergrowths may preclude a mining operation. The McArthur River deposit in the Northern Territory of Australia contains 200 Mt grading 10% zinc, 4% lead, 0.2% copper and 45 ppm silver with high grade sections running up to 24% zinc and 12% lead. This enormous deposit of base metals has remained unworked since its discovery in 1956 because of the ultra-fine grain size and despite years of mineral processing research on the 'ore'.

(f) *Undesirable substances*. Deleterious substances may be present in both ore and gangue minerals. For example, tennantite ($Cu_{12}As_4S_{13}$) in copper ores can introduce unwanted arsenic and sometimes mercury into copper concentrates. These, like phosphorus in iron concentrates and arsenic in nickel concentrates, will lead to custom smelters imposing financial penalties. The ways in which gangue minerals may lower the value of an ore are very varied. For example, an acid leach is normally employed to extract uranium from the crushed ore, but if calcite is present, there will be excessive acid consumption and the less effective alkali leach method may have to be used. Some primary tin deposits contain appreciable amounts of topaz which, because of its hardness, increases the abrasion of crushing and grinding equipment, thus raising the operating costs.

(g) *Size and shape of deposits*. The size, shape and nature of ore deposits also affects the workable grade. Large, low grade deposits which occur at the surface can be worked by cheap open pit methods, whilst thin tabular vein deposits will necessitate more expensive underground methods of extraction, although they can generally be worked in much smaller volumes so that a relatively small initial capital outlay is required. Although the initial capital outlay for larger deposits may be higher, open pitting, aided by the savings from bulk handling of large daily tonnages of ore (say $>$ 30 kt), has led to a trend towards the large scale mining of low grade orebodies. As far as shape is concerned, orebodies of regular shape can generally be mined more cheaply than those of irregular shape particularly when they include barren zones. For an open pit mine the shape and attitude of the orebody will also determine how much waste has to be removed during mining, which is quoted as the waste-to-ore or stripping ratio. The waste will often include not only overburden (waste rock above the orebody) but waste rock around and in the orebody, which has to be cut back to maintain a safe overall slope to the side of the pit.

(h) *Ore character*. A loose unconsolidated beach sand deposit can be mined cheaply by dredging and does not require crushing. Hard compact ore must be drilled, blasted and crushed. In hard rock mining operations a related aspect is the strength of the country rocks. If these are badly sheared or fractured they will be weak and require roof supports in underground working, and in open pitting a gentler slope to the pit sides will be required, which in turn will affect the waste-to-ore ratio adversely.

(i) *Cost of capital.* Big mining operations have now reached the stage, thanks to inflation, where they require enormous initial capital investments. For example, to develop the large Roxby Downs Project in South Australia, Western Mining Corporation and British Petroleum have estimated that a capital investment of $1200 million will be necessary. This means that the stage has been reached where few companies can afford to develop a mine with their own financial resources. They must borrow the capital from banks and elsewhere, capital which has to be repaid with interest. Thus the revenue from the mining operation must cover the payment of taxes, royalties, the repayment of capital plus interest on it, and provide a profit to shareholders who have risked their capital to set up or invest in the company.

(j) *Location.* Geographical factors may determine whether or not an orebody is economically viable. In a remote location there may be no electric power supply, roads, houses, schools, hospitals, etc. The cost of transporting the mine product may thus be prohibitively high and wages will have to be high to attract skilled workers.

(k) *Environmental considerations.* Conflicts over land use are common in the more developed and more populous countries. The resolution of such conflicts may involve the payment of compensation and the eventual cost of rehabilitating mined out areas.

(l) *Taxation.* Greedy governments may demand so much tax that mining companies cannot make a reasonable profit. On the other hand, some governments have encouraged mineral development with taxation incentives such as a waiver on tax during the early years of a mining operation. This proved to be a great attraction to mining companies in the Irish Republic in the 1960s and brought considerable economic gains to that country.

When a company only operates one mine, then it is particularly true that dividends to shareholders should represent in part a return of capital, for once an orebody is under exploitation it has become a *wasting asset* and one day there will be no ore, no mine and no further cash flow. The company will be wound up and its shares will have no value. In other words, all mines have a limited life and for this reason should not be taxed in the same manner as other commercial undertakings. When this fact is allowed for in the taxation structure of a country, it can be seen to be an important incentive to investment in mining in that country.

(m) *Political factors.* Many large mining houses will not now invest in politically unstable countries. Fear of nationalization with perhaps very inadequate or even no compensation is perhaps the main factor. Nations with a history of nationalization generally have poorly developed mining industries. Possible political turmoil, civil strife and currency controls may all combine to greatly increase the financial risks of investing in certain countries.

ORE RESERVE CLASSIFICATION

In delineating and working an orebody the mining geologist often has to classify his ore reserves into three classes: proved, probable and possible. Proved ore has been so thoroughly sampled that we can be certain of its outline, tonnage and average grade, within certain limits. Elsewhere in the orebody, sampling from drilling and development workings may not have been so thorough, but there may be enough information to be reasonably sure of its tonnage and grade; this is probable ore. On the fringes of our exploratory workings we may have enough information to infer that ore extends for some way into only partially explored ground and that it may amount to a certain volume and grade of possible ore. In most countries, these, or equivalent, words have nationally recognized definitions and legal connotations. The practising geologist must therefore know the local definitions thoroughly and make sure that he uses them correctly.

MINERAL RESOURCES

These represent the total amount of a particular commodity (e.g. tin) and they are usually estimated for a nation as a whole and not for a company. They consist of ore reserves; known but uneconomic deposits; and hypothetical deposits not yet discovered. The estimation of the undiscovered potential of a region can be made by comparison with well explored areas of similar geology.

Theoretically, world resources of most metals are enormous. Taking copper as an example, there are large amounts of rock running 0.1–0.3% and enormous volumes containing about 0.01%. The total quantity of copper in such deposits probably exceeds that in proven reserves by a factor of from 10^3 to 10^4. Nevertheless, the enormous amount of such material does not at present imply a virtually endless resource of metals. As grades approach low values, a concentration (the mineralogical limit) is reached, below which an element no longer forms a distinct physically recoverable mineral phase.

An interesting and provocative discussion of mineral resources and some of the elements of mineral economics can be found in Wolfe (1984).

Geochemical considerations

It is traditional in the mining industry to divide metals into groups with special names. These are as follows:

(a) *precious metals*: gold, silver, platinum group (PGM);
(b) *non-ferrous metals*: copper, lead, zinc, tin, aluminium (the first four being commonly known as the *base metals*);
(c) *iron and ferroalloy metals*: iron, manganese, nickel, chromium, molybdenum, tungsten, vanadium, cobalt;
(d) *minor metals and related non-metals*: antimony, arsenic, beryllium, bismuth, cadmium, magnesium, mercury, REE, selenium, tantalum, tellurium, titanium, zirconium, etc.;

(e) *fissionable metals*: uranium, thorium, (radium).

For the formation of an orebody the element or elements concerned must be enriched to a considerably higher level than their normal crustal abundance. The degree of enrichment is termed the concentration factor and typical values are shown in Table 1.1.

Table 1.1. Concentration factors.

	Average crustal abundance (%)	Average minimum exploitable grade (%)	Concentration factor
Aluminium	8	30	3.75
Iron	5	25	5
Copper	0.005	0.4	80
Nickel	0.007	0.5	71
Zinc	0.007	4	571
Manganese	0.09	35	389
Tin	0.0002	0.5	2500
Chromium	0.01	30	3000
Lead	0.001	4	4000
Gold	0.000 000 4	0.000 1	250

2

The Nature and Morphology of the Principal Types of Ore Deposit

A good way to start an argument among mining geologists is to suggest that a deposit held by common consensus to be syngenetic is in fact epigenetic! These words are clearly concerned with the manner in which deposits have come into being and, like all matters of genesis, they are fraught with meaning and are frequently heard on the lips of mining geologists wherever they gather. What do these magic words mean? A syngenetic deposit is one which has formed at the same time as the rocks in which it occurs and it is sometimes part of a stratigraphic succession like an iron-rich sedimentary horizon. An epigenetic deposit, on the other hand, is one believed to have come into being after the host rocks in which it occurs. A good igneous analogy is a dyke; an example among ore deposits is a vein. Before discussing their nature we must learn some of the terms used in describing orebodies.

If an orebody viewed in plan is longer in one direction than the other we can designate this long dimension as its strike (Fig. 2.1). The inclination of the orebody perpendicular to the strike will be its dip and the longest dimension of the orebody its axis. The plunge of the axis is measured in the vertical plane ABC but its pitch or rake can be measured in any other plane, the usual choice being the plane containing the strike, although if the orebody is fault controlled then the pitch may be measured in the fault plane. The meanings of other terms are self-evident from the figure.

It is possible to classify orebodies in the same way as we divide up igneous intrusions according to whether they are discordant or concordant with the lithological banding (often bedding) in the enclosing rocks. Considering discordant orebodies first, this large class can be subdivided into those orebodies which have an approximately regular shape and those which are thoroughly irregular in their outlines.

Discordant orebodies

REGULARLY SHAPED BODIES

(a) *Tabular orebodies*. These bodies are extensive in two dimensions, but have a restricted development in their third dimension. In this class we have veins (sometimes called fissure-veins) and lodes (Fig. 2.2). In the past, some workers have made a genetic distinction between these terms; veins were considered to have resulted mainly from the infilling of pre-existing open spaces, whilst the

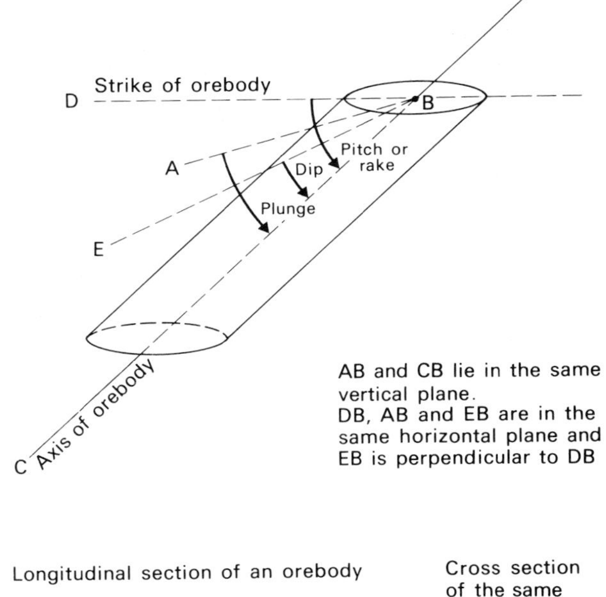

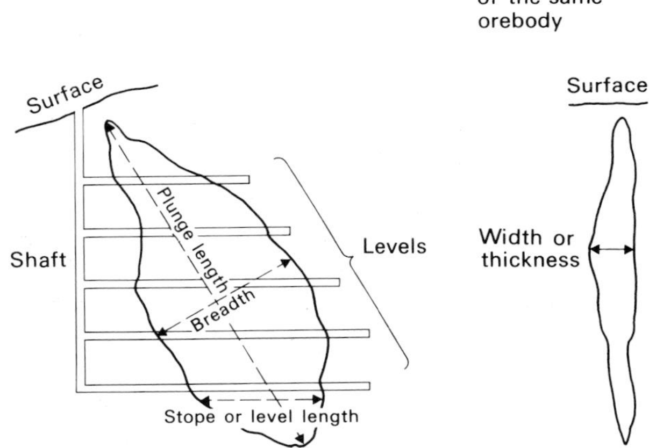

Fig. 2.1. Diagrams illustrating terms used in the description of orebodies.

formation of lodes was held to involve the extensive replacement of pre-existing host rock. Such a genetic distinction has often proved to be unworkable and the writer advises that all such orebodies be called veins and the term lode be dropped.

Veins are often inclined, and in such cases, as with faults, we can speak of the hanging wall and the footwall. Veins frequently pinch and swell out as they are followed up or down a stratigraphic sequence (Fig. 2.2). This pinch-and-swell structure can create difficulties during both exploration and mining often because only the swells are workable and if these are imagined in a section at right angles to that in Fig. 2.2 it can be seen that they form ribbon ore shoots. The origin of pinch-and-swell structure is shown in Fig. 2.3. An initial fracture in rocks changes its attitude as it crosses them according to the

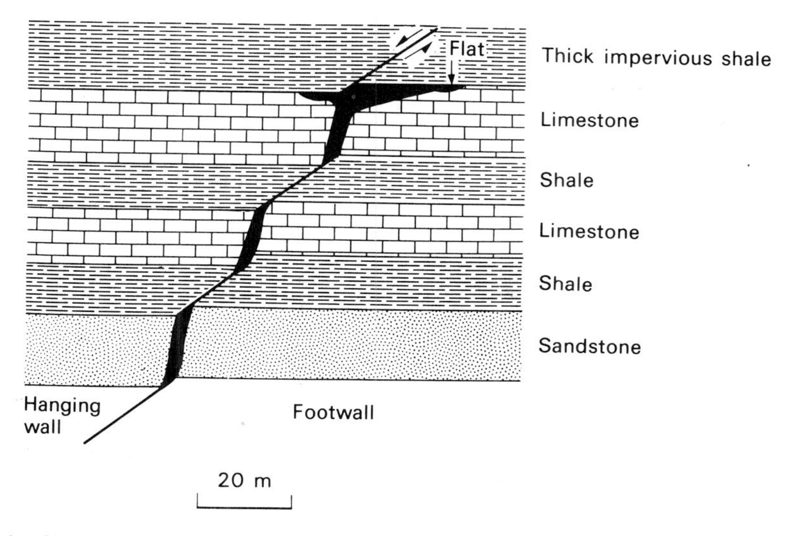

Fig. 2.2. Vein occupying a normal fault and exhibiting pinch-and-swell structure, giving rise to ribbon ore shoots. The development of a flat beneath impervious cover is also shown.

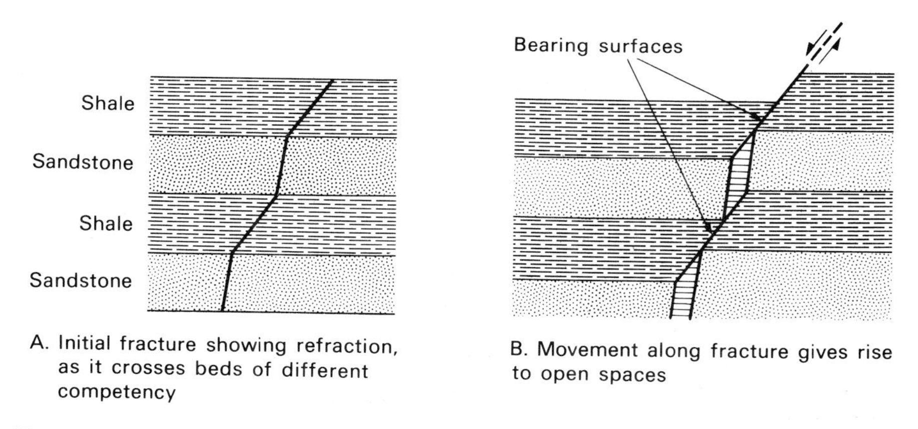

Fig. 2.3. Formation of pinch-and-swell structure in veins.

changes in physical properties of the rocks and these properties are in turn governed by changes in lithology (Fig. 2.3A). When movement occurs producing a normal fault then the less steeply dipping sections are held against each other to become bearing surfaces, and open spaces (dilatant zones) form in the more steeply dipping sections. Then, should minerals be deposited in these cavities, a vein will be formed. If the reader carries out the experiment of reversing the movement on the initial fracture, he will find that the steeper parts of the fault now act as bearing surfaces and the dilatant zones are formed in the less steeply dipping sections. Veins are usually developed in fracture systems and therefore show regularities in their orientation (Figs. 2.4, 17.2, 17.3).

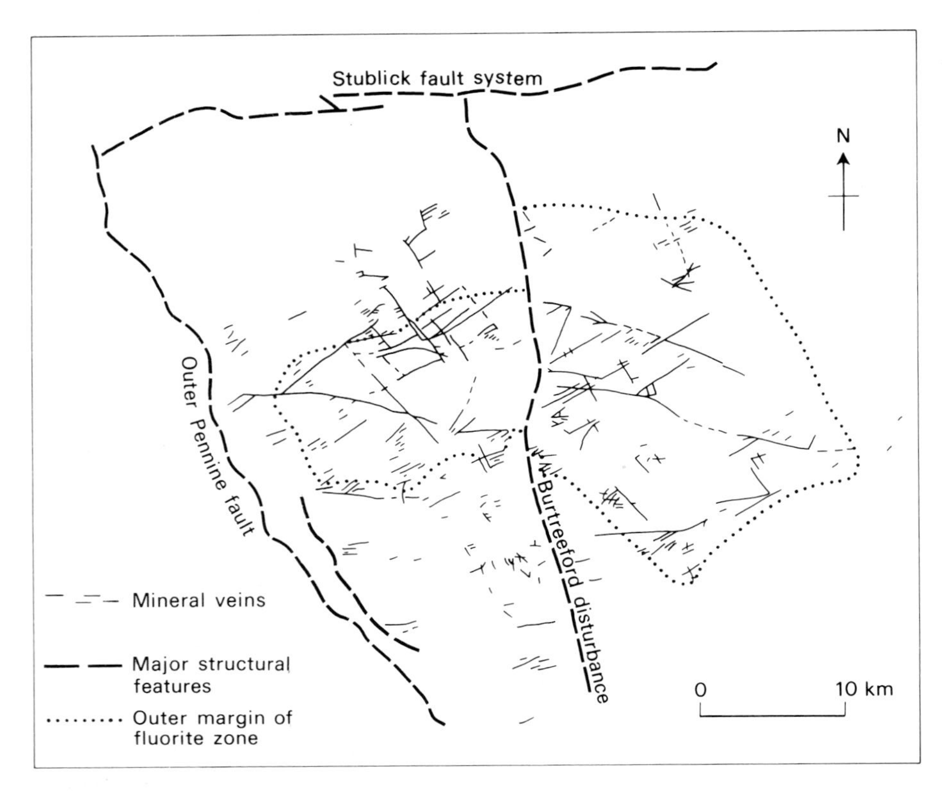

Fig. 2.4. Vein system of the Alston block of the Northern Pennine Orefield, England. Note the three dominant vein directions. (Modified from Dunham 1959).

The infilling of veins may consist of one mineral but more usually it consists of an intergrowth of ore and gangue minerals. The boundaries of vein orebodies may be the vein walls or they can be assay boundaries within the veins.

(b) *Tubular orebodies*. These bodies are relatively short in two dimensions but extensive in the third. When vertical or subvertical they are called pipes or chimneys, when horizontal or subhorizontal, 'mantos'. The Spanish word manto is inappropriate in this context for its literal translation is blanket: it is, however, firmly entrenched in the English geological literature. The word has been and is employed by some workers for flat-lying tabular bodies, but the perfectly acceptable word 'flat' is available for these; therefore the reader must look carefully at the context when he encounters the term 'manto'.

In eastern Australia, along a 2400 km belt from Queensland to New South Wales, there are hundreds of pipes in and close to granite intrusions. Most have quartz fillings and some are mineralized with bismuth, molybdenum, tungsten and tin; an example is shown in Fig. 2.5. Pipes may be of various types and origins (Mitcham 1974). Infillings of mineralized breccia are particularly common, a good example being the copper-bearing breccia pipes of Messina in South Africa (Jacobsen & McCarthy 1976).

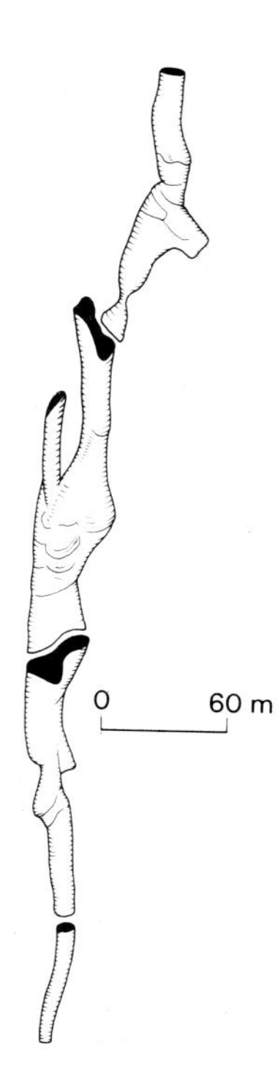

Fig. 2.5. Diagram of the Vulcan tin pipe, Herberton, Queensland. The average grade was 4.5% tin. (After Mason 1953).

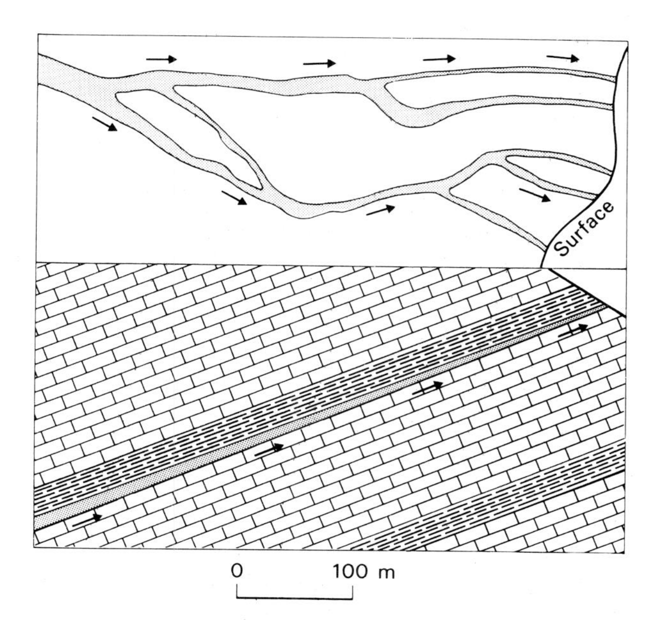

Fig. 2.6. Plan and section of part of the Hidden Treasure manto, Ophir mining district, Utah. (After Gilluly 1932).

Mantos and pipes may branch and anastomose; an example of a branching manto is given in Fig. 2.6. Mantos and pipes are often found in association, the pipes frequently acting as feeders to the mantos. Sometimes mantos pass upwards from bed to bed by way of pipe connections, often branching as they go, an example being the Providencia mine in Mexico where a single pipe at depth feeds into twenty mantos nearer the surface.

In some tubular deposits formed by the sub-horizontal flow of mineralizing fluid, ore grade mineralization may be discontinuous, thus creating pod-shaped orebodies (Figs. 2.7, 2.8) as with the McClean deposits, Saskatchewan. These pods undulate along the unconformity between a regolith and the overlying Proterozoic Athabaska Group sediments, and their position appears to have been controlled by a vertically disposed fracture system (Wallis *et al.* 1984). The mineralizing fluid removed much quartz from both the regolith and the overlying sediments and deposited new minerals including a considerable amount of pitchblende in its place.

IRREGULARLY SHAPED BODIES

(a) *Disseminated deposits.* In these deposits, ore minerals are peppered throughout the body of the host rock in the same way as accessory minerals are disseminated through an igneous rock; in fact, they often *are* accessory minerals. A good example is that of diamonds in kimberlites; another is that of some orthomagmatic nickel-copper deposits such as the La Perouse Layered Gabbro, Alaska (Czamanske *et al.* 1981), which contains disseminated sulphide mineralization throughout its entire thickness of about 6 km. This deposit has over 100 Mt grading about 0.5% nickel and 0.3% copper. Given the economic

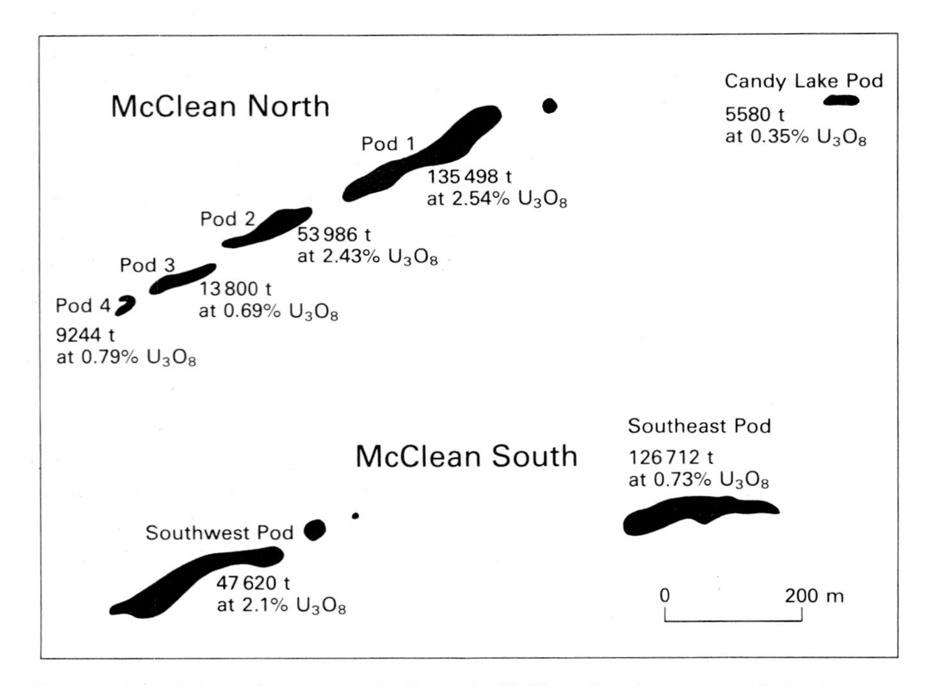

Fig. 2.7. Distribution of uranium orebodies at the McClean deposits, northern Saskatchewan. (After Wallis *et al.* 1984).

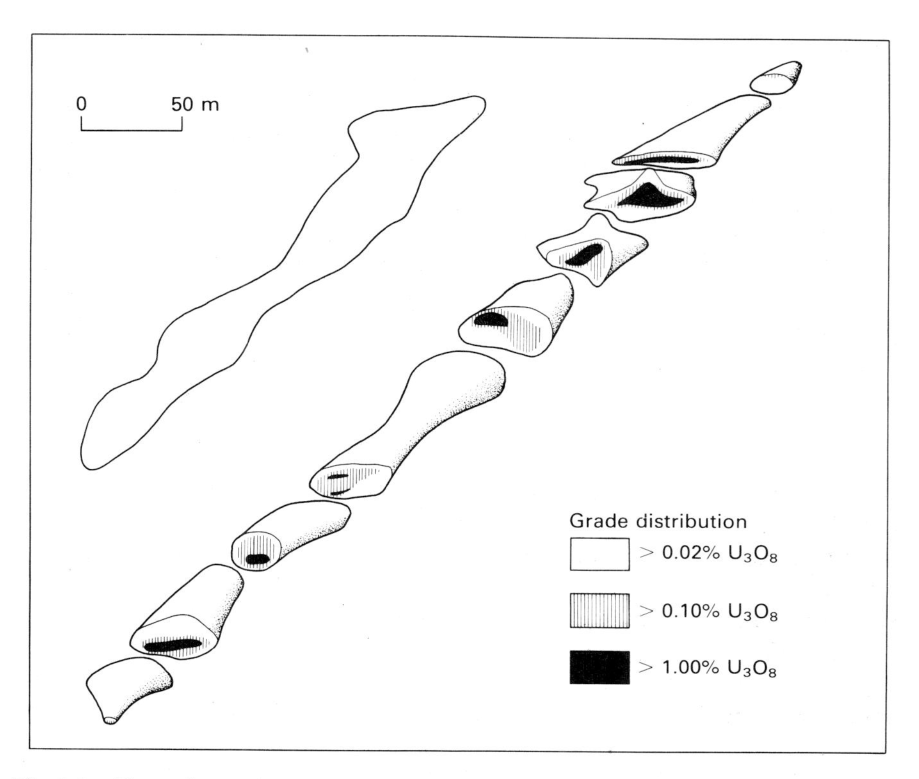

Fig. 2.8. Plan and expanded views of McClean North Pod 1. The pods are arbitrarily defined by mineralization which averages a minimum of 0.15 m%, i.e. 0.1% U_3O_8 over a thickness of at least 1.5 m. (After Wallis *et al.* 1984).

climate of the seventies and a suitable geographical location, such a body might have constituted ore. However, in its remote position and at an elevation of over 1000 m, where much of it is covered by up to 200 m of ice, it must remain for the moment an item to be grouped among mineral resources and not ore reserves (see Chapter 1), despite the 1.2–1.5 ppm of PGM in flotation concentrates which one day could form a lucrative by-product. In other deposits, the disseminations may be wholly or mainly along close-spaced veinlets cutting the host rock and forming an interlacing network called a stockwork (Fig. 15.1) or the economic minerals may be disseminated through the host rock and along veinlets (Fig. 15.2). Whatever the mode of occurrence, mineralization of this type generally fades gradually outwards into subeconomic mineralization and the boundaries of the orebody are assay limits. They are, therefore, often irregular in form and may cut across geological boundaries. The overall shapes of some are cylindrical (Fig. 15.5) and others are caplike (Fig. 15.13). The mercury-bearing stockworks of Dubník in Slovakia are sometimes pear-shaped.

Stockworks most commonly occur in acid to intermediate plutonic igneous intrusions, but they may cut across the contact (Fig. 2.9) into the country

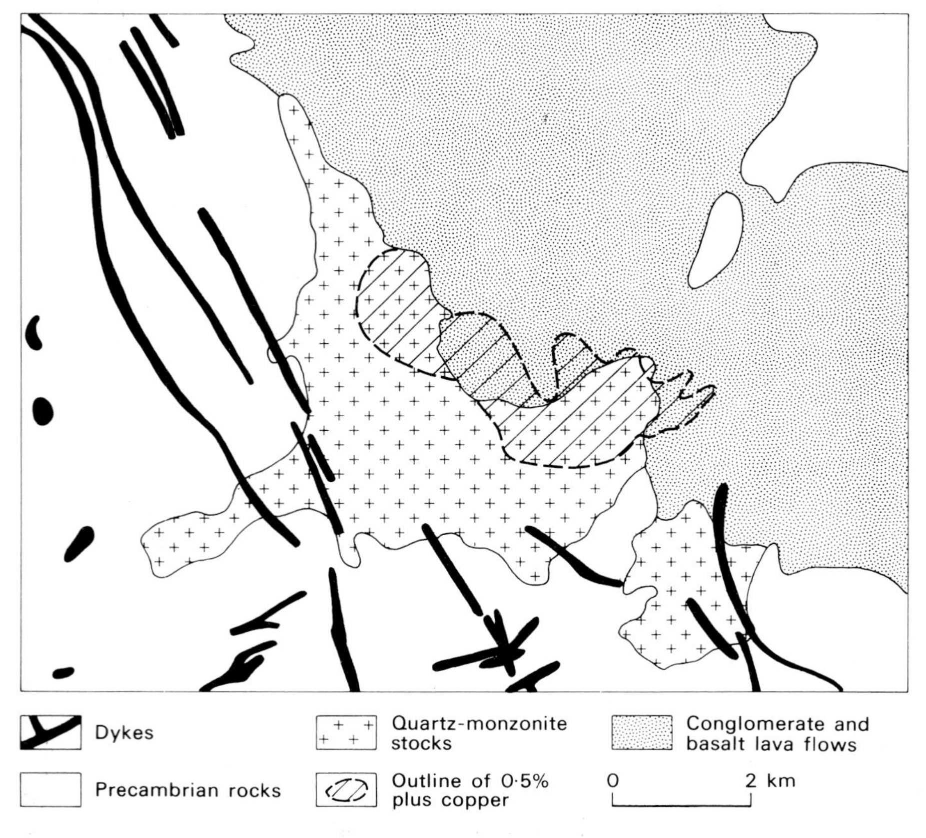

Fig. 2.9. Generalized geological map of the Bagdad mine area, Arizona. (Modified from Anderson 1948).

rocks, and a few are wholly or mainly in the country rocks. Disseminated deposits produce most of the world's copper and molybdenum and they are also of some importance in the production of tin, gold, silver, mercury and uranium.

(b) *Irregular replacement deposits*. Many ore deposits have been formed by the replacement of pre-existing rocks, particularly carbonate-rich sediments. These replacement processes often occured at high temperatures, at contacts with medium-sized to large igneous intrusions. Such deposits have therefore been called contact metamorphic or pyrometasomatic; however, *skarn* is now the preferred and more popular term. The orebodies are characterized by the development of calc-silicate minerals such as diopside, wollastonite, andradite garnet and actinolite. These deposits are extremely irregular in shape (Fig. 2.10); tongues of ore may project along any available planar structure — bedding, joints, faults, etc. and the distribution within the contact aureole is often apparently capricious. Structural changes may cause abrupt termination of the orebodies. The principal materials produced from pyrometasomatic

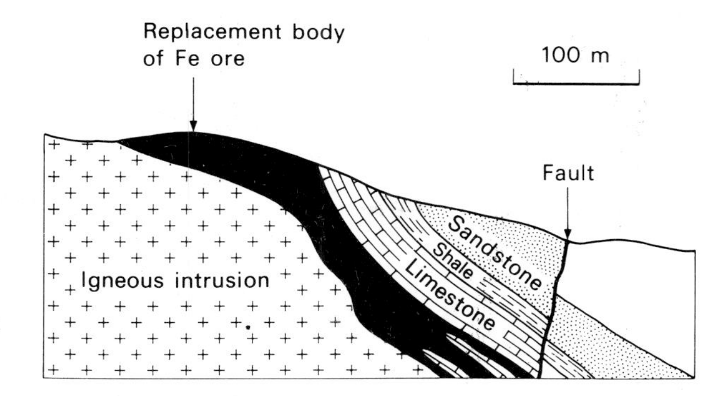

Fig. 2.10. Skarn deposit at Iron Springs, Utah. (After Gilluly *et al.* 1959).

deposits are: iron, copper, tungsten, graphite, zinc, lead, molybdenum, tin and uranium.

Other replacement deposits occur which do not belong to the pyrometasomatic class; examples of these include flats. These are horizontal or sub-horizontal bodies of ore which commonly branch out from veins and lie in carbonate host rocks beneath an impervious cover such as shale (Fig. 2.11).

Concordant orebodies

SEDIMENTARY HOST ROCKS

Concordant orebodies in sediments are very important producers of many different metals, being particularly important for base metals and iron, and are of course concordant with the bedding. They may be an integral part of the stratigraphic sequence, as is the case with Phanerozoic ironstones (syngenetic ores formed by the exhalation of mineralizing solutions at the sediment-water interface) — or they may be epigenetic infillings of pore spaces or replacement orebodies. Usually these orebodies show a considerable development in two dimensions, i.e. *parallel* to the bedding and a limited development *perpendicular* to it (Figs. 2.12, 2.13, 2.15) and for this reason such deposits are referred to as stratiform. This term must not be confused with strata-bound, which refers to any type or types of orebody, concordant or discordant, which are restricted to a particular part of the stratigraphic column. Thus the veins, pipes and flats of the Southern Pennine Orefield of England can be designated as strata-bound, as they are virtually restricted to the Carboniferous limestone of that region. A number of examples of concordant deposits which occur in different types of sedimentary rocks will be considered.

(a) *Limestone hosts.* Limestones are very common host rocks for base metal sulphide deposits. In a dominantly carbonate sequence ore is often developed in a small number of preferred beds or at certain sedimentary interfaces. These are often zones in which the permeability has been increased by dolomitization or fracturing. When they form only a minor part of the stratigraphical succession, limestones, because of their solubility and reactivity, can become favour-

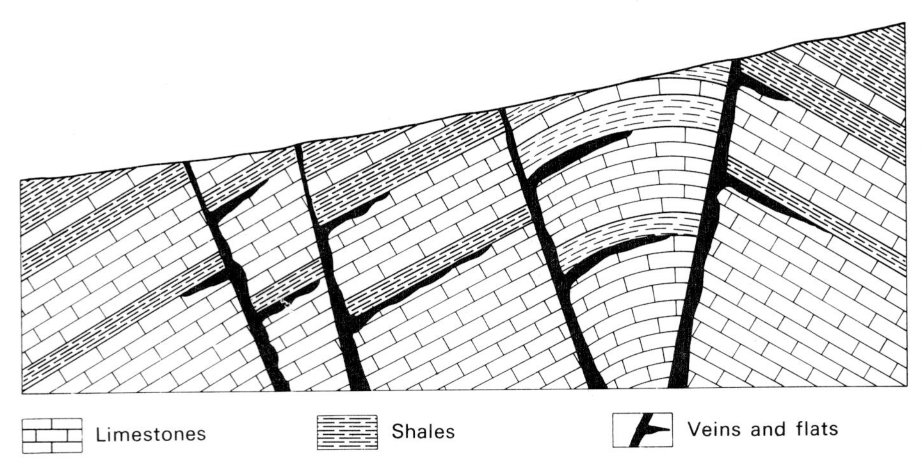

Fig. 2.11. Lead-zinc orebodies in a faulted anticline in Devonian rocks, Gyumushlug, Transcaucasia, USSR. Note the development of veins along the faults with flats branching off them beneath impervious beds of shale. (After Malyutin & Sitkovskiy 1968).

able horizons for mineralization. For example, the lead-zinc ores of Bingham, Utah, occur in limestones which make up 10% of a 2300 m succession mainly composed of quartzites.

At Silvermines in Ireland, lead-zinc mineralization occurred (mining ceased in 1982) as syngenetic stratiform orebodies in a limestone sequence (Fig. 2.12), as fault-bounded epigenetic strata-bound orebodies in the basal Carboniferous, and as structurally controlled vein or breccia zones in the Upper Devonian sandstones (Taylor & Andrew 1978, Taylor 1984). The larger stratiform orebody shown in Fig. 2.12 occurred in massive, partly brecciated pyrite at the base of a thick sequence of dolomite breccias. The maximum thickness of the ore was 30 m, and, at the base, there was usually an abrupt change to massive pyrite from a footwall of nodular micrite, shale biomicrite and other limestone units of the Mudbank Limestone, although sometimes the contact was gradational. The upper contact was always sharp. Pyrite and marcasite made up 75% of the ore, sphalerite formed 20% and galena 4%.

(b) *Argillaceous hosts.* Shales, mudstones, argillites and slates are important host rocks for concordant orebodies which are often remarkably continuous and extensive. In Germany, the Kupferschiefer of the Upper Permian is a prime example. This is a copper-bearing shale a metre or so thick which, at Mansfeld, occurred in orebodies which had plan dimensions of 8, 16, 36 and 130 km^2. Mineralization occurs at exactly the same horizon in Poland, where it is being worked extensively, and across the North Sea in north-eastern England, where it is subeconomic.

The world's largest, single lead-zinc orebody occurs at Sullivan, British Columbia. The host rocks are late Precambrian argillites. Above the main orebody (Fig. 2.13) there are a number of other mineralized horizons with concordant mineralization. This deposit appears to be syngenetic and the lead,

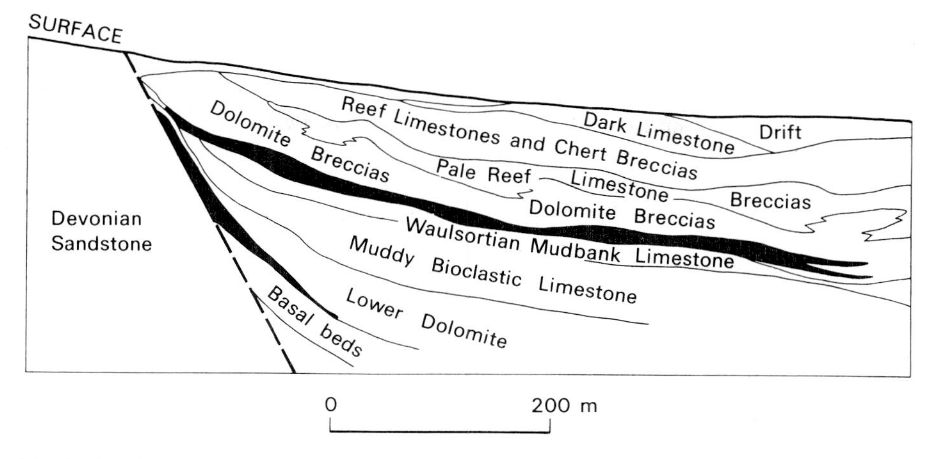

Fig. 2.12. Vertical section through the G zone at Silvermines, Co. Tipperary, Ireland. The orebodies are shown in black. (After Taylor & Andrew 1978).

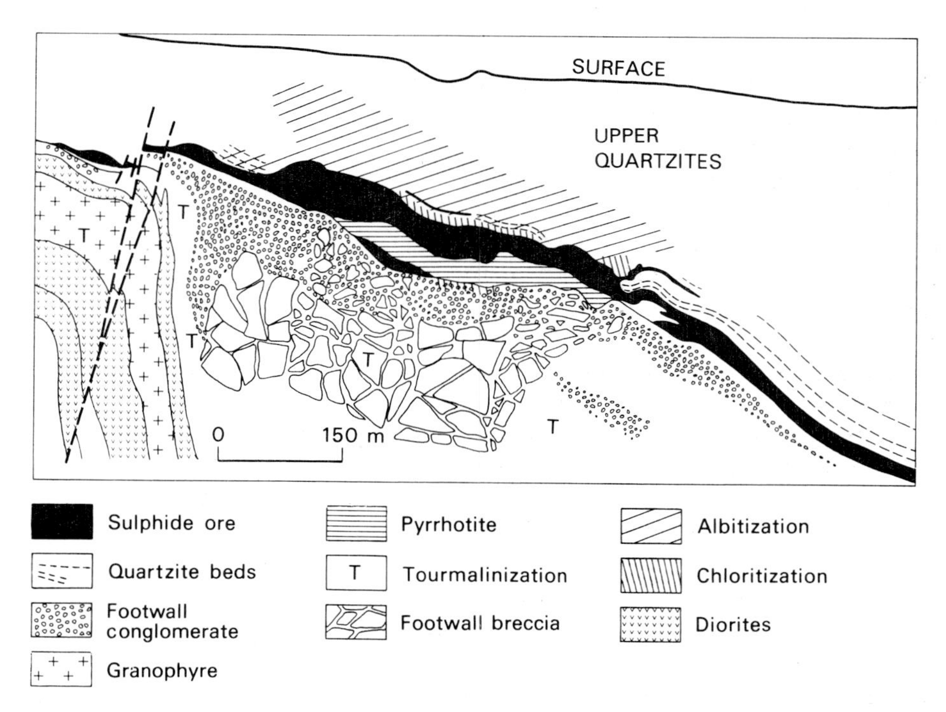

Fig. 2.13. Cross section through the ore zone, Sullivan Mine, British Columbia. (After Sangster & Scott 1976).

zinc and other metal sulphides form an integral part of the rocks in which they occur. They are affected by sedimentary deformation such as slumping, pull apart structures, load casting, etc., in a manner identical to that in which poorly consolidated sand and mud respond. (Fig. 2.14).

The orebody occurs in a single generally conformable zone between 60 and 90 m thick and runs 6.6% lead and 5.9% zinc. Other metals recovered are silver, tin, cadmium, antimony, bismuth, copper and gold. Before mining commenced in 1900, the orebody contained at least 155 million tonnes of ore, and at the current rate of mining (2 Mt p.a.), the mine has a remaining life of about twenty years. The footwall rocks consist of graded impure quartzites and argillites and, in places, conglomerate. The hanging wall rocks are more thickly bedded and arenaceous. The ore zone is a mineralized argillite in which the principal sulphide-oxide minerals are pyrrhotite, sphalerite, galena, pyrite and magnetite with minor chalcopyrite, arsenopyrite and cassiterite. Beneath the central part of the orebody there are extensive zones of brecciation and tourmalinization which extend downwards for at least 100 m. In places, the matrix of the breccias is heavily mineralized with pyrrhotite and occasionally with galena, sphalerite, chalcopyrite and arsenopyrite. This zone may have been a channelway up which solutions moved to debouch on to the sea floor, to precipitate the ore minerals among the accumulating sediment; if this was the case, then Sullivan could be called a sedimentary-exhalative deposit (Hamilton *et al.* 1983).

Other good examples of concordant deposits in argillaceous rocks, or slightly metamorphosed equivalents, are the lead-zinc deposits of Mount Isa, Queensland, many of the Zambian Copperbelt deposits and the copper shales of the White Pine Mine, Michigan.

(c) *Arenaceous hosts*. Not all the Zambian Copperbelt deposits occur in shales and metashales. Some orebodies occur in altered feldspathic sandstones (Fig.

Fig. 2.14. Sulphiditic siltstone, Sullivan Mine, British Columbia. The lighter grey material is sulphide. Note the pull-apart and load structures which affect both the sulphide and silicate layers.

2.15). The Mufulira copper deposit occurs in Proterozoic rocks on the eastern side of an anticline (Fleischer *et al.* 1976) and lies just above the unconformity with an older, strongly metamorphosed Precambrian basement. The gross ore reserves in 1974 stood at 282 Mt assaying 3.47% copper and the largest orebody stretches for 5.8 km along the strike and for several kilometres down dip. Chalcopyrite is the principal sulphide mineral, sometimes being accompanied by significant amounts of bornite. Fluviatile and aeolian arenites form the footwall rocks. The ore zone consists of feldspathic sandstones which, in places, contain carbon-rich lenses with much sericite. The basal portion is coarse-grained and characterized by festoon cross-bedding in which bornite is concentrated along the cross-bedding together with well rounded, obviously detrital zircon; whilst in other parts of the orebody, concentrations of sulphides occur in the hollows of ripplemarks and in desiccation cracks. These features suggest that some of the sulphides are detrital in origin. Mineralization ends abruptly at the hanging wall, suggesting a regression and at this sharp cut-off the facies changes from an arenaceous one to dolomites and shallow-water muds.

Conformable deposits of copper occur in some sandstones which were laid down under desert conditions. As these rocks are frequently red, the deposits are known as Red Bed Coppers. Dune sands are frequently porous and permeable and the copper minerals are generally developed in pore spaces. Examples of such deposits occur in the Permian of the Urals and the Don Basin in the USSR in the form of sandstone layers 10–40 cm thick running 1.5–1.9% copper. They are also found in the Trias of central England, in

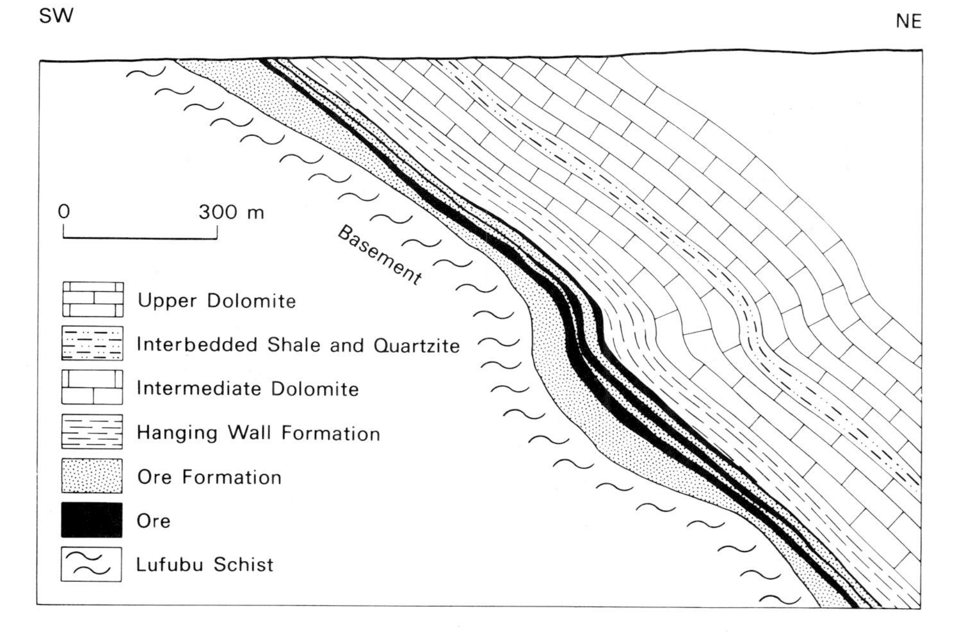

Fig. 2.15. Cross section through the Mufulira orebodies, Zambia. (After Fleischer *et al.* 1976).

Nova Scotia, in Germany and in the south-western USA. At the Nacimiento Mine in New Mexico a deposit of 11 Mt averaging 0.65% copper is being worked by open pit methods. Like other red bed coppers, this deposit has a high metal/sulphur ratio as the principal mineral is chalcocite. This yields a copper concentrate low in sulphur which is very acceptable to present-day custom smelters faced with stringent anti-pollution legislation.

Copper is not the only base metal which occurs in such deposits. Similar lead ores are known in Germany and silver deposits in Utah. Another important class of pore-filling deposits are the uranium-vanadium deposits of Colorado Plateau or western states-type which occur mainly in sandstones of continental origin but also in some siltstones and conglomerates. The orebodies are very variable in form, and pods and irregularly shaped deposits occur, although large concordant sheets up to 3 m thick are also present. The orebodies follow sedimentary structures and depositional features.

Many mechanical accumulations of high density minerals such as magnetite, ilmenite, rutile and zircon occur in arenaceous hosts, usually taking the form of heavy mineral rich layers in Pleistocene and Holocene sands. As the sands are usually unlithified, the deposits are easily worked and no costly crushing of the ore is required. These orebodies belong to the group called placer deposits —

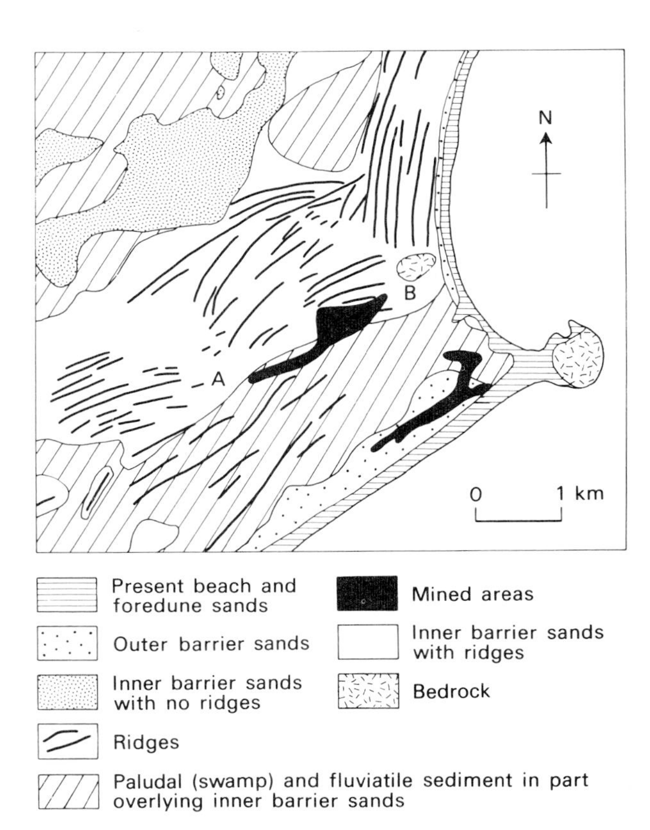

Fig. 2.16. Geology and mining areas of the beach sand deposits of Crowdy Head, New South Wales, Australia. (After Winward 1975).

beach sand placers are a good example (Fig. 2.16). Beach placers supply much of the world's titanium, zirconium, thorium, cerium and yttrium. They occur along present-day beaches or ancient beaches where longshore drift is well developed and frequent storms occur. Economic grades can be very low and sands running as little as 0.6% heavy minerals are worked along Australia's eastern coast. The deposits usually show a topographic control, the shapes of bays and the position of headlands often being very important; thus in exploring for buried orebodies a reconstruction of the palaeogeography is invaluable.

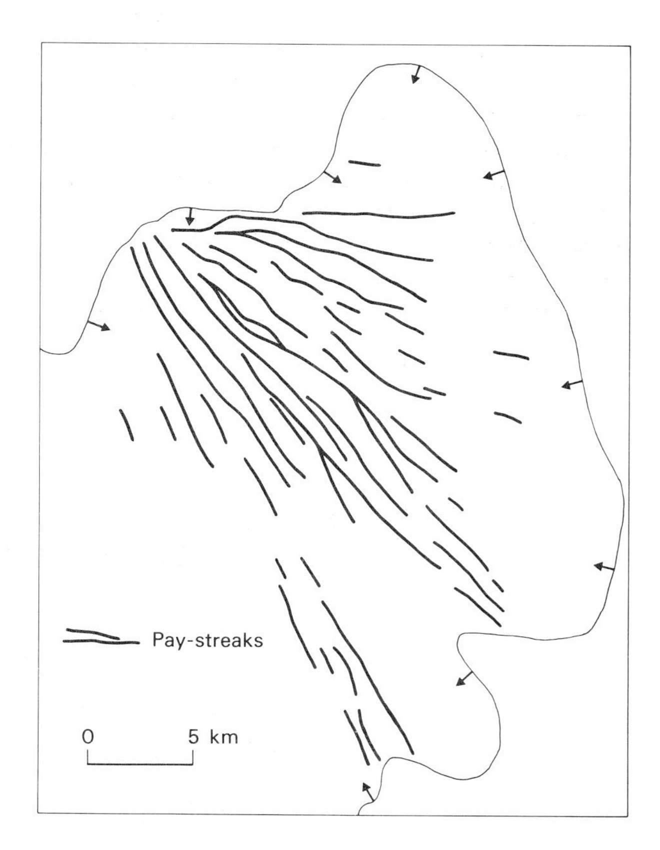

Fig. 2.17. Distribution of pay-streaks (gold orebodies) in the Main Leader Reef of the East Rand Basin of the Witwatersrand Goldfield of South Africa. The arrows indicate the direction of dip at the outcrop or sub-outcrop. (After Du Toit 1954).

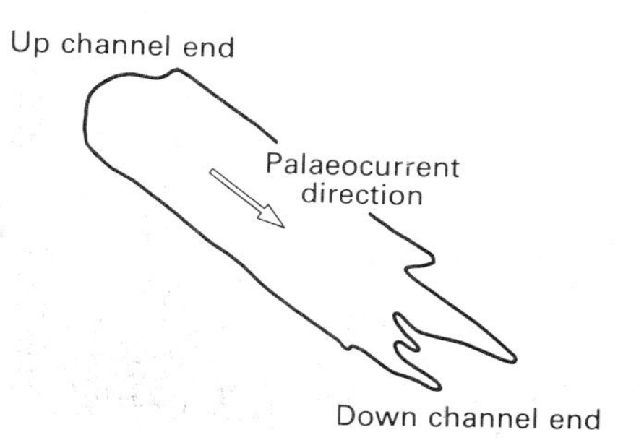

Fig. 2.18. Plan of an ore sheet, Blind River, Ontario, showing blunt up channel end and fingering down channel end. (After Robertson 1962).

(d) *Rudaceous hosts*. Alluvial gravels and conglomerates also form important recent and ancient placer deposits. Alluvial gold deposits are often marked by 'white runs' of vein quartz pebbles as in the White Channels of the Yukon, the White Bars of California and the White Leads of Australia. Such deposits form one of the few types of economic placer deposits in fully lithified rocks, and indeed the majority of the world's gold is won from Precambrian deposits of this type in South Africa. Fig. 2.17 shows the distribution of the gold orebodies in the East Rand Basin where the vein quartz pebble conglomerates occur in quartzites of the Upper Witwatersrand System. Their fan-shaped distribution strongly suggests that they occupy distributary channels. Uranium is recovered as a by-product of the working of the Witwatersrand goldfields. In the very similar Blind River area of Ontario uranium is the only metal produced. In this field the conglomeratic orebodies lie in elongate south-easterly trending sheets (Fig. 2.18) that are composed of layers of braided, interfingering channels and beds. These ore sheets, carrying the individual orebodies, have dimensions measured in kilometres (Robertson 1962, Theis 1979). Similar mineralized conglomerates occur elsewhere in the Precambrian.

(e) *Chemical sediments*. Sedimentary iron and manganese formations occur scattered through the stratigraphical column where they form very extensive beds conformable with the stratigraphy. They are described in Chapter 16.

IGNEOUS HOST ROCKS

(a) *Volcanic hosts*. There are two principal types of deposit to be found in volcanic rocks, vesicular filling deposits and volcanic-associated massive sulphide deposits. The first deposit type is not very important but the second type is a widespread and important producer of base metals often with silver and gold as by-products.

The first type forms in the permeable vesicular tops of basic lava flows whose permeability may have been increased by autobrecciation. The mineralization is normally in the form of native copper and the best examples occurred in late Precambrian basalts of the Keweenaw Peninsula of northern Michigan. Mining commenced in 1845 and the orebodies, which are now virtually worked out, were very large and were mined down to nearly 2750 m. There were six main producing horizons, five in the tops of lava flows and one in a conglomerate. The orebodies averaged 4 m in thickness and 0.8% copper. Occasionally, so-called veins of massive copper were found, one such mass weighing 500 t. Similar deposits occur around the Coppermine River in northern Canada where 3 Mt running 3.48% copper has been found but, although the copper content was worth some £100 million at 1979 prices, the deposit is unlikely to be worked because of its remote location. Other uneconomic deposits of this type are known in many other countries.

Volcanic-associated massive sulphide deposits often consist of over 90% iron sulphide usually as pyrite, although pyrrhotite is well developed in some

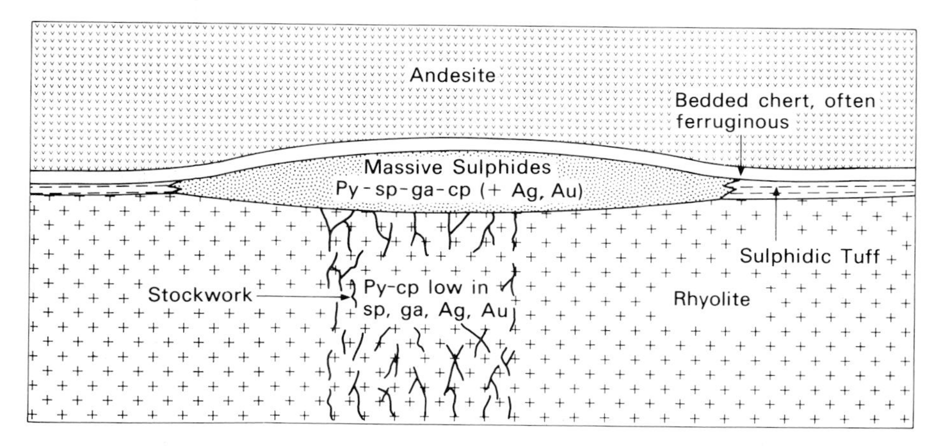

Fig. 2.19. Schematic cross section through an idealized volcanic-associated massive sulphide deposit showing the underlying feeder stockwork and typical mineralogy. Py = pyrite, sp = sphalerite, ga = galena, cp = chalcopyrite.

deposits. They are generally stratiform bodies, lenticular to sheetlike (Fig. 2.19), developed at the interfaces between volcanic units or at volcanic-sedimentary interfaces. With increasing magnetite content, these ores grade to massive oxide ores of magnetite and/or hematite such as Savage River in Tasmania, Fosdalen in Norway and Kiruna in Sweden (Solomon 1976). They can be divided into three classes of deposit: (a) zinc-lead-copper, (b) zinc-copper, and (c) copper. Typical tonnages and copper grades are given in Fig. 15.3.

The most important host rock is rhyolite and lead-bearing ores are only associated with this rock-type. The copper class is usually, but not invariably, associated with mafic volcanics. Massive sulphide deposits commonly occur in groups and in any one area they are found at one or a restricted number of horizons within the succession. These horizons may represent changes in composition of the volcanic rocks, a change from volcanism to sedimentation, or simply a pause in volcanism. There is a close association with volcaniclastic rocks and many orebodies overlie the explosive products of rhyolite domes. These ore deposits are usually underlain by a stockwork that may itself be ore grade and which appears to have been the feeder channel up which mineralizing fluids penetrated to form the overlying massive sulphide deposit.

(b) *Plutonic hosts*. Many plutonic igneous intrusions possess rhythmic layering and this is particularly well developed in some basic intrusions. Usually the layering takes the form of alternating bands of mafic and felsic minerals, but sometimes minerals of economic interest such as chromite, magnetite and ilmenite may form discrete mineable seams within such layered complexes (Fig. 11.2). These seams are naturally stratiform and may extend over many kilometres as is the case with the chromite seams in the Bushveld Complex of South Africa (Fig. 11.1).

Another form of orthomagmatic deposit is the nickel-copper sulphide orebody formed by the sinking of an immiscible sulphide liquid to the bottom of a magma chamber containing ultrabasic or basic magma. These are known to the learned as liquation deposits and they may be formed at the base of lava flows as well as in plutonic intrusions. The sulphide usually accumulates in hollows in the base of the igneous body and generally forms sheets or irregular lenses conformable with the overlying silicate rock. From the base upwards, massive sulphide gives way through disseminated sulphides in a silicate gangue to lightly mineralized and then barren rock (Figs. 12.6, 12.9).

METAMORPHIC HOST ROCKS

Apart from some deposits of metamorphic origin such as the irregular replacement deposits already described, metamorphic rocks are mainly important for the metamorphosed equivalents of deposits that originated in sedimentary and igneous rocks and which have been discussed above.

RESIDUAL DEPOSITS

These are deposits formed by the removal of non-ore material from protore. For example, the leaching of silica and alkalis from a nepheline-syenite may leave behind a surface capping of hydrous aluminium oxides (bauxite). Some residual bauxites occur at the present surface, others have been buried under younger sediments to which they form conformable basal beds.

Other examples of residual deposits include some laterites sufficiently high in iron to be worked and nickeliferous laterites formed by the weathering of peridotites (Figs. 20.1, 20.2, 20.3).

SUPERGENE ENRICHMENT

This is a process which may affect most orebodies to some degree. After a deposit has been formed, uplift and erosion may bring it within reach of circulating ground waters which may leach some of the metals out of that section of the orebody above the water table. These dissolved metals may be redeposited in that part of the orebody lying beneath the water table and this can lead to a considerable enrichment in metal values. Supergene processes are discussed in Chapter 20.

3

Textures and Structures of Ore and Gangue Minerals. Fluid Inclusions. Wall Rock Alteration

The study of textures can tell us much about the genesis and subsequent history of orebodies. The textures of orebodies vary according to whether their constituent minerals were formed by deposition in an open space from a silicate or aqueous solution, or by replacement of pre-existing rock or ore minerals. Subsequent metamorphism may drastically alter primary textures. The interpretation of mineral textures is a very large and difficult subject and only a few important points can be touched on here; for further information the reader should consult Edwards (1952, 1960), Ramdohr (1969), Stanton (1972) and Craig & Vaughan (1981).

Open space filling

PRECIPITATION FROM SILICATE MELTS

Critical factors in this situation are the time of crystallization and the presence or absence of simultaneously crystallizing silicates. Oxide ore minerals such as chromite often crystallize out early and thus may form good euhedral crystals, although these may be subsequently modified in various ways. Chromites deposited with interstitial silicate liquid may suffer corrosion and partial resorption to produce atoll textures (Fig. 3.1) and rounded grains, whereas

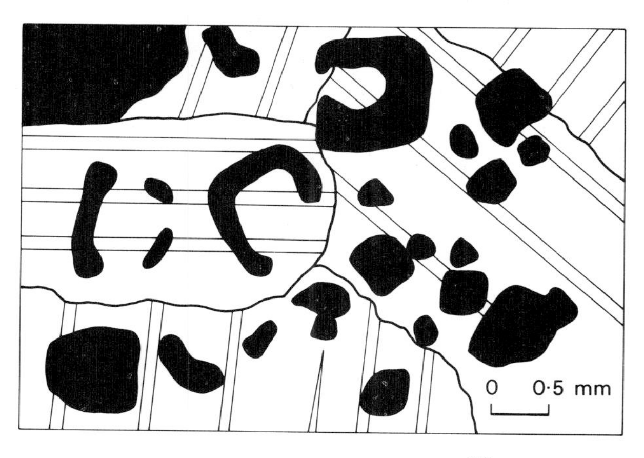

Fig. 3.1. Chromite grains in anorthosite, Bushveld Complex, South Africa. The chromites are euhedral crystals which have undergone partial resorption, producing rounded grains of various shapes, including atoll texture.

those developed in monomineralic bands (Figs. 3.2, 11.2) may, during the cooling of a large parent intrusion, suffer auto-annealing and develop foam texture (see Chapter 21).

When oxide and silicate minerals crystallize simultaneously, xenomorphic to hypidiomorphic textures similar to those of granitic rocks develop, due to mutual interference during the growth of the grains of all the minerals. A minor development of micrographic textures involving oxide ore minerals may also occur at this stage.

Sulphides, because of their lower melting points, crystallize after associated silicates and, if they have not segregated from the silicates, they will either be present as rounded grain aggregates representing frozen globules of immiscible sulphide liquid (Fig. 4.1), or as anhedral grains or grain aggregates which have crystallized interstitially to the silicates and whose shapes are governed by those of the enclosing silicate grains.

Fig. 3.2. Chromite bands in anorthosite, Dwars River Bridge, Bushveld Complex, South Africa. The central band is 1.3 cm thick at the right-hand end.

Open spaces such as dilatant zones along faults, solution channels in areas of karst topography, etc., may be permeated by mineralizing solutions. If the prevailing physico-chemical conditions induce precipitation then crystals will form. These will grow as the result of spontaneous nucleation within the solution, or, more commonly, by nucleation on the enclosing surface. This leads to the precipitation and outward growth of the first formed minerals on vein walls. If the solutions change in composition, there may be a change in mineralogy and crusts of minerals of different composition may give the vein filling a banded appearance (Fig. 3.3), called crustiform banding. Its development in some veins demonstrates that mineralizing solutions may change in composition with time and shows us the order in which the minerals were precipitated, this order being called the paragenetic sequence.

In an example of simple opening and filling of a fissure as depicted in Fig. 3.3 the banding is symmetrical about the centre of the vein. With repeated opening and mineralization this symmetry will of course be disturbed, but it may still be present among the constituents deposited after the last phase of opening.

Open space deposition also occurs at the surface at sediment-water or rock-water interfaces during, for example, the formation of volcanic-associated massive sulphide deposits. Under such situations rapid flocculation of material occurs and a common primary texture which results is colloform banding. This is a very fine scale banding involving one or more sulphides very like the banding in agate. Some workers believe that it results from colloidal deposition and

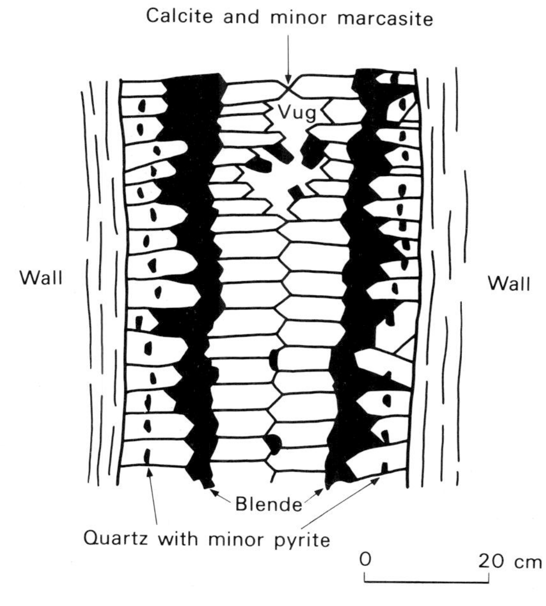

Fig. 3.3. Section across a vein showing crustiform banding.

that the banding is analogous to the Liesegang rings formed in some gels. This banding is very susceptible to destruction by recrystallization and may be partially or wholly destroyed by diagenesis or low grade metamorphism, producing a granular ore. The textures of sedimentary iron and manganese ores are discussed in Chapter 19.

Replacement

Edwards (1952) defined replacement as 'the dissolving of one mineral and the simultaneous deposition of another mineral in its place, without the intervening development of appreciable open spaces, and commonly without a change of volume'. Replacement has been an important process in the formation of many ore deposits, particularly the pyrometasomatic class. This process involves not only the minerals of the country rocks, but also the ore and gangue minerals. Nearly all ores, including those developed in open spaces, show some evidence of the occurrence of replacement processes.

The most compelling evidence of replacement is pseudomorphism. Pseudomorphs of cassiterite after orthoclase have been recorded from Cornwall, England, and of pyrrhotite after hornblende from Sullivan, British Columbia. Numerous other examples are known. The preservation of delicate plant cells by marcasite is well known and crinoid ossicles replaced by cassiterite occur in New South Wales. An overall view of ore deposits suggests that there is no limit to the direction of metasomatism. Given the right conditions, any mineral may replace any other mineral, although natural processes often make for unilateral reactions. Secondary (supergene) replacement processes, leading to sulphide enrichment by downward percolating meteoric waters, are sometimes most dramatic and fraught with economic importance. They can be every bit as important as primary (hypogene) replacement brought about by solutions emanating from crustal or deeper sources.

Fluid inclusions

The growth of crystals is never perfect and as a result samples of the fluid in which the crystals grew may be trapped in tiny cavities usually $< 100\ \mu m$ in size. These are called fluid inclusions and can be divided into various types. *Primary* inclusions formed during the growth of crystals provide us with samples of the ore-forming fluid. They also yield crucial geothermometric data and tell us something about the physical state of the fluid, e.g. whether it was boiling at the time of entrapment. They are common in all rocks and veins.

The principal matter in most fluid inclusions is water. Second in abundance is carbon dioxide. The commonest inclusions in ore deposits fall into four groups (Nash 1976). Type I, moderate salinity inclusions, are generally two phase, consisting principally of water and a small bubble of water vapour which forms 10–40% of the inclusion (Fig. 3.4). The presence of the bubble indicates trapping at an elevated temperature with formation of the bubble on

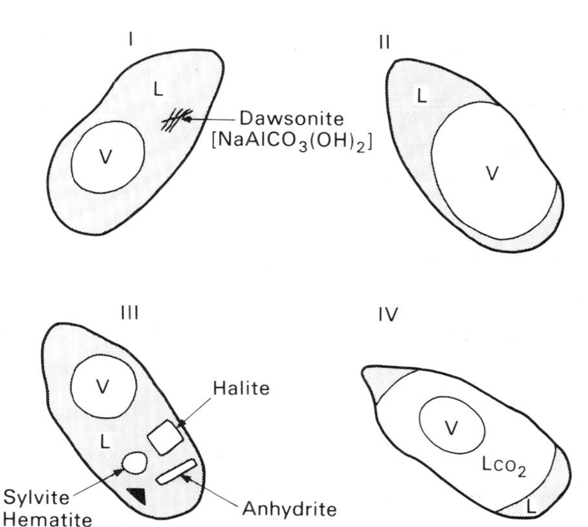

Fig. 3.4. Sketches of four important types of fluid inclusions. L = aqueous liquid, V = vapour, L_{CO_2} = liquid CO_2. (After Nash 1976). For explanation see text.

cooling. Heating on a microscope stage causes rehomogenization to one liquid phase and the homogenization temperature indicates the temperature of growth of that part of the containing crystal (provided the necessary pressure correction can be made). Sodium, potassium, calcium and chlorine occur in solution and salinities range from 0–23 wt % NaCl equivalent. In some of these inclusions small amounts of daughter salts have been precipitated during cooling, among them carbonates and anhydrite.

Type II, gas-rich inclusions, generally contain more than 60% vapour. Again they are dominantly aqueous but CO_2 may be present in small amounts. They often appear to represent trapped steam. The simultaneous presence of gas-rich and gas-poor aqueous inclusions is good evidence that the fluids were boiling at the time of trapping. Type III, halite-bearing inclusions, have salinities ranging up to more than 50%. They contain well-formed, cubic halite crystals and generally several other daughter minerals, particularly sylvite and anhydrite. Type IV, CO_2-rich inclusions, have CO_2:H_2O ratios ranging from 3 to over 30 mol %. They grade into type II inclusions and indeed there is a general gradation in many situations, e.g. porphyry copper deposits (Chivas & Wilkins 1977), between the common types of fluid inclusion.

Perhaps one of the most surprising results of fluid inclusion studies is the evidence of the common occurrence of exceedingly strong brines in nature, brines more concentrated than any now found at the surface (Roedder 1972). These are not only present in mineral deposits but are common in igneous and metamorphic rocks. Many, but not all, strong brine inclusions are secondary features connected with late-stage magmatic and metamorphic phenomena such as the genesis of greisens, pegmatites and ore deposits as well as wall rock alteration processes such as sericitization and chloritization. They are compelling evidence that many ore-forming fluids are hot, saline aqueous solutions. They form a link between laboratory and field studies, and it should be noted that there is strong experimental and thermodynamic evidence which shows that chloride in hydrothermal solutions is a potent solvent for metals through the formation of metal-chloride complex ions, and indeed inclusions that carry more than 1% weight of precipitated sulphides are known. An excellent up-to-date review of fluid inclusion studies, including those on mineral deposits, is that by Roedder (1984).

Wall rock alteration

Frequently alongside veins or around irregularly shaped orebodies we find alteration of the country rocks. This may take the form of colour, textural, mineralogical or chemical changes, or any combination of these. Alteration is not always present, but when it is, it may vary from minor colour changes to extensive mineralogical transformations and complete recrystallization. Generally speaking, the higher the temperature of deposition of the ore minerals the more intense is the alteration, but it is not necessarily more widespread. This alteration, which shows a spatial and usually a close temporal relationship to ore deposits, is called wall rock alteration.

The areal extent of the alteration can vary considerably, sometimes being limited to a few centimetres on either side of a vein, at other times forming a thick halo around an orebody and then, since it widens the drilling target, it may be of considerable exploration value. Hoeve (1984) estimated that the drilling targets in the uranium field of the Athabaska Basin in Saskatchewan are enlarged by a factor of ten to twenty times by the wall rock alteration. The spatial and temporal relationships suggest that wall rock alteration is due to reactions caused by the mineralizing fluid permeating parts of the wall rocks. Many alteration haloes show a zonation of mineral assemblages resulting from the changing nature of the hydrothermal solution as it passes through the wall rocks. The associated orebodies in certain deposits (e.g. porphyry coppers, Chapter 15) may show a special spatial relationship to the zoning, knowledge of which may be invaluable in probing for the deposit with a diamond drill.

The different wall rock alteration mineral assemblages can be compared with metamorphic facies as, like these, they are formed in response to various pressure, temperature and compositional changes. They are not, however, generally referred to as alteration facies but as types of wall rock alteration. It can be very difficult in some areas to distinguish wall rock alteration, such as chloritization, from the effects of low-grade regional or contact metamorphism, but it is, however, essential that the two effects are separately identified, otherwise a considerable degree of exploration effort may be expended in vain.

There are two main divisions of wall rock alteration: hypogene and supergene. Hypogene alteration is caused by ascending hydrothermal solutions, and supergene alteration by descending meteoric water reacting with previously mineralized ground. A third mechanism giving rise to the formation of wall rock alteration is the metamorphism of sulphide orebodies. In this chapter we will be concerned mainly with hypogene alteration.

The study of hydrothermal fluids has shown that they are commonly weakly acidic, but may become neutral or slightly alkaline by reaction with wall rocks (or by mixing with other waters, e.g. ground water). The solutions contain dissolved ions that are important in ion exchange reactions, and the composition of a particular hydrothermal solution will have an important bearing on the nature of the wall rock alteration it may give rise to. Since the chemistry of wall rocks can also vary greatly according to their petrography, it is clear that predictions as to the course of wall rock alteration reactions are fraught with difficulties. Nevertheless, there is, despite a variety of controls, a considerable uniformity in the types of wall rock alteration which facilitates their study and classification. The controls of wall rock alteration fall into two groups governed respectively by the nature of the host rocks and the nature of the ore-forming solution. Besides the chemistry of the host rocks, other factors of importance are their grain size, physical state (e.g. sheared or unsheared) and permeability, and for the hydrothermal solution important properties are the pressure, temperature, chemistry, pH and Eh.

Although most rock-forming minerals are susceptible to attack by acid solution, carbonates, zeolites, feldspathoids and calcic plagioclase are least resistant; pyroxenes, amphiboles and biotite are moderately resistant, and sodic plagioclase, potash feldspar and muscovite are strongly resistant. Quartz is often entirely unaffected.

Modern geothermal fields provide a natural laboratory for the study of wall rock alteration particularly at temperatures up to about 300°C. Boreholes in such fields yield samples of the aqueous solutions responsible for the alteration so that we can study their chemistry, measure their physical properties and determine the mineral assemblages in the wall rocks. Studies of this sort are described by Capuano & Cole (1982), Cavarretta *et al.* (1982), Henley & Ellis (1983) and many other workers. Such investigations have given us *inter alia* data on the temperature ranges over which alteration minerals have been observed. Some of this data is shown in Fig. 3.5 for a selection of minerals commonly developed in the alteration zones around orebodies. The actual presence or absence of a particular mineral depends on a number of other factors such as CO_2 and H_2S activities, solution pH, etc.

Studies of wall rock alteration are important because they (a) contribute to our knowledge of the nature and evolution of ore-forming solutions, (b) are often valuable in exploration, and (c) produce minerals such as phyllosilicates which can be used to obtain radiometric dates on the wall rock alteration and, by inference, on the associated mineralization.

TYPES OF WALL ROCK ALTERATION

These have been extensively described by Meyer & Hemley (1967) and by Rose & Burt (1979) from whose work much of the following is drawn.

(a) *Advanced argillic alteration.* This alteration is characterized by dickite, kaolinite [both $Al_2Si_2O_5(OH)_4$], pyrophyllite [$Al_2Si_4O_{10}(OH)_2$] and quartz.

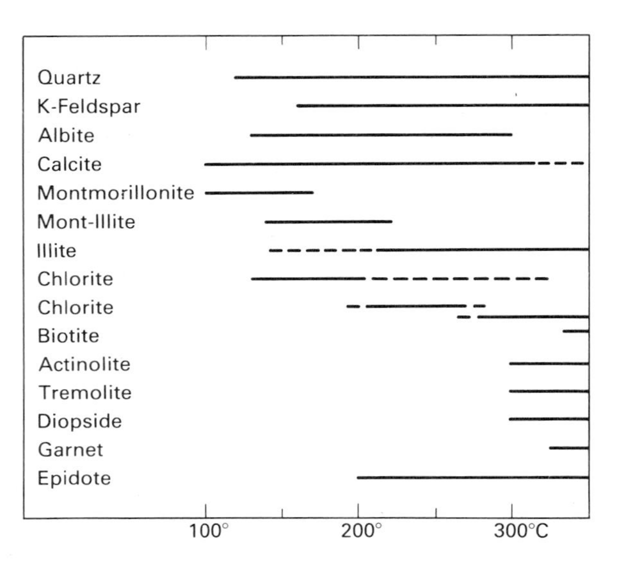

Fig. 3.5. Generalized summary of the temperature ranges over which some alumino-silicate alteration minerals have been observed in geothermal systems. (Solid lines give an indication of the most commonly observed temperature ranges). The three chlorite stability ranges indicate the transition from swelling through mixed layer to non-swelling chlorite with increasing temperature. (After Henley & Ellis 1983).

Sericite is usually present and frequently alunite, pyrite, tourmaline, topaz, zunyite and amorphous clay minerals. This is one of the more intense forms of alteration, often present as an inner zone adjoining many base metal vein or pipe deposits associated with acid plutonic stocks as at Butte, Montana, and Cerro de Pasco in Peru. It is also found in hot spring environments and in telescoped shallow precious metal deposits. The associated sulphides of the orebodies are generally sulphur-rich; covellite, digenite, pyrite and enargite are most common.

This alteration involves extreme leaching of bases (alkalies and calcium) from all aluminous phases such as feldspars and micas, but is only present if aluminium is not appreciably mobilized. When aluminium is also removed it grades into silicification and, with increasing sericite, it grades outwards into sericitization. The generation of advanced argillic alteration can be very important in developing the high permeability necessary for the circulation of enormous quantities of hydrothermal fluids and vein growth (Brimhall & Ghiorso 1983).

(b) *Sericitization.* In orefields the world over this is one of the commonest types of alteration in aluminium-rich rocks such as slates, granites, etc. The dominant minerals are sericite and quartz, pyrite often accompanying them. Care should be taken to ensure that the sericite is muscovite because illite, paragonite, phlogopite, talc and pyrophyllite can be mistaken for sericite. Muscovite is stable over a wide pressure-temperature field and this accounts for its common occurrence as an alteration mineral. If potassium is introduced into the wall rocks then rocks low in this element, such as diorites, can be sericitized. The reader must not assume that during this and other wall rock alteration processes the wall rocks necessarily become solid sericite or clay minerals, as the case may be. What we see is the appearance in significant amounts, or an increase in quantity, of the mineral or minerals concerned. Sometimes the new mineral(s) will develop, to the exclusion of all other minerals, but this is not necessarily the case. During the sericitization of granite, the feldspars and micas may be transformed to sericite, with secondary quartz as a reaction by-product, but the primary quartz may be largely unaffected except for the development of secondary fluid inclusions. Wall rock alteration is progressive, with some minerals reacting and being altered more rapidly than others. If shearing accompanies this, or any of the other types of alteration carrying phyllosilicates, then a schistose rock may result, otherwise hornfelsic-like textures develop.

With the appearance of secondary potash feldspar and/or secondary biotite, sericitization grades into potassium silicate alteration which is very common in the central deeper portions of porphyry copper deposits as at Bingham Canyon, Utah. In fluorine-rich environments, topaz together with zunyite and quartz may accompany the sericite to form greisen. Outside the sericitization zone lower grade intermediate argillic alteration may occur. Thus sericitization may grade into three types of higher grade alteration and one of lower grade.

(c) *Intermediate argillic alteration*. The principal minerals are now kaolin- and montmorillonite-group minerals occurring mainly as alteration products of plagioclase. The intermediate argillic zone may itself be zoned with montmorillonite minerals dominant near the outer fringe of alteration and kaolin minerals nearer the sericitic zone. Sulphides are generally unimportant. Outwards from the intermediate argillic alteration zone propylitic alteration may be present before fresh rock is reached.

(d) *Propylitic alteration*. This is a complex alteration generally characterized by chlorite, epidote, albite and carbonate (calcite, dolomite or ankerite). Minor sericite, pyrite and magnetite may be present, less commonly zeolites and montmorillonites. The term propylitic alteration was first used by Becker in 1882 for the alteration of diorite and andesite beside the Comstock Lode, Nevada (a big gold-silver producer in the boom days of the last century). Here the main alteration products are epidote, chlorite and albite. The propylitic alteration zone is often very wide and therefore, when present, is a useful guide in mineral exploration. For example at Telluride, Colorado, narrow sericite zones along the veins are succeeded outwards by a wide zone of propylitization.

With the intense development of one of the main propylitic minerals we have what are sometimes considered as subdivisions of propylitization: chloritization, albitization and carbonatization. Albitization will be dealt with under feldspathization.

(e) *Chloritization*. Chlorite may be present alone or with quartz or tourmaline in very simple assemblages; however, other propylitic minerals are usually present, and anhydrite may also be in evidence. Hydrothermal chlorites often show a change in their Fe:Mg ratio with distance from the orebody, usually being richer in iron adjacent to the sulphides, although the reverse has been reported. The change in this ratio can be recorded by simple refractive index measurements so that this offers the possibility of a cheap exploration tool.

The development of secondary chlorite may result from the alteration of mafic minerals already present in the country rocks or from the introduction of magnesium and iron, though of course both processes can occur together, as at Ajo, Arizona. Chloritization is common alongside tin veins in Cornwall where progressive alteration of the country rocks occurs:

Unaltered porphyritic granite	Pinking of feldspars	Chlorite in groundmass developed from biotite and small feldspars	Rim, core and cleavage replacement of phenocrysts	Quartz-chlorite rock carrying tin values	Quartz vein with tin

(f) *Carbonatization.* Dolomitization is a common accompaniment of low to medium temperature ore deposition in limestones, and dolomite is probably the commonest of the carbonates formed by hydrothermal activity. Dolomitization is most commonly associated with low temperature lead-zinc deposits of 'Mississippi Valley-type'. These can have wall rocks of pure dolomite; usually this rock is coarser and lighter in colour than the surrounding limestone. Dolomitization generally appears to have preceded sulphide deposition, but the relationship and timing have been much debated in the various fields where dolomitization occurs. Nevertheless, dolomitization often appears to have preceded mineralization, to have increased the permeability of the host limestones and thus to have prepared them for mineralization. Of course not all dolomitized limestone contains ore and sometimes in such fields ore may occur in unaltered limestone. Often the limestone is then recrystallized to a coarse white calcite rock, as at Bisbee, Arizona.

Other carbonates may be developed in silicate rocks, especially where iron is available and ankerite may then be common, particularly in the calcium-iron environment of carbonatized basic igneous rocks and volcaniclastics. This is particularly the case with many Precambrian and Phanerozoic vein gold deposits, for example the Mother Lode in California where ankerite, sericite, albite, quartz, pyrite and arsenopyrite are well developed in the altered wall rocks. At Larder Lake in Ontario, dolomite and ankerite have replaced large masses of greenstone. As with the chlorites, there may be a chemical variation in the Fe:Mg ratio with proximity to ore.

Most of the alteration types we have dealt with so far have involved hydrolysis, i.e. the introduction of hydrogen ion for the formation of hydroxyl-bearing minerals such as micas and chlorite, often accompanied by the removal of bases (K^+, Na^+, Ca^{2+}). Thus the H^+/OH^- ratio of the mineralizing solutions will decrease and concomitantly they will be enriched in bases. The chemical significance of carbonatization may take two different forms. Dolomitization of limestone involves very extensive magnesium metasomatism, but only as a base exchange process. This is called cation metasomatism and in this case the mineralizing (or pre-mineralization) solutions must have carried

abundant magnesium. Carbonatization of silicate rocks involves *inter alia* anion metasomatism with the introduction of CO_3^{2-} rather than bases.
(g) *Potassium silicate alteration*. Secondary potash feldspar and/or biotite are the essential minerals of this alteration. Clay minerals are absent but minor chlorite may be present. Anhydrite is often important especially in porphyry copper deposits, e.g. El Salvador, Chile, and it can form up to 15% of the altered rock. It is, however, easily hydrated and removed in solution, even at depths of up to 1000 m, so that it has probably been removed from many deposits by groundwater solution. Magnetite and hematite may be present and the common sulphides are pyrite, molybdenite and chalcopyrite, i.e. there is an intermediate sulphur:metal ratio.
(h) *Silicification*. This involves an increase in the proportion of quartz or crypto-crystalline silica (i.e. cherty or opaline silica) in the altered rock. The silica may be introduced from the hydrothermal solutions, as in the case of chertified limestones associated with lead-zinc-fluorite-baryte deposits, or it may be the by-product of the alteration of feldspars and other minerals during the leaching of bases. Silicification is often a good guide to ore, e.g. the Black Hills, Dakota. At the Climax porphyry molybdenum deposit in Colorado, intensive and widespread silicification accompanied the mineralization.

The silication of carbonate rocks leading to the development of skarn is dealt with under pyrometasomatic (skarn) deposits (see Chapters 4 and 14).
(i) *Feldspathization*. This leads to the development of either potash feldspar or albite. Secondary orthoclase or microcline results from the introduction of potassium as in the deeper zones of porphyry copper deposits. Albitization, on the other hand, may result from the introduction of sodium or from removal of calcium from plagioclase-bearing rocks. Albitization is found adjacent to some gold deposits, often replacing potash feldspar, e.g. Treadwell, Alaska.
(j) *Tourmalinization*. This is associated with medium to high temperature deposits, e.g. many tin and some gold veins have a strong development of tourmaline in the wall rocks and often actually in the veins as well. The Sigma Gold Mine in Quebec has veins which in places are massive tourmaline and this mineral is well developed in the adjacent wall rock. This is also the case in the granodiorite wall rock of the Siscoe Mine, Quebec. At Llallagua, Bolivia, the world's largest primary tin mine, the porphyry host is altered to a quartz-sericite-tourmaline rock.

If the altered country rocks are lime-rich then axinite rather than tourmaline may be formed.
(k) *Other alteration types*. There are many other types of alteration; among these may be mentioned *alunitization*, which may be of either hypogene or supergene origin; *pyritization*, due to the introduction of sulphur which may attack both iron oxides and mafic minerals; *hematitization*, an alteration type often associated with uranium (particularly pitchblende deposits); *bleaching*, due in many cases to the reduction of hematite; *greisenization*, a frequent form of alteration alongside tin-tungsten and beryllium deposits in granitic rocks or gneisses; *fenitization*, which is associated with carbonatite hosted deposits and

which is characterized by the development of nepheline, aegirine, sodic amphiboles and alkali feldspars in the aureoles of the carbonatite masses; *serpentinization,* and the allied development of talc, can occur in both ultrabasic rocks and limestones; it is associated with some gold and nickel deposits but where serpentine and talc are developed in limestones there is generally an introduction of SiO_2 and H_2O and frequently some Mg; finally *zeolitization* is marked by the development of stilbite, natrolite, heulandite, etc., and often accompanies native copper mineralization in amygdaloidal basalts — calcite, prehnite, pectolite, apophyllite and datolite are generally also present.

INFLUENCE OF ORIGINAL ROCK-TYPES

A survey of world literature on the subject shows that wall rock alteration exhibits a certain regularity with respect to the nature of the host rock (Boyle 1970). There are of course exceptions, but certain generalizations can be made. Thus, for example, the most prevalent types of alteration in acidic rocks are sericitization, argillization, silicification and pyritization. Intermediate and basic rocks generally show chloritization, carbonatization, sericitization, pyritization and propylitization. In carbonate rocks the principal high temperature alteration is skarnification, whereas normal shales, slates and schists are frequently characterized by tourmalinization, especially when hosting tin and tungsten deposits. A more detailed discussion of this subject can be found in Boyle's work.

CORRELATION WITH TYPE OF MINERALIZATION

Boyle has also shown that certain types of mineralization tend to be accompanied by characteristic types of alteration but space only permits the mention here of a few examples. Red bed uranium, vanadium, copper, lead and silver deposits are generally accompanied by bleaching. Vein deposits of native silver are usually characterized by carbonatization and chloritization and molybdenum-bearing veins by silicification and sericitization. Other examples have been given above.

TIMING OF WALL ROCK ALTERATION

The problem of disentangling age relationships between different assemblages of wall rock alteration minerals in a given deposit can be extremely difficult. Their presence, together with evidence such as crustiform banding, has suggested to many workers that the mineralizing solutions came in pulses of different compositions (polyascendant solutions). In some deposits, however, such as Butte, Montana, there is a considerable body of data suggesting a single long-lasting phase of mineralization and accompanying alteration during the formation of the main stage veins. Both mechanisms have probably taken place and some deposits may have been formed from monoascendant solutions whilst others are the result of polyascendant mineralization. This important aspect is discussed at length by Meyer & Hemley (1967).

THE NATURE OF ORE-FORMING SOLUTIONS AS DEDUCED FROM WALL ROCK ALTERATION

Studies of wall rock alteration indicate that aqueous solutions played a large part in the formation of epigenetic deposits. Clearly, in some cases these solutions carried other volatiles such as CO_2, S, B and F. The pH of ore-forming solutions is difficult to assess from wall rock alteration studies, but in cases of hydrogen metasomatism the pH value must have been low; it would have increased in value during reactions with the wall rocks, so that in some cases solutions may have become neutral or even slightly alkaline. A very instructive study of the use of wall rock alteration studies in throwing a light on the nature of ore-forming solutions is to be found in Fournier (1967).

4

Some Major Theories of Ore Genesis

Some indication has been given in Chapter 2 of the very varied nature of ore deposits and their occurrence. This variety of form has given rise over the last hundred years or more to an equally great variety of hypotheses of ore genesis. The history of the evolution of these ideas is an interesting study in itself and one which has been well documented by Stanton (1972). There is no room for such a discussion here and the reader is referred to Stanton's work and to references given by him. This chapter will, therefore, be concerned only with major theories of ore genesis current at the present time and these will be divided for the sake of convenience into internal and surface processes. The reader should be warned, however, that very often several processes contribute to the formation of an orebody. Thus, where we have rising hot aqueous solutions forming an epigenetic stockwork deposit just below the surface and passing on upwards through it to form a contiguous syngenetic deposit under, say, marine conditions, even the above simple classification is in difficulties. This is the reason why ore geologists, besides producing a plethora of ore genesis theories, have also created a plethora of orebody classifications! A summary of the principal theories of ore genesis is given in Table 4.1.

Origin due to internal processes

MAGMATIC SEGREGATION

The terms magmatic segregation deposit or orthomagmatic deposit are used for those ore deposits, apart from pegmatites, that have crystallized direct from a magma. Those formed by fractional crystallization are usually found in plutonic igneous rocks; those produced by liquation (separation into immiscible liquids) may be found associated with both plutonic and volcanic rocks. Magmatic segregation deposits may consist of layers within or beneath the rock mass (chromite layers, subjacent copper-nickel sulphide ores).

(a) *Fractional crystallization.* This includes any mechanical process by which early formed crystals are prevented from equilibrating with the melt from which they grew. The important processes until recently were thought to be gravity fractionation, flowage differentiation, filter pressing and dilatation (Carmichael *et al.* 1974) but the simple hypotheses of separation and mechanical sorting by magmatic currents invoked by various workers as being at least partly responsible for stratiform chromite accumulations (e.g. Cameron &

Table 4.1. Simple classification of ore genesis theories.

Theory	Nature of process	Typical deposits
ORIGIN DUE TO INTERNAL PROCESSES		
Magmatic segregation	Separation of ore minerals by fractional crystallization and related processes during magmatic differentiation	Chromite layers in the Great Dyke of Rhodesia and the Bushveld Complex, South Africa
	Liquation, liquid immiscibility. Settling out from magmas of sulphide, sulphide-oxide or oxide melts which accumulated beneath the silicates or were injected into wall rocks or in rare cases erupted on the surface	Copper-nickel orebodies of Sudbury, Canada; Pechenga, USSR and the Yilgarn Block, Western Australia
Pegmatitic deposition	Crystallization as disseminated grains or segregations in pegmatites	Lithium-tin-caesium pegmatites of Bikita, Zimbabwe. Uranium pegmatites of Bancroft, Canada, Rössing, Namibia
Hydrothermal	Deposition from hot aqueous solutions which may have had a magmatic, metamorphic, surface or other source	Tin-tungsten-copper veins and stockworks of Cornwall, UK. Molybdenum stockworks of Climax, USA. Porphyry copper deposits of Panguna, PNG and Bingham, USA
Lateral secretion	Diffusion of ore- and gangue-forming materials from the country rocks into faults and other structures	Yellowknife gold deposits, Canada. Mother Lode, USA
Metamorphic processes	Pyrometasomatic (skarn) deposits formed by replacement of wall rocks adjacent to an intrusion	Copper deposits of Mackay, USA and Craigmont, Canada. Magnetite bodies of Iron Springs, USA
	Initial or further concentration of ore elements by metamorphic processes, e.g. granitization, alteration processes	Some gold veins, and disseminated nickel deposits in ultramafic dykes
ORIGIN DUE TO SURFACE PROCESSES		
Mechanical accumulation	Concentration of heavy durable minerals into placer deposits	Rutile-zircon sands of New South Wales, Australia, and Trail Ridge, USA. Tin placers of Malaysia. Gold placers of the Yukon, Canada
Sedimentary precipitates	Precipitation of particular elements in suitable sedimentary environments, with or without the intervention of biological organisms	Banded iron formations of the Precambrian shields. Manganese deposits of Chiaturi, USSR
Residual processes	Leaching from rocks of soluble elements leaving concentrations of insoluble elements in the remaining material	Nickel laterites of New Caledonia. Bauxites of Hungary, France, Jamaica and Arkansas, USA
Secondary or supergene enrichment	Leaching of valuable elements from the upper parts of mineral deposits and their precipitation at depth to produce higher concentrations	Many gold and silver bonanzas. The upper parts of a number of porphyry copper deposits
Volcanic-exhalative (=sedimentary exhalative)	Exhalations of hydrothermal solutions at the surface, usually under marine conditions and generally producing stratiform orebodies	Meggan, Germany; Sullivan, Canada; Mount Isa, Australia; Rio Tinto, Spain; Kuroko deposits of Japan; black smoker deposits of modern oceans

Emerson 1959, Irvine & Smith 1969), were questioned as long ago as 1961 by Jackson, following his work on layering in the Stillwater Complex. The critical evidence and the new explanations of deposition from density currents and *in situ* bottom crystallization were succinctly summarized by Best (1982). The second of these two processes is favoured by Eales & Reynolds (1985) to explain evidence from the Bushveld Complex.

Whatever the formative processes may be, their products are the rocks called cumulates, which often display conspicuous lithological alternations called rhythmic layering, due to their frequent repetition in vertical sections of the plutonic bodies in which they occur. Usually, olivine-, pyroxene-, or plagioclase-rich layers are formed. However, when oxides such as chromite are precipitated, layers of this mineral may develop, as in the Bushveld Complex of South Africa. This enormous layered intrusion is characterized by cumulus magnetite in the upper zone. The chromite layers have been mined for decades, the magnetite now being exploited for its high vanadium content. Another mineral which may be concentrated in this way is ilmenite. Whilst chromite accumulations are nearly all in ultrabasic rocks and to a lesser extent in gabbroic or noritic rocks, ilmenite accumulations show an association with anorthosites or anorthositic gabbros. These striking rock associations are strong evidence for the magmatic origin of the minerals.

(b) *Liquation*. A different form of segregation results from liquid immiscibility. In exactly the same way that oil and water will not mix but form immiscible globules of one within the other, so in a mixed sulphide-silicate magma the two liquids will tend to segregate. Sulphide droplets separate out and coalesce to form globules which, being denser than the magma, sink through it to accumulate at the base of the intrusion or lava flow (Fig. 4.1). Iron sulphide is the principal constituent of these droplets which are associated with basic and ultrabasic rocks, because sulphur and iron are both more abundant in these rocks than in acid or intermediate rocks. Chalcophile elements such as copper and nickel also enter ('partition into' is the pundits' phrase) these droplets and sometimes the platinum group metals.

A basic or ultrabasic magma is generated by partial melting in the mantle and it may acquire its sulphur at this time, or later by assimilation in the crust. For the formation of an ore deposit the timing of the liquation is critical (Naldrett *et al*. 1984). If it is too early then the sulphides may settle out in the mantle or the lower crust; if it is late then the crystallization of silicates may be in full swing and they will dilute any sulphide accumulations. The processes that can promote sulphide immiscibility are: cooling, silication (increase of silica content by assimilation), sulphur assimilation and magma mixing.

The accumulation of Fe-Ni-Cu sulphide droplets beneath the silicate fraction can produce massive sulphide orebodies. These are overlain by a zone with subordinate silicates enclosed in a network of sulphides — net-textured ore, sometimes called disseminated ore. This zone is, in turn, overlain by one of weak mineralization which grades up into overlying peridotite, gabbro or komatiite, depending on the nature of the associated silicate fraction. To

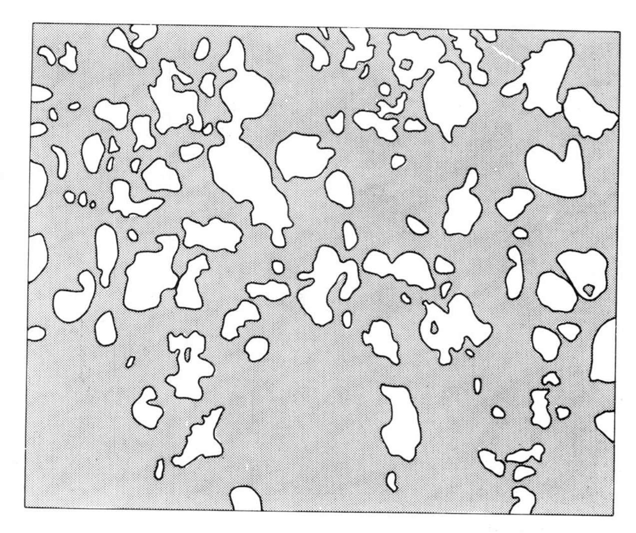

Fig. 4.1. Tracing of an ore specimen from Sudbury, Ontario. Sulphides, mainly pyrrhotite with minor pentlandite and chalcopyrite, are shown white; surrounding silicates are shown grey. Note the rounded discontinuous nature of the sulphide globules. They appear to have formed as a result of liquid immiscibility from a silicate-sulphide melt. Note especially the rounded silicate blebs within the sulphide bodies, and that many of the sulphide globules appear to have formed from the coalescence of smaller bodies of sulphide liquid.

explain the mechanism of formation of these zones Naldrett (1973) proposed his 'billiard ball' model (Fig. 4.2).

Imagine a large beaker partly filled with billiard balls and water (Fig. 4.2A). These represent olivine grains and interstitial silicate liquid. Then consider the effect of adding mercury to represent the immiscible sulphide liquid. This will sink to the bottom and the balls will tend to float on it with the lower balls being forced down into the mercury by the weight of the overlying ones. If the contents of the beaker are frozen before all the mercury has had a chance to percolate to the bottom then the situation shown in Fig. 4.2B will exist. There is an obvious analogy between the massive mercury and the zone of massive sulphides, between the overlying zone of balls immersed in mercury and net-textured ore, and between the zone of scattered globules of mercury and the zone of weak mineralization.

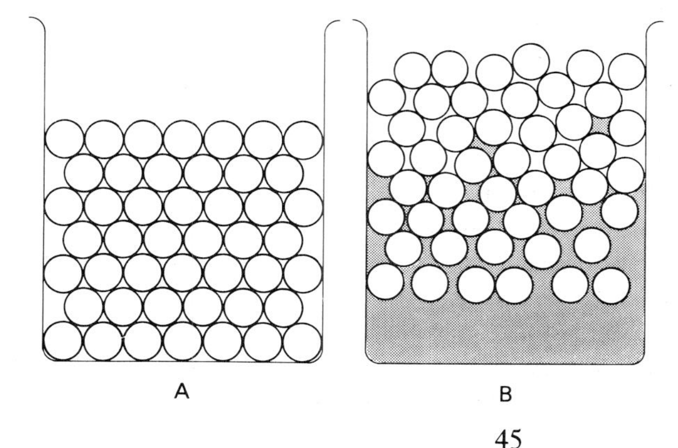

Fig. 4.2. The billiard ball model illustrating the formation of Fe-Ni-Cu sulphide deposits by liquation. For explanation see text. (After Naldrett 1973).

Hot aqueous solutions have played a part in the formation of many different types of mineral and ore deposit, for example veins, stockworks of various types, volcanic-exhalative deposits and others. Such fluids are usually called hydrothermal solutions and many lines of evidence attest to their important role as mineralizers. The evidence from wall rock alteration and fluid inclusions has been discussed in Chapter 3. Homogenization of fluid inclusions in minerals from hydrothermal deposits and other geothermometers has shown that the depositional range for all types of deposit is approximately 50–650°C. Analysis of the fluid has shown water to be the common phase and usually it has salinities far higher than that of sea water. Hydrothermal solutions are believed to be capable of carrying a wide variety of materials and of depositing these to form minerals as diverse as gold and muscovite showing that the physical chemistry of such solutions is complex and very difficult to imitate in the laboratory. Our knowledge of their properties and behaviour is still somewhat hazy and there are many ideas about the origin of such solutions and the materials they carry (see relevant chapters in Barnes 1979a)

It should always be remembered that hydrothermal ore deposits are small compared with most geological features — the largest are only a few km^3 in volume. At first sight they may appear to be random accidents with little or no control over the position or geological environment in which they occur within the crust. Nevertheless deposits can be classified into families and individual family members occur more frequently in some areas of the crust than others. Moreover, despite their variation in occurrence and the large number of minerals known in nature, hydrothermal deposits display a chemical consistency that is best expressed by the limited and repetitive ranges of minerals, mostly sulphides and oxides, that are found concentrated within them. This chemical consistency suggests that relatively few chemical processes are important in their genesis (Skinner 1979).

Ore geology, like most branches of geology, is what might be termed a forensic science. We study the evidence collected in the field and attempt to reconstruct the crime. But many decades of study have revealed that the evidence is always partial and open to various interpretations, which implies that the best investigated deposits are still worthy of further research and our ideas concerning their modes of genesis will be continually modified. The principal problems are the source and nature of the solutions, the sources of the metals and sulphur in them and the driving force that moved the solutions through the crust, the means of transport of these substances by the solutions, and the mechanisms of deposition.

(a) *Sources of the solutions and their contents.* As explained in the section on fluid inclusions in Chapter 3, there is much evidence that saline hydrothermal solutions are, and have been, very active and widespread in the crust. In some present day geothermal systems the circulation of hydrothermal solutions is under intensive study. Whence the water of these solutions? Data from water in mines, tunnels, drill holes, hot springs, fluid inclusions, minerals and rocks suggest that there are five sources of subsurface hydrothermal waters. The five are:

(1) surface water, including ground water, commonly referred to by geologists as meteoric water;
(2) ocean (sea) water;
(3) connate and deeply penetrating meteoric water;
(4) metamorphic water;
(5) magmatic water.

Most connate water may have been originally meteoric water but long burial in sediments and reactions with the rock minerals give it a different character.

Measurements of the relative abundances of oxygen and hydrogen isotopes give us information on the source or sources of water, but there are problems of interpretation of the data that we obtain. Both connate and metamorphic water (produced by dehydration of minerals during metamorphism) may once have been meteoric water, but subsurface rock-water reactions may change the isotopic compositions and, if these reactions are incomplete, then a range of isotopic compositions will result. Another mechanism that may produce intermediate isotopic compositions is the mixing of waters, e.g. magmatic and meteoric.

Present evidence appears to show that similar deposits can be formed from detectably different types of water and, on the other hand, that waters of at least two parentages have played an important role in the formation of some orebodies. As White (1974) has warned us, we must not try to make our mineralization models too simple but be prepared to accept diversity of sources for the water in hydrothermal solutions. We must also keep in mind the possibility of overprinting, as for example when a post-mineralization flow of meteoric water, through rock-water reactions (and the formation of new fluid inclusions) partially or wholly changes the isotopic character of the orebody and its immediate wall rocks. In many cases where an intrusion, acting as a heat engine, provided the driving force for the movement of solutions, such an overprinting may be expected to have taken place, and evidence for this should be sought (e.g. Scratch *et al.* 1984). Finally it must be said that in putting forward a mineralization model for a particular deposit or class of deposit, thought should be given to the amount of water that will be required for its formation. Can the postulated source provide a sufficient volume?

Let us now turn our thoughts to the source of the dissolved contents. Firstly, what are they? Secondly, where do we look for this information? The answer to the second question is fluid inclusions (which provide most of our data), modern geothermal systems, hot springs, and waters encountered during deep drilling operations in oilfields — oilfield brines. Some data from these sources are given in Table 4.2. They indicate that the major constituents are sodium, potassium, calcium and chlorine. Other elements and radicles are usually present in amounts less than 1000 ppm. Little can be said about the source of any of these except lead, which always contains a number of stable isotopes developed by the radioactive decay of ^{232}Th, ^{235}U and ^{238}U plus

Table 4.2. Concentrations in ppm of *some* of the elements in some modern and ancient hydrothermal solutions. The data are from Skinner (1979) in which the original literature sources are to be found. (1) Salton Sea geothermal brine (California); (2) Cheleken geothermal brine (USSR); (3) oilfield brine, 3350 m depth (Michigan); (4) fluid inclusion in fluorite (Illinois); (5) fluid inclusion in sphalerite, Creede (Colorado); (6) fluid inclusions, Core Zone, Bingham (Utah).

	Modern solutions			Ancient solutions		
Element	1	2	3	4	5	6
Cl	155 000	157 000	158 200	87 000	46 500	295 000
Na	50 400	76 140	59 500	40 400	19 700	152 000
Ca	28 000	19 708	36 400	8 600	7 500	4 400
K	17 500	409	538	3 500	3 700	67 000
Mg	54	3 080	1 730	5 600	570	—
B	390	—	—	<100	185	—
Br	120	526.5	870	—	—	—
F	15	—	—	—	—	—
NH_4	409	—	39	—	—	—
HCO_3^-	>150	31.9	—	—	—	—
H_2S	16*	0	—	—	—	—
SO_4^{2-}	5	309	310	1 200	1 600	11 000
Fe	2 290	14.0	298	—	—	8 000
Mn	1 400	46.5	—	450	690	—
Zn	540	3.0	300	10 900	1 330	—
Pb	102	9.2	80	—	—	—
Cu	8	1.4	—	9 100	140	—

*Sulphide present; all S reported as H_2S.

non-radiogenic lead. As with hydrogen and oxygen isotopes, these have been used as tracers to seek the source of the lead and these studies have yielded some interesting results. For example, hydrothermal fluids responsible for depositing lead in the Old Lead Belt of south-east Missouri appear to have leached it from a sandstone underlying the orefield (Doe & Delavaux 1972). Stacey *et al.* (1968) showed that much of the lead in some leadfields of Utah was derived from associated igneous intrusions; Marcoux (1982) demonstrated that lead in the Pentivy district of France was derived directly from the mantle, and Plimer (1985), citing lead isotopic and other evidence, also postulated a mantle source for the lead and other metals in the huge Broken Hill orebodies of New South Wales. The evidence from lead isotopic studies thus suggests that ore fluids may collect their metals from a magma, if that was their source, Alternatively, they may collect more metals from the rocks they pass through, or obtain all their metallic content form the rocks they traverse, which can contain (in trace amounts) all the metals required to form an orebody.

Our present knowledge indicates that most rocks can act as a source of geochemically scarce elements which can be leached out under suitable conditions by hydrothermal solutions. In laboratory experiments, Bischoff *et al.* (1981) have shown that heavy metals present in trace quantities will fractionate from greywacke into sea water or natural brine at 350°C. Similar experiments have revealed that basalt will also yield up much of its heavy metal content under similar conditions, and both these experiments produced sufficiently high concentrations for the solutions to be ranked as ore-forming fluids. High trace concentrations of most heavy metals in source rocks do not seem to be a *sine qua non* for the formation of ore fluids. The greywacke of the above experiments contained only 15 ppm of lead. Differences in source rocks and differences in leaching conditions will of course produce different concentrations of economic metals in the reacting solutions. Very scarce elements such as tin, mercury and silver may require pre-enrichment.

Because of the spatial relationship that exists between many hydrothermal deposits and igneous rocks, a strong school of thought holds that consolidating magmas are the source of many, if not all, hydrothermal solutions. The solutions are considered to be low temperature residual fluids left over after pegmatite crystallization, and containing the base metals and other elements which could not be accommodated in the crystal lattices of the silicate minerals precipitated by the freezing magma. This model derives not only the water, the metals and other elements from a hot body of igneous rock, but also the heat to drive the mineralization system. The solutions are assumed to move upwards along fractures and other channelways to cooler parts of the crust where deposition of minerals occurs. Laboratory (e.g. Khitarov *et al.* 1982, Manning 1984) and field studies (Strong, 1981), backed by thermodynamic considerations, show that this is a perfectly feasible process of mineralization, and it has been discussed at length by Burnham (1979). There are many recent papers in which the authors, who studied particular deposits, favoured a magmatic origin for the hydrothermal solutions that formed them; see for example Eadington (1983), Kay & Strong (1983), Norman & Trangcotchasan (1982), Lehman (1985) and Wilton (1985).

In many orefields, however, such as the Northern Pennine Orefield of England, there are no acid or intermediate plutonic intrusions which might be the source of the ores. Some workers have therefore postulated a more remote magmatic source such as the lower crust or, more frequently, magmatic processes in the mantle, whilst an important body of opinion has favoured deposition from connate solutions — that is water which was trapped in sediments during deposition and which has been driven up dip by the rise in temperature and pressure caused by deep burial. Such burial might occur in sedimentary basins, and solutions from this source are often called basinal brines. With a geothermal gradient of 1°C per 30 m, temperatures around 300°C would be reached at a depth of 9 km. Hot solutions from this source are believed to leach metals, but not necessarily sulphur, from the rocks through which they pass, ultimately precipitating them near the surface in shelf facies carbonates on the fringe of the basin, and far from any igneous intrusion. This, too, is a model favoured at present by many workers, particularly as an explanation for the genesis of low temperature, carbonate-hosted lead-zinc-fluorite-baryte deposits (Mississippi Valley-type). Useful discussions are Hanor (1979), Cathles & Smith (1983) and Sverjensky (1984). It has been suggested, however, that the available volumes of connate water are insufficient to carry the amount of metal that is present in such deposits (Duhovnik 1967). If this is a serious objection to the model as depicted in the simple terms above, then there are various hypotheses with which it may be refuted. Hsü (1984) has suggested one and applied it in particular to deposits of this nature in the southern Alps. Other workers favour a comparable flow of water, under a hydrostatic head, passing through sedimentary basins to produce the ore fluid. There is no problem of a lack of water for this model and a recent statement of it is to be found in Garven (1985).

An important paper by White in 1955 engendered a fertile field of research into geothermal systems as possible generators of orebodies and a number of epigenetic deposits are now considered by many researchers to have formed in ancient geothermal systems. Useful reading on this subject includes Ellis (1979), Weissberg *et al.* (1979) White (1981), Henley & Ellis (1983), Henley *et al.* (1984), Höll (1985) and Henley (1986). Geothermal systems form where a heat engine (usually magmatic) at depths of a few km sets deep ground waters in motion (Fig. 4.3A). These waters are usually meteoric in origin but in some systems connate or other saline waters (Salton Sea) may be present. Systems near the coast may be fed by sea water or both sea water and meteoric water (Svartsengi, Iceland). Magmatic water may be added by the heat engine, and some ancient systems appear to have been dominated by magmatic water, at least in their early stages, e.g. porphyry copper and molybdenum deposits. Dissolved constituents (see Table 4.2, columns 1 and 2) may be derived by the circulating waters from a magmatic body at depth, or from the country rocks which contain the system. These may be altered by the solutions to mineral assemblages identical with those found in some wall rock alteration zones associated with orebodies. Common sulphides, such as galena and sphalerite, occur in a number of modern systems, and at Broadlands, an amorphous Sb-As-Hg-Tl sulphide precipitate enriched to ore grade in gold and silver has been formed.

The principal features of a geothermal system are shown in Fig. 4.3A. Meteoric water sinking to several km depth (A) enters a zone of high heat flow, absorbs heat and rises into one or a succession of permeable zones (BD). There may be outflow at an appreciable rate along a path such as (C) or much slower outflow by permeation of the cap rock (mudstone, tuff, etc.). Outflow through (C) depends on the permeability of the rocks and the pressure at the top of the zone (BD). If the outflow rate does not exceed that of the inflow, an all liquid system will prevail. With a higher outflow rate, a steam phase will form in (BD) and the steam pressure will decrease until the mass outflow through (C) is reduced to equal the mass inflow. A dynamic balance than obtains with a lowered water level in the permeable horizon, boiling water and, most important from our point of view, the development of convection currents in the water.

Fig. 4.3B illustrates the structure of a geothermal system in volcanic terrane like those of the Taupo Volcanic Zone, New Zealand. Note that the hot waters are circulating through, reacting with and probably obtaining dissolved constituents from both the magmatic intrusion and the country rocks. In Figs.

4.3C and D geothermal systems are postulated to explain vein tin and copper mineralization in and adjacent to the Land's End Granite in south-west England. In Fig. 4.3E we have a broader picture, with geothermal systems being invoked to explain some of the different types of mineralization in south-west England and the zoning of metals that is one of the well known features of this orefield. A similar model, to explain the epigenetic uranium mineralization of the Variscan Metallogenic Province of western Europe, has been proposed by Cathelineau (1982). An important and detailed application of this model to epithermal, precious metal deposits is to be found in Hayba *et al.* (1986). Although hydrothermal deposits are only small geological features, fossil geothermal systems can be very large. One of the largest so far documented is the Casto Ring Zone in central Idaho, which occupies an area of over 4500 km^2 (Criss *et al.* 1984).

(b) *Means of transport.* Sulphides and other minerals have such low solubilities in pure water that it is now generally believed that the metals were transported as complex ions. A few simple figures will illustrate this. The amount of zinc in a saturated zinc sulphide solution at a pH of 5 and a temperature of 100°C (possible mineralizing conditions) is about 1×10^{-5} $g\,l^{-1}$. A small orebody containing 1 Mt of zinc could have been formed from a solution of this strength (assuming all the zinc was precipitated) provided 10^{17}l of solution. passed through the orebody. This is equal to the volume of a tank having an area of 10 000 km^2 and sides 10 km high — an impossible quantity of solution. This difficulty is further illustrated in Fig. 4.4 where the calculated lead ion (line AB) and H_2S concentrations in water in equilibrium with galena at 80°C are shown. This indicates that (in the absence of ion complexing) concentrations of lead and H_2S high enough to form an orebody can only be achieved in very acid solutions, about pH = 0–3 (Anderson 1977). These pH values are most unlikely to hold for hydrothermal solutions except those in contact with a relatively insoluble rock such as quartzite. With other rocks, hydrogen ion would be consumed by wall rock alteration reactions until a pH nearer neutral was achieved. Probable values of pH for hydrothermal solutions lie to the right of the thick vertical line, particularly when the solutions are passing through limestones. So we can rule out the possibility that pure acidified water could be the transporting medium. What we require is a mechanism of transportation which will operate in the upper right-hand portion of such a diagram.

Laboratory, thermodynamic studies and the examination of modern geothermal systems have led geochemists to conclude that metals are transported in hydrothermal solutions as complex ions, i.e. the metals are joined to complexing groups (ligands). The most important are HS^- or H_2S, Cl^- and OH^-; other liqands including organic ligands may also contribute to complexing in ore fluids Barnes (1979b). Bisulphide complexes can exist stably in near neutral solutions containing abundant H_2S. Ions such as $PbS(HS)^-$ are formed and these have much higher solubilities than pure ionic solutions. The main objection to what is a very useful and promising hypothesis is the high concentration of H_2S and HS^- required to keep the complexes stable, a concentration much higher than that usually found in hot springs, fluid inclusions and geothermal systems (Table 4.2). For this and other reasons many workers favour the idea of metal transport in chloride complexes such as $AgCl_2^-$ and $PbCl^{3-}$. It is likely however, that both, and other complexes, play a part in metal transportation. Barnes (1979b) has pointed out that available data show that the geologically improbable requirement of very alkaline solutions is

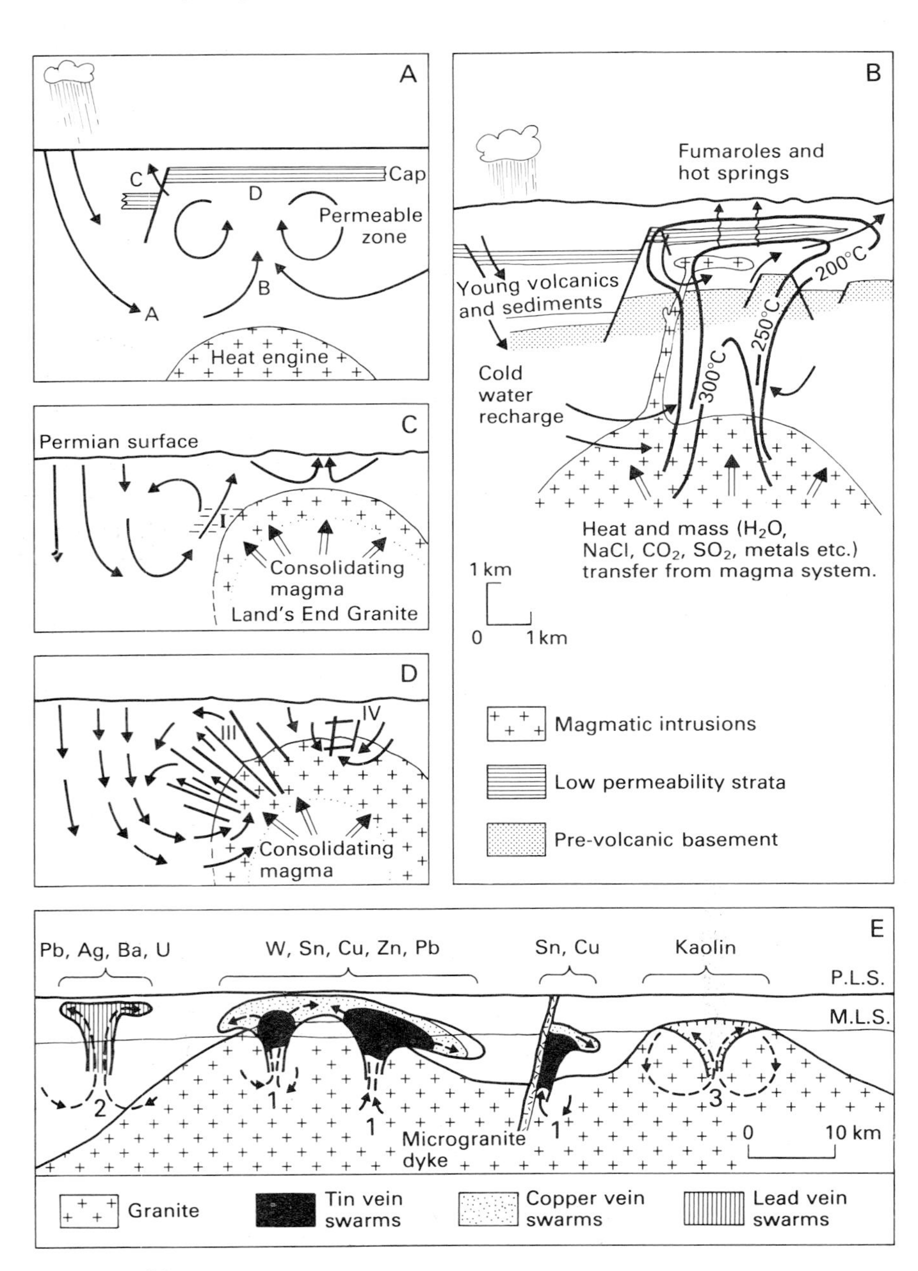

necessary above approximately 300°C for HS^- to become a dominant species. For this reason bisulphide complexes may only be important for ore transportation at temperatures below about 350°C. Henley *et al.* (1984) suggested that gold probably travels as a bisulphide complex up to about 300°C but that chloride complexing may be important at higher temperatures. It is interesting to note that Weissberg *et al.* (1979) indicated that in the Broadlands (NZ) geothermal system, gold is probably in solution as a bisulphide complex at 260°C, whilst lead is probably travelling as a chloride complex. One particular advantage of the bisulphide hypothesis is that it can be used to account for the zonal distribution of minerals in epigenetic deposits, as will be discussed later. A successful explanation for zoning using the chloride complex hypothesis has not yet been produced.

The origin of the sulphur at the site of deposition is also a problem; did it originate at this site or was it carried there with the metals in solution? Some mineralization situations, e.g. structurally deep sulphide veins in granites and quartzites, seem to demand that sulphur and the metals travelled together. Isotopic evidence, evidence from modern geothermal systems (Weissberg *et al.* 1979), ocean floor fissures venting hydrothermal solutions, and evidence from some fluid inclusions, many of which carry daughter sulphides in amounts often implying quite high concentrations of metals and sulphur in the ore fluid (Sawkins & Scherkenbach 1981), also favour this interpretation for many deposits. It must be emphasized that the apparent absence of H_2S from many fluid inclusions may be due to analytical difficulties of detection. The advent of Raman spectroscopy as an analytical tool may result in many more reports of the presence of H_2S in fluid inclusions (Touray & Guilhaumou 1984).

If the ore metals are transported as bisulphide complexes then abundant sulphur will be available for the precipitation of sulphides at the site of deposition. On the other hand, the postulate that chloride complexes are the

Fig. 4.3.
A. Schema showing some of the features of a geothermal system.
B. Schema showing the structure of a geothermal system like that of the Taupo Volcanic Zone, N.Z.; after Henley & Ellis (1983).
C. and D. are schemata illustrating the evolution of some of the mineralization in a flank of the Land's End Granite; in detail, these show:
C. Initial emplacement of the pluton with the development of an H_2O-saturated carapace enclosing still-consolidating magma. It also shows formation of tin- and magnetite-bearing skarns (I) in aureole rocks, by aqueous solutions of a dominantly magmatic origin. Time: about 290–270 Ma ago.
D. Further crystallization of the pluton has taken place, and joints and fractures have formed in the crystallized carapace. With the formation of a water-rich phase that has separated from the H_2O-saturated melt, an extensive geothermal system has come into being. This has produced the main stage mineralization (III & IV) of tin- and copper-bearing quartz veins. Time: about 270 Ma. (Type II mineralization is that of pegmatites). After Jackson *et al.* (1982).
E. Schema (after Moore 1982) of possible fossil geothermal systems associated with the granite batholith of south-west England, illustrating the different types and settings of mineralization in that region, and the district zoning developed there.
(1) Dines (1956) type emanative centres.
(2) Cross course mineralization (succinctly described by Alderton 1978).
(3) Kaolin deposits (weathering may have played a part in their formation).

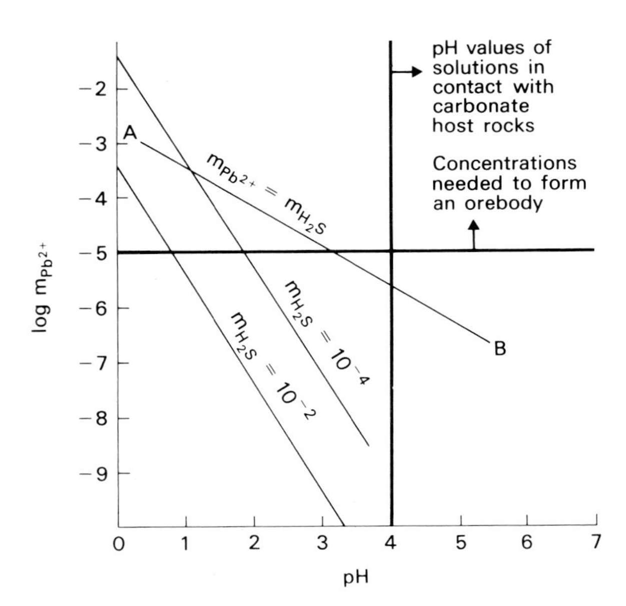

Fig. 4.4. Calculated lead-ion concentrations in water in equilibrium with galena at 80°C:line AB. Two lines showing H_2S concentrations of 10^{-2} and 10^{-4} m are plotted. (After Anderson 1977).

metal transporters, which is the most favoured hypothesis of those studying carbonate-hosted base metal deposits, creates difficulties as far as sulphur supply and metal transport are concerned. These difficulties may be explained in this way: Anderson (1977) has shown that, if the molality of H_2S is 10^{-5}, a 3 molal NaCl solution will transport almost 10^{-5} m (about 2 ppm) Pb at a pH of 4 at 80°C. At temperatures up to 150°C and otherwise similar conditions a little more than 10^{-5} m Pb would dissolve. This means that ore transport and deposition in equilibrium with carbonate rocks could occur at these temperatures. pH 4 is at the acid end of the possible pH range. With pH values in the more probable range 5–6, chloride solutions at up to 150°C cannot carry more than 10^{-5} m of Pb and H_2S at the same time. This implies that transportation and precipitation conditions would be in the very restricted field labelled *Galena precipitation* (Fig. 4.5), which is too acid to co-exist with limestone or dolomite. Thus, as Anderson (1983) remarks, we have a transport problem. Lead (and zinc) solubilities appear to be too low, even at 100–150°C, to allow metal transport in realistic amounts at realistic pH values to form an orebody. Two of the possible ways round this problem for low temperature carbonate-hosted deposits are as follows. The first way is to accept the hypothesis of Beales & Jackson (1966) that sulphur is added from another solution to the ore fluid at the site of deposition — the mixing model. We can then propose transport under relatively oxidizing and alkaline conditions well above the line AB in Fig. 4.5, which will permit the transport of significantly greater amounts of metal, but of almost no H_2S. A second way out of the problem, and one which allows us to keep a single solution model with metal and sulphur travelling together, has been put forward by Giordano & Barnes (1981), Barnes (1983) and Giordano (1985), who have proposed that organometallic

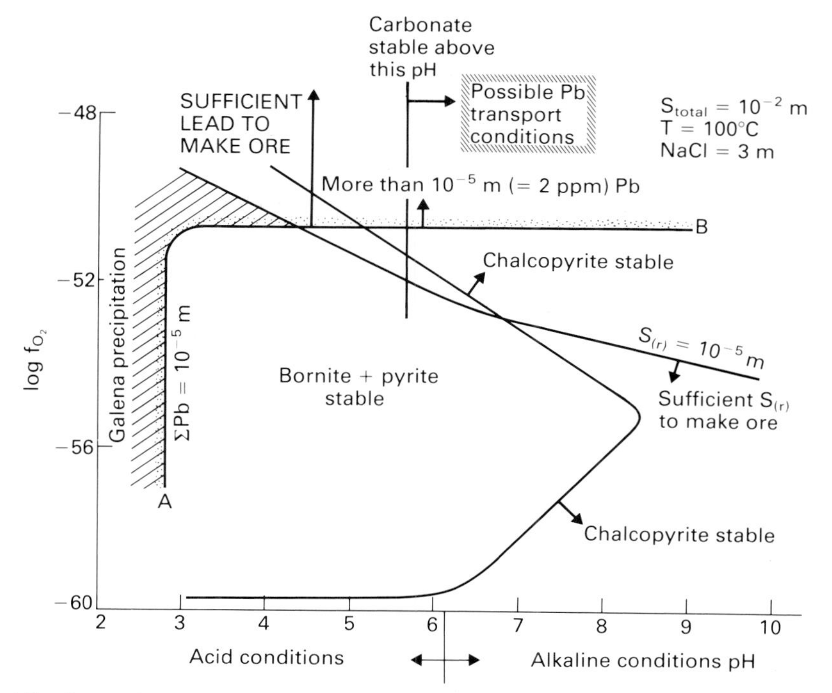

Fig. 4.5. Some mineral stabilities as a function of oxygen fugacities and pH. Other conditions as indicated. The stippled line is the 10^{-5} contour for total lead in equilibrium with galena. Larger solubilities lie to the left and above this line. (Adapted from Anderson 1975). $S_{(r)}$ = reduced sulphur (i.e. H_2S). For further discussion see text.

complexes were the metal carriers. Such complexes could carry many thousand ppm lead and zinc. The presence of these complexing agents would have the effect of shifting the solubility contours to lower f_{O_2} values, possibly permitting metals and H_2S to travel together.

In connexion with the idea of oilfield brines as potential ore fluids, particularly in the formation of carbonate-hosted lead-zinc deposits (Mississippi Valley-type), it must be pointed out that there are both single solution models as postulated by Sverjensky (1984), which may be criticized as having too dilute a solution to form an orebody and as pushing the chemical parameters of the ore fluid to the fringes of what might be expected under natural conditions (Giordano 1985), and the mixing model as proposed by Beales & Jackson (1967).

(c) *Sulphide deposition*. There is a useful concise illustration of various factors involved in ore mineral deposition in Henley *et al.* (1984) which we will follow here. If $PbCl_2$ is the dominant lead complex in solution then we may write:

$$PbS + 2H^+ + 2Cl^- \rightleftharpoons PbCl_2 + H_2S.$$

From this equation we can see that *dilution or addition of H_2S* would precipitate galena from an initially saturated solution. Dilution can occur in various ways, the most important being when a rising ore fluid enters a zone saturated with ground water. Addition of H_2S may take place when another solution carrying H_2S is encountered or if sulphate *in the ore fluid* is reduced by organic material. (It should be stressed that it is only reduced sulphur in ore solutions that causes precipitation and Anderson's 'transport problem' discussed above. Much sulphate ion is usually present in geothermal systems and sulphate values in fluid inclusions are usually comparable with base metal concentrations — see Table 4.2.) An *increase in pH* due to *boiling* can induce precipitation and *cooling* of course also reduces solubility. What is the relative effectiveness of these factors? We can illustrate this by reference to the above solution, initially saturated with respect to galena, when it is cooled from 300°C to 280°C. A temperature change of this amount causes the solubility to drop to less than one fifth of that at 300°C, whereas the solubility change due to dilution is much less. On the other hand, if boiling occurs (how could this come about?), accompanied by a temperature drop to 280°C, then the solubility is one hundred times less than that of the original solution. Of the three processes leading to galena precipitation, dilution, cooling and boiling, boiling is clearly the most efficient. If the initial solution was considerably undersaturated then dilution alone might lead to saturation and nothing more.

If, after reading the above sections, the reader feels baffled, and uncertain of his own opinion on the difficult subject of hydrothermal ore genesis, then let him take consolation from the recent statement of Giordano and Barnes (1981) concerning the vast number of geological and geochemical studies of Mississippi Valley-type deposits, 'In spite of this wealth of information, the mechanisms of base metal transport and deposition remain poorly understood'. And if that is not consolation enough let him turn to Henley *et al.* (1984, pp. 119–200) where the authors demonstrate that we could use our present state of geochemical knowledge to prove that it would be impossible for Nature to form Mississippi Valley-type deposits. To quote from them, 'It is a good thing that someone had already found these deposits because geochemists might have proven that they couldn't exist!'

In the foregoing, some ideas concerning the genesis and nature of hydrothermal solutions have been dealt with briefly by particular reference to lead-zinc deposits in limestone host rocks. The subject is, however, vast and the above discussion must only be considered as a short introduction to some of the principles involved. The reader who wishes to further his study of the chemistry of hydrothermal solutions should turn to Henley *et al.* (1984) and then progress onwards from that excellent work which is well endowed with pertinent references.

LATERAL SECRETION

It has been accepted for many years that quartz lenses and veins in metamorphic rocks commonly result from the infilling of dilatational zones and open fractures by silica which has migrated out of the enclosing rocks, and that this silica may be accompanied by other constituents of the wall rocks including metallic components and sulphur. This derivation of materials *from the immediate neighbourhood* of the vein is called lateral secretion. A very interesting example of deposits formed in this way has been described by Boyle (1959) from the Yellowknife Goldfield of the North-west Territories of Canada; but before discussing it we must consider the probable behaviour of element levels in rocks adjacent to veins forming under different conditions. In Fig. 4.6A we have a vein forming from an uprising hydrothermal solution supersaturated in silica. Some of this diffuses into the wall rocks and causes some silicification. The curve showing the level of silica decreases away from the source (i.e. the vein). In Fig. 4.6B we have the opposite situation where silica is being supplied to the vein from the wall rocks. The curve now climbs as it leaves the vein, indicating a zone of silica depletion in the rocks next to the vein. Clearly silica has been abstracted from the wall rocks and has presumably accumulated in the vein.

The principal economic deposits of the Yellowknife field occur in quartz-carbonate lenses in extensive chloritic shear zones cutting amphibolites (metabasites). The deposits represent concentrations of silica, carbon dioxide, sulphur, water, gold, silver and other metallic elements. The principal minerals are quartz, carbonates, sericite, pyrite, arsenopyrite, stibnite, chalcopyrite, sphalerite, pyrrhotite, various sulphosalts, galena, scheelite, gold and aurostibnite. The regional metamorphism of the host rocks varies from amphibolite to greenschist facies. Alteration haloes of carbonate-sericite-schist and chlorite-carbonate-schist occur in the host rocks adjoining the deposits.

It is very instructive to remember that the dominant mineral of the veins is quartz. The profile of silica alongside the lenses is shown in Fig. 4.7. This demonstrates that a very substantial amount of silica has been subtracted from both the alteration zones and this has occurred of course *on both sides* of the vein. Clearly more silica has been subtracted from the wall rocks than is present in the lenses and the problem is not, where has the silica in the lenses come from, but where has the surplus silica gone to? Some subtraction of magnesia, iron oxides, lime, titania and manganese oxide has also occurred and

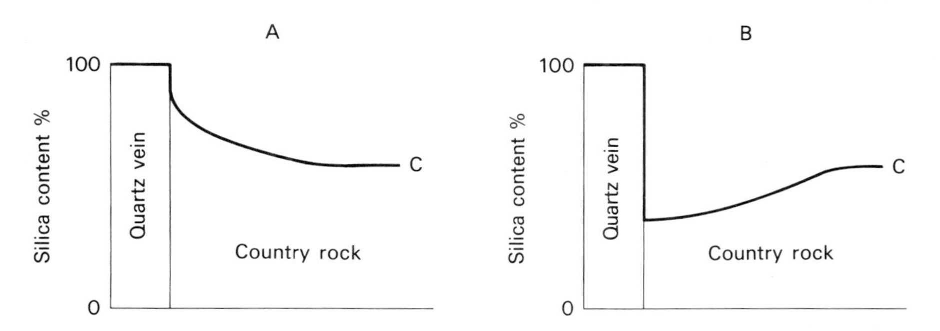

Fig. 4.6. Comparison of hypothetical profiles of silica. In case A, silica is added to the wall rocks from the hydrothermal solution which is depositing quartz in the vein. In B, silica is abstracted from the wall rocks and deposited as quartz in the vein. C indicates the normal level of silica in the country rocks.

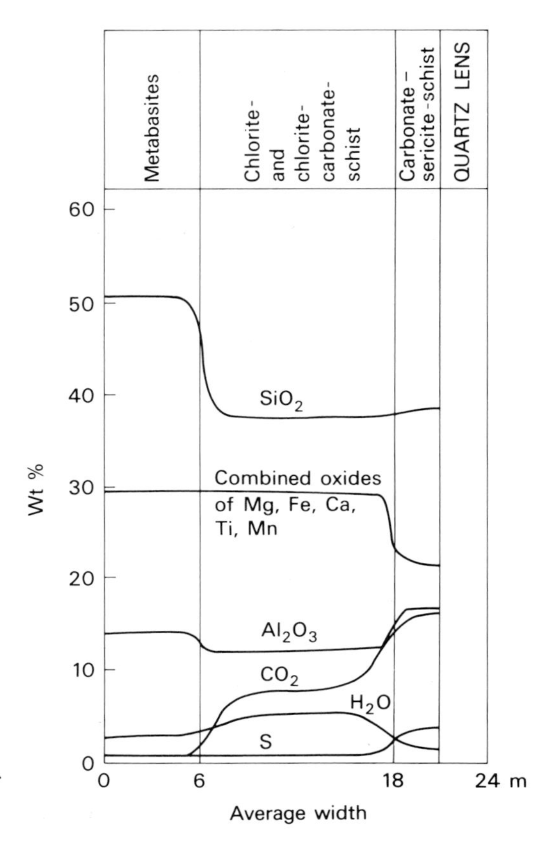

Fig. 4.7. Chemical changes produced by alteration of metabasites. Yellowknife gold deposits, Canada. (Modified from Boyle 1959).

doubtless this is the source of iron in such minerals as pyrite, pyrrhotite and chalcopyrite in the lenses. Alumina shows a depletion in the outer zone of alteration and a concentration in the inner zone where it has collected for the formation of sericite. As a result of the extensive development of carbonates in the alteration zones carbon dioxide develops a much higher level than in the unaltered country rocks. Water shows a similar but not identical behaviour. Boyle produced good evidence that these two oxides were passing through the rocks in considerable quantity, being mobilized by regional metamorphism and migrating down the metamorphic facies. They passed into the shear zones to form the chlorite, carbonate and sericite.

It appears highly probable that the major constituents of the shear zones resulted from rearrangement and introduction of material from the country rocks; the remaining question is whether the metabasites could have been the source of the sulphur and metallic elements in the deposits. The metabasites consist of metamorphosed basic volcanic lavas and tuffs. These rocks are richer in elements such as gold, silver, arsenic, copper, etc., than other igneous rocks, and for the unsheared metabasites of the Yellowknife area Boyle obtained the following values (all in ppm): S = 1500; As = 12; Sb = 1;

Cu = 50; Zn = 50; Au = 0.01; Ag = 1. For purposes of calculation the rock system was taken to be: length = 16 km; width = 152 m; depth = 4.8 km. The amount of ore in the system was assumed to be 6 × 10^6 t with average grades of S = 2.34%; As = 1.35%; Sb = 0.15%; Cu = 0.07%; Zn = 0.28%; Au = 0.654 oz/ton and Ag = 0.139 oz/ton. The total contents of these elements in the shear system prior to shearing and alteration, and in the deposits is shown in Table 4.3. It is apparent from these figures that all the

Table 4.3. Contents of chalcophile elements in shear zones and deposits, Yellowknife gold deposit, Canada.

Element	Total content in shear system before shearing and alteration (millions of tons)	Total content in deposits (millions of tons)
S	62	0.14
As	0.5	0.081
Sb	0.04	0.009
Cu	2.0	0.004
Zn	2.0	0.017
Au	12.2 × 10^6 oz	3.9 × 10^6 oz
Ag	1219 × 10^6 oz	0.834 × 10^6 oz

elements considered could have been derived solely from the sheared rock of the shear zone and there is no need to postulate another source. Indeed, there is such a difference between the values in columns two and three that it may well be that significant quantities of chalcophile elements accompanied the surplus silica to higher zones in the crust to form deposits which have now been eroded away.

Other applications of the lateral secretion theory in very different geological settings include Brimhall (1979) and Dejonghe & de Walque. Brimhall has proposed that the main veins at Butte were formed by the concentration of ore-forming material already present in the host granodiorite intrusion in the form of protore (see Chapter 16). Dejonghe & de Walque (1981) have discussed the formation of a lead-zinc-copper-bearing vein in Carboniferous sediments in Belgium in terms of this theory.

METAMORPHIC PROCESSES

(a) *Skarn deposits*. These deposits have been termed hydrothermal metamorphic, igneous metamorphic, contact metamorphic and pyrometasomatic. The last term neatly summarizes the origin of these deposits which are formed at high temperatures (pyro) with the addition and subtraction of material (metasomatic). Their general morphology and nature have been summarized in Chapter 2. They are developed most often, but not invariably, at the contact of intrusive plutons and carbonate country rocks. The latter are converted to marbles, calc-silicate hornfelses and/or skarns by contact metamorphic effects.

The calc-silicate minerals such as diopside, andradite and wollastonite, which are often the principal minerals in these ore-bearing skarns, attest to the high temperatures involved, and various lines of evidence suggest a range of 650–400°C for initial skarn formation (Einaudi *et al.* 1981). The pressures at the time of formation were very variable as the depths of formation were probably from one to several km. Some of the classic pyrometasomatic deposits of the United States are associated with porphyry copper intrusions, indicating a relatively shallow depth of emplacement.

During the genesis of these deposits, the first stage was that of recrystallization of the country rocks. (In any carbonate rocks impurities led to the formation of calc-silicate hornfelses.) The next stage was the introduction of calc-silicate-forming materials such as silica, iron, magnesia and alumina. This was the stage of large-scale metasomatism and Lindgren (1924) showed that at Morenci, Arizona, vast quantities of material had been added *and subtracted*. He pointed out that if all the CaO in 1 cm^3 of calcite is converted into andradite [$Ca_3Fe_2(SiO_4)_3$] skarn then the volume would increase to 1.4 cm^3. All the evidence at Morenci, as in similar deposits, suggests a volume for volume replacement process with no concomitant expansion. In that case, for every m^3 of altered limestone 460 kg of CaO and 1190 kg of CO_2 were removed, and 1330 kg of SiO_2 and 1180 kg Fe_2O_3 were added.

The next stage in the development of many deposits was the formation of volatile-bearing minerals such as amphiboles, epidote, idocrase, fluorite, tourmaline and axinite. The introduction of ore minerals may have overlapped stages two and three. Oxide minerals appear to have crystallized before sulphides.

The origin of all this introduced material is much debated. In the past, it was held that the pluton responsible for the contact metamorphism was also the source of the metasomatizing solutions. Whilst it is conceivable that a granitic pluton might supply much silica, it might be thought unlikely that it could have supplied the amount of iron that is present in some deposits. However, Whitney *et al.* (1985) have shown that it is probable that in natural magmatic systems, the concentration of iron in chloride solutions coexisting with magnetite or biotite is very high. This high solubility may explain the large quantities of iron in some skarns associated with granitic intrusions. On the other hand, where the pluton concerned is basic, the supply of iron does not present such great problems. These difficulties do become insurmountable, however, for the small class of pyrometasomatic deposits such as the Ausable Magnetite District, New York State, which have no associated intrusions. Perhaps the main function of the intrusion is that of a heat engine. For Ausable, Hagner & Collins (1967) suggested migration of iron from accessory magnetite in granite-gneiss, with concentration in shear zones to form magnetite-rich bodies, together with the release of iron during the recrystallization of clinopyroxene- and hornblende-gneisses.

The experimental replacement of marble by sulphides has been achieved by Howd & Barnes (1975) who found that ore-bearing bisulphide solutions at 400–450°C and 500 MPa when oxidized, produced acid solutions which dissolved the marble and provided sites for sulphide deposition. These experiments and the proximity of many skarn deposits to porphyry copper intrusions suggest that circulating hydrothermal solutions may have played a part in the ore genesis.

(b) *The role of other metamorphic processes in ore formation*. Some examples of lateral secretion are clearly the result of metamorphism. This subject has, however, already been covered above and will not be discussed further. In this section we are concerned with those metamorphic changes which involve recrystallization and redistribution of materials by ionic diffusion in the solid state or through the medium of volatiles, especially water. Under such conditions relatively mobile ore constituents may be transported to sites of lower pressure such as shear zones, fractures or the crests of folds. In this way, the occurrence of quartz-chalcopyrite-pyrite veins in amphibolites and schists and many gold veins in greenstone belts (Saager *et al.* 1982) may have come about.

The behaviour of trace amounts of ore minerals in large volumes of rock undergoing regional metamorphism is uncertain and is a field for more extensive research. It might be thought that with the progressive expulsion of large volumes of water and other volatiles during prograde metamorphism, natural hydrothermal systems might evolve which would carry away elements such as copper, zinc or uranium which are enriched in trace amounts in pelites. Shaw (1954), however, in a study of the progressive metamorphism of pelitic sediments from clays through to gneisses showed that such changes are but slight. Taylor (1955) in a study of greywackes reported similar results. On the other hand, De Vore (1955) calculated that during the transformation of one cubic mile of epidote-amphibolite facies hornblendite into the granulite facies there may be a release of 9 Mt of Cr_2O_3, 4.5 Mt of NiO and 900 000 tonnes of CuO. Similarly, retrograde metamorphism can release large quantities of zinc, lead and manganese. In most cases, the liberated elements are probably dispersed rather than concentrated, but with the diversion of hydrothermal fluids also expelled by the metamorphism into suitable structural situations, ore concentration may occur. Fyfe & Henley (1973) have suggested just such a mechanism.

They envisage a situation where a volcanic-sedimentary pile is being metamorphosed under amphibolite facies conditions. It would be losing about 2% water, and if salt is present and oxygen buffered by magnetite-ferrous silicate assemblages, then gold solubilities of the order of 0.1 ppm at 500°C would be achieved. This gold would either be dispersed through greenschist facies rocks or concentrated into a favourable structure. This could happen if the solution flow was focused into a large vein or fault system where seismic pumping could have forced the ore fluid upwards (Sibson *et al.* 1975).

Fyfe & Henley show that a source region of 30 km^3 could have provided all the gold, silica and water required to form a deposit as large as Morro Velho, Brazil. Their figures are as follows. The orebody of auriferous quartz occupies 0.01 km^3 and contains about 3×10^8 g of gold. With an average crustal gold

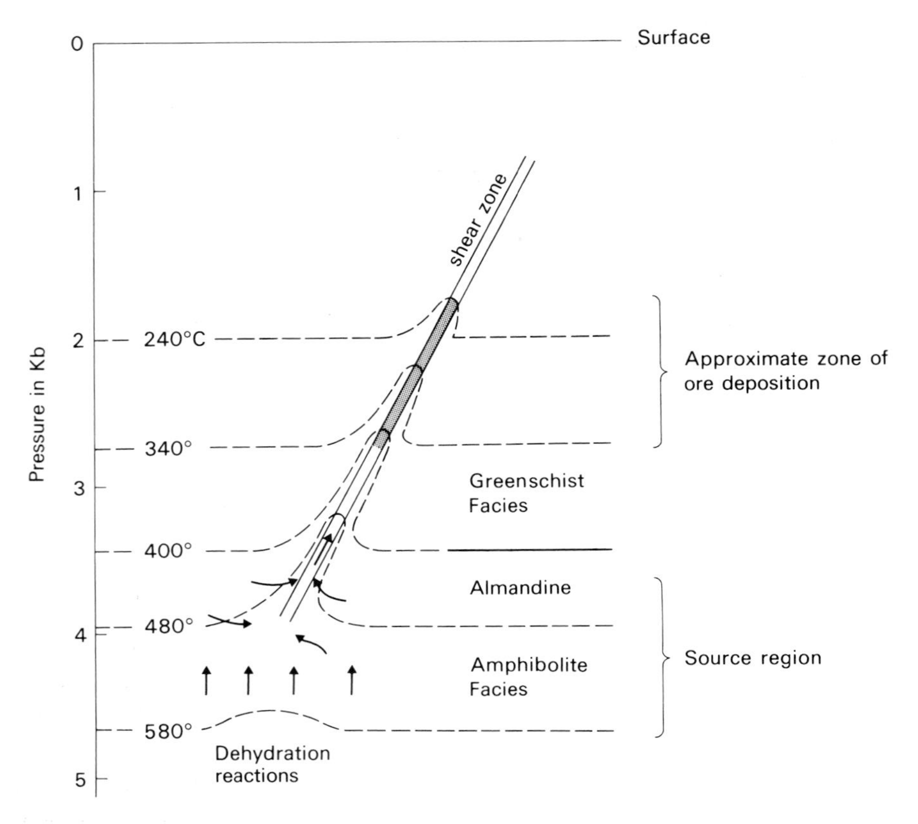

Fig. 4.8. Diagram of a shear zone where metamorphic water from a large volume of rock is rising to higher levels. (After Fyfe & Henley 1973).

content of about 3 parts per billion (ppb), approximately 30 km^3 of volcanics or sediments are required to form the source region. About 2×10^{15} g of water would be released which at 0.1 ppm could transport 2×10^8 g of gold. At 5 kb and 500°C this volume of water could also transport 2×10^{13} g silica (solubility under these conditions 10 g kg^{-1}). This silica is about 0.01 km^3 in volume, i.e. that of the orebody. This model of gold mineralization is similar to that proposed by Boyle (see earlier), but now the major constituents are derived from a deep source region and not by lateral secretion.

This model, and variations of it, have been used to explain the formation of a number of gold deposits — see for example Kerrich & Fryer (1979), Phillips, Groves & Martin (1984), Shepherd & Allen (1985). In developing this model for exploration purposes the comments of Phillips & Groves (1983 p. 36) should be carefully considered. A recent counterblast to this metamorphic model for the genesis of hydrothermal, gold depositing solutions has come from Burrows *et al.* (1986) who have used a great deal of carbon isotope data from the world's two largest Archaean gold vein and shear zone systems, Hollinger-McIntyre in Canada and the Golden Mile in Western Australia, to support a return to the once orthodox hypothesis of a magmatic-hydrothermal origin of the mineralizing solutions!

Uranium deposits formed from metamorphic fluids have been described by Adamek & Wilson (1979), and Nisbet *et al.* (1983) have described cobalt-tungsten mineralization formed during metamorphism.

Origin due to surface processes

Processes involving mechanical and chemical sedimentation will not be dealt with here. The reader is referred to texts on sedimentology for the general principles involved, e.g. Reading (1986) and Tucker (1981). However, certain aspects will be touched upon in Chapter 18. Residual processes and supergene enrichment will be dealt with in Chapter 19. The remaining space in this chapter will be devoted to a consideration of exhalative processes. These, it should be noted, are a surface expression of the activity of hydrothermal solutions.

VOLCANIC-EXHALATIVE (SEDIMENTARY-EXHALATIVE) PROCESSES

We are concerned here with a group of deposits often referred to as exhalites and including massive sulphide ores. Some of the characteristics of these ores have been dealt with on pp. 26–27. They frequently show a close spatial relationship to volcanic rocks but this is not the case with all the deposits, e.g. Sullivan, Canada (Fig. 2.11). They are conformable and frequently banded; the principal constituent is usually pyrite with varying amounts of copper, lead, zinc and baryte; precious metals together with other minerals may be present. For many decades they were considered to be epigenetic hydrothermal replacement orebodies (Bateman 1950). In the 1950s, however, they were recognized as being syngenetic, submarine-exhalative, sedimentary orebodies, and deposits of this type have been observed in the process of formation from hydrothermal vents (black smokers) at a number of places in the Pacific Ocean (Anon. 1984, Hekinian & Fouquet 1985). They are now often referred to by one or other of the two terms in this section heading, or as volcanic-associated massive sulphide deposits.

The ores with a volcanic affiliation show a progression of types. Associated with basic volcanics, usually in the form of ophiolites and presumably formed at oceanic or back arc spreading ridges, we find the Cyprus types. These are essentially cupriferous pyrite bodies. They are exemplified by the deposits of the Troodos Massif in Cyprus and the Ordovician Bay of Islands Complex in Newfoundland. Associated with the early part of the main calc-alkaline stage of island arc formation are the Besshi-type deposits. These occur in successions of mafic volcanics in complex structural settings characterized by thick greywacke sequences. They commonly carry zinc as well as copper and are exemplified by the Palaeozoic Sanbagawa deposits of Japan, and the Ordovician deposits of Folldal in Norway. The more felsic volcanics, developed at a later stage in island arc evolution, have a more varied metal association. They are copper-zinc-lead ores often carrying gold and silver. Large amounts of baryte, quartz and gypsum may be associated with them. They are called Kuroko deposits after the Miocene ores of that name in Japan. All these different types are normally underlain in part by a stockwork up which the generating hydrothermal solutions appear to have passed (Fig. 2.19).

There is today wide agreement that these deposits are submarine-hydrothermal in origin, but there is a divergence of opinion as to whether the solutions responsible for their formation are magmatic in origin or whether they represent circulating sea water. Here, there is only room to deal briefly with this controversy.

The general correlation of ore types with basic to acid volcanics favours a magmatic origin. In the case of the Kuroko deposits, Sato (1977) has argued strongly for a magmatic origin. In particular he notes the close correlation of the Kuroko deposits with rhyolite domes rather than with the preceding andesitic and dacitic vulcanism, together with supporting isotopic and fluid inclusion evidence. The lead isotopic ratios are almost identical with the rock leads of the Neogene volcanics which, in turn, show a uniform variation across the island arc. This suggests a common direct magmatic source for the volcanics and the ore lead, because the volcanics beneath the deposits are too thin for circulating brines to have leached the lead from them. Stanton (1978), using petrological and geochemical evidence, argued cogently for a magmatic origin. Aye (1982) postulated a mantle or magmatic source on sulphur isotope and other grounds for some Brittany deposits for which $\delta^{34}S$ values are close to zero (see Chapter 12 for significance of this value).

The circulating sea water model has been put forward mainly by workers on stable isotopes who argue that sea water is the main source of the water and sulphur, and that the metals were leached out of the rocks through which the fluids flowed. This model has received great support from the recent discoveries of black smokers along oceanic rifts in the Pacific Ocean, of the remains of ancient vents in the Cretaceous deposits of Cyprus and of the peculiar fauna associated with these vents in similar deposits of Oman (Oudin & Constantinou 1984, Haymon *et al.* 1984).

As was mentioned earlier in this chapter, different types of water have characteristic hydrogen (D/H) and oxygen ($^{18}O/^{16}O$) isotopic ratios (Sheppard 1977). Using these ratios it has been shown that various types of water were involved in general mineralization processes. Magmatic fluids were dominant in some cases, and in others initiated the mineralization and wall rock alteration only to be swamped by convective meteoric water set in motion by hot intrusions or other heat sources. In massive sulphide deposits, the isotopic evidence favours sea water as the principal or only fluid. The possibility must be borne in mind, however, that the sea water was a late addition to the hydrothermal system and that it has overprinted the pre-existing magmatic values (Sato 1977). Let us examine the evidence a little more deeply, as isotopic studies are now being applied to all types of mineral deposits and they will be mentioned again in later chapters.

Variations in the isotopic ratios of hydrogen and oxygen are given in the δ notation in parts per thousand (per mil, ‰) where:

$$\delta_x = \frac{R_{sample}}{R_{standard}} - 1 \times 1000$$

In the above formula for hydrogen, $\delta_x = \delta D$ and $R = D/H$; for oxygen, $\delta_x = \delta^{18}O$ and $R = {}^{18}O/{}^{16}O$.

The standard for both hydrogen and oxygen is standard mean ocean water or SMOW. In nature, D/H is about 1/7000 and ${}^{18}O/{}^{16}O$ is about 1/500. These values are measured directly on natural substances such as thermal waters, connate waters in sediments and fluid inclusions, or they are determined indirectly using minerals after removal of all the absorbed water. In the latter case the isotopic composition of the mineral is not that of the fluid with which it was in contact at the time of crystallization or recrystallization. The δ values for the fluid have to be calculated from the mineral values using equilibrium fractionation factors determined by laboratory experiment or from studies of active geothermal systems. A temperature fractionation effect also occurs, so the temperature must be known (from fluid inclusion studies, etc.) to determine the isotopic composition of the water in equilibrium with the mineral. For example, in Fig. 4.9, raising the temperature from 10 to 200°C gives rise to isotope exchange and re-equilibration such that the isotopic composition of the water changes from Y_1 to X_1 whilst that of the coexisting kaolinite changes

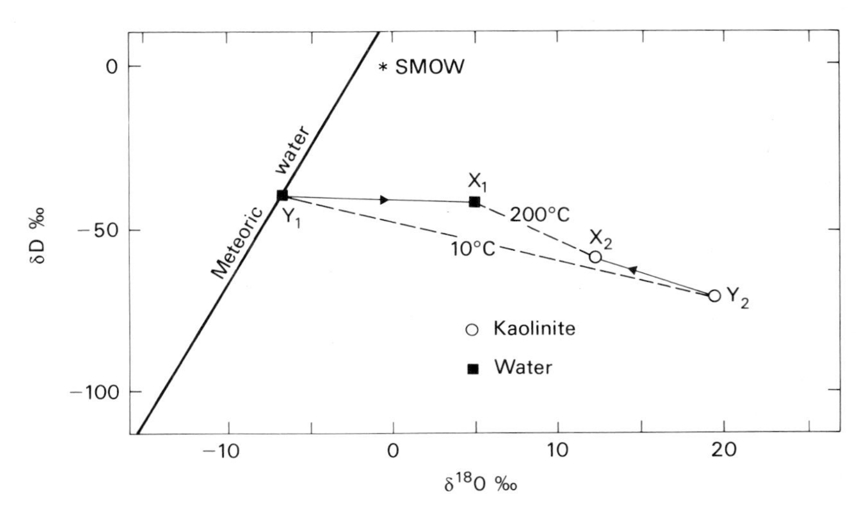

Fig. 4.9. Hydrogen and oxygen isotopic fractionations between water and kaolinite at two different temperatures. From a knowledge of the system kaolinite-water, and given the values X_2 and 200°C, the value of X_1 can be calculated. (Modified from Sheppard 1977).

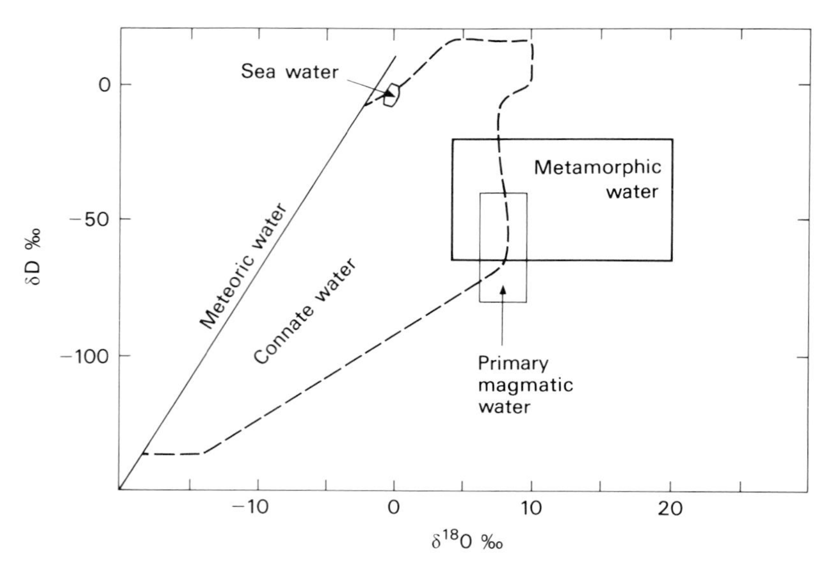

Fig. 4.10. Fields of isotopic composition of sea water, connate water, metamorphic water and magmatic water. (Modified from Sheppard 1977).

from Y_2 to X_2. In other words rock-water reactions cause a shift in the delta values of both the circulating meteoric water and the rock with which it is in contact, with the result that the water is enriched in ^{18}O as the temperature rises.

The isotopic compositions of the various types of water show useful differences. Sea water in general plots very close to SMOW (Fig. 4.10) and shows very little variation. Meteoric water varies fairly systematically with latitude along the line shown in Fig. 4.10. Values for metamorphic and magmatic waters have been deduced from measurements on minerals. Connate (formational) water can be measured directly and plots as shown. Since many of the connate waters are richer in ^{18}O than SMOW they cannot have resulted from simple mixing of meteoric and sea water. There must have been isotopic exchange with the sediments at elevated temperatures (shown by Fig. 4.9 to result in ^{18}O enrichment of the water, e.g. the change from Y_1 to X_1), addition of rising metamorphic water or some other process. We are now in a position to compare the results obtained from a study of the Cyprus and Kuroko deposits.

These results are plotted in Fig. 4.11. Those for the Cyprus stockwork deposits coincide exactly with the values for sea water, and Heaton & Sheppard (1977) suggest a model involving deeply circulating sea water as sketched in Fig. 4.12. They also present evidence that the associated country rocks were thoroughly permeated by sea water during their metamorphism into the greenschist and zeolitic facies. The Kuroko fluids show $\delta^{18}O$ values commensurate with a sea water origin but δD is depleted by 11–26‰ relative to

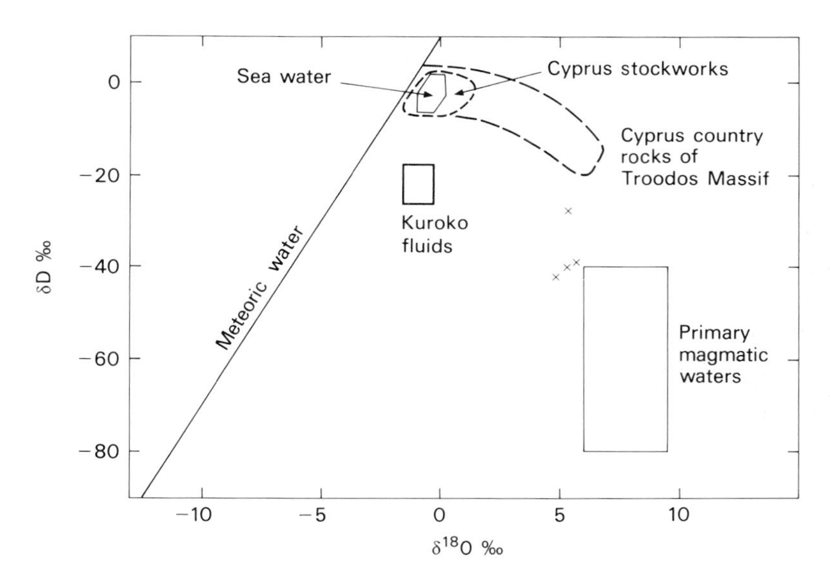

Fig. 4.11. Plot of δD against $\delta^{18}O$ for Cyprus stockworks, associated country rocks and Kuroko fluids. (Modified from Sheppard 1977). Crosses mark values for sericites from the Kosaka deposit, Japan.

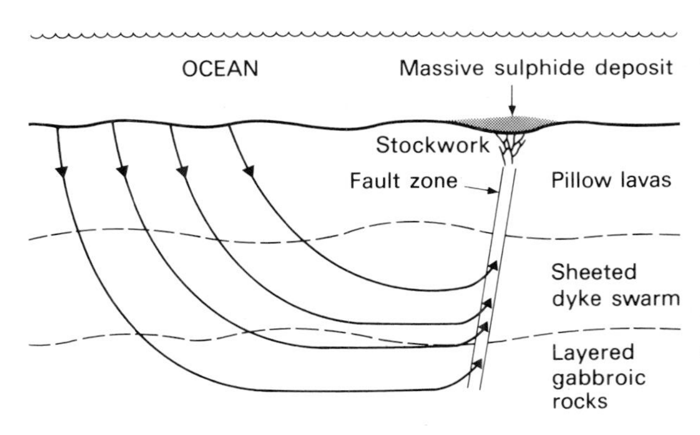

Fig. 4.12. Diagram showing how sea water circulation through oceanic crust might give rise to the formation of an exhalative massive sulphide deposit.

sea water. Ohmoto & Rye (1974) concluded that sea water was the dominant source of the hydrothermal fluid but that it contained a small meteoric and/or magmatic contribution. As Kuroko deposits belong to the Island Arc environment, meteoric water could be involved. This would suggest a model, similar to that in Fig. 4.12, of circulating sea water becoming a concentrated brine at depth, dissolving copper and other metals from the rocks it traversed and carrying these up to the surface where they were precipitated as sulphides with sulphur derived from the sea water. This leads us into another field of fruitful research in isotopic studies — sulphur.

Sulphur isotopic data for the Cyprus ores have been reviewed by Spooner (1977) who has pointed out that $\delta^{34}S$ for the pyrite is higher than that of the country rocks from which the sulphur may have been partly derived. This suggests an additional source of isotopically heavy sulphur. Spooner felt that this was probably the circulating Cretaceous sea water which would have had a value of $\delta^{34}S = +16‰$. Spooner *et al.* (1977) also showed that the mineralized zones are enriched in ^{87}Sr relative to the initial magmatic $^{87}Sr/^{86}Sr$ of the igneous country rocks. The values obtained range up to the value for Upper Cretaceous sea water but no higher. Many other examples of similar, more recent studies support this view of circulating sea water being the ore fluid.

Sato's point (Sato 1977) that the sea water circulation can overprint an earlier magmatic-hydrothermal mineralization event must, however, be borne in mind. Consistent negative δD values obtained in a number of investigations of Kuroko deposits indicate the involvement of water of another origin during the ore genesis. The data of Hattori and Muehlenbachs (1980) from the Kosaka deposit (Fig. 4.11), for example, strongly support this view and plot very close to the magmatic water box. In their paper on Palaeozoic deposits of this type in northern Queensland, Gregory & Robinson (1984) have used sulphur isotope analyses to infer that a magmatic ore fluid was responsible for the formation of one deposit, but that in a nearby one the magmatic ore fluid was diluted with sea water. The reader must beware of taking such studies too much at their face value; sulphur isotope values in sea water passing through mantle-derived rocks may, through water-rock reactions, become close to or virtually the same as magmatic (mantle) values. For further discussion on this point see Skirrow & Coleman (1982).

One of the difficulties concerning the isotope studies outlined above is that they have been largely confined to mineralized areas. One non-mineralized area where such studies have been carried out in depth is Mull in Scotland. The isotopic evidence from this district shows without doubt that vast quantities of meteoric water can be flushed through igneous intrusive and extrusive rocks without the genesis of any significant mineralization. This fact should be remembered when considering the evidence of the circulation of meteoric water through some porphyry copper deposits (Chapter 14).

The most important factor in controlling the behaviour of exhalative hydrothermal solutions when they reach the sea floor is probably the density difference between the ore solution and sea water. Solomon & Walshe (1979) deduced that the ore solutions were buoyant on entering sea water (all the evidence from black smokers bears out their conclusion) and sometimes rose from the sea floor as conical plumes, especially in shallow water, distributing their load of sulphides over a wide area. As a result, massive ores may not necessarily be underlain by a feeder vent or stockwork. The location of ore deposition is, however, mainly controlled by the site of discharge of the hydrothermal solutions and this may lead to deposition in gravitationally unstable positions. The unconsolidated ores may therefore be reworked by submarine sliding, leading perhaps to turbidity current transportation and deposition. The common occurrence of graded fragmental ore is a characteristic feature of many Kuroko deposits (Sato 1977), and Thurlow (1977) and Badham (1978) have described the results of transportation by slumping and sliding at Buchans, Newfoundland, and Avoca, Ireland.

5

Geothermometry, Geobarometry, Paragenetic Sequence, Zoning and Dating of Ore Deposits

Geothermometry and geobarometry

Ores are deposited at temperatures and pressures ranging from very high, at deep crustal levels, to atmospheric, at the surface. Some pegmatites and magmatic segregation deposits have formed at temperatures around 1000°C and under many kilometres of overlying rock, whilst placer deposits and sedimentary ores have formed under surface conditions. Most orebodies were deposited between these two extremes. Clearly, knowledge of the temperatures and pressures obtaining during the precipitation of the various minerals will be invaluable in assessing their probable mode of genesis and such knowledge will also be of great value in formulating exploration programmes. In this small volume it is only possible to touch on a few of the methods that can be used.

FLUID INCLUSIONS

The nature of fluid inclusions and the principle of this method have been outlined in Chapter 3. Clearly, primary inclusions are those which must be examined. Secondary inclusions produced after the mineral was deposited and commonly formed by the healing of fractures will not give us data on the mineral depositional conditions, but their study can be very important in the investigation of certain deposits, e.g. porphyry coppers (Chapter 15).

The filling (homogenization) temperatures of aqueous inclusions (i.e. the temperature at which the inclusion becomes a single phase fluid) indicate the depositional temperature of the enclosing mineral if a correction can be made for the confining pressure and salinity of the fluid. The salinity, in terms of equivalent weight per cent NaCl, can be determined by studying the depression of the freezing point using a freezing stage. Frequently, confining pressures have to be estimated by reconstructing the stratigraphical and structural succession above the point of mineral depositional. This clearly leads to a degree of uncertainty. Pressure corrections increase the homogenization temperatures obtained in the laboratory, so these are still of great value even when uncorrected as they record minimum temperatures of deposition. Where CO_2-rich and H_2O-rich inclusions coexist, the pressure can be estimated from the filling temperature of the aqueous inclusions and the density of the CO_2 inclusions (Groves & Solomon 1969). If the fluid inclusion assemblage indicates boiling at the time of trapping then the depth of formation can be estimated (Haas 1971). However, if boiling did not occur, then the confining pressure must have exceeded the vapour pressure of the fluid, which can be calculated from the salinity and temperature data, giving us a minimum pressure value. Studies on the use of fluid inclusions in geobarometry were reviewed by Roedder & Bodnar (1980).

INVERSION POINTS

Some natural substances exist in various mineral forms (polymorphs) and some use of their inversion temperatures can be made in geothermometry. For example, β-quartz inverts to α-quartz with falling temperature at 573°C. We can determine whether quartz originally crystallized as β-quartz by etching with hydrofluoric acid and thus decide whether it was deposited above or below 573°C. Examples among ore minerals include:

$$\underset{\text{monoclinic}}{\text{acanthite}} \overset{177^\circ\text{C}}{\rightleftharpoons} \underset{\text{cubic}}{\text{argentite}}$$

$$\text{orthorhombic chalcocite} \overset{104^\circ\text{C}}{\rightleftharpoons} \text{hexagonal chalcocite}$$

This second low inversion temperature is important in distinguishing hypogene chalcocite from low temperature near surface supergene chalcocite, a distinction which can be of great economic importance in evaluating many copper deposits, for if the near surface ore is only just of economic grade and much of the mineralization is supergene, then the ore below the zone of supergene enrichment may be uneconomic. Fortunately, the distinction can be made, for relict cleavage remains after chalcocite has inverted from the high temperature hexagonal form, and this cleavage is revealed by etching polished sections with nitric acid. A comprehensive list of invariant points (including inversion points) is included in Barton & Skinner (1979).

EXSOLUTION TEXTURES

As a result of restricted solid solution at lower temperatures between various pairs of oxide and sulphide minerals, exsolution bodies of the minor phase segregate from the host solid solution on cooling (Fig. 5.1). Their presence indicates the former existence of a solid solution of the two minerals which was deposited at an elevated temperature. An idea of that temperature can be obtained by reference to laboratory work on the sulphide system concerned (Edwards 1960), or by resolution of natural exsolution bodies in the host grain by heating samples in the laboratory. An example of the latter approach was reported by Edwards & Lyon (1957) who performed resolution experiments on samples from the Aberfoyle tin mine, Tasmania. After the geologically short annealing time of one week they obtained the following temperatures for the onset of resolution:

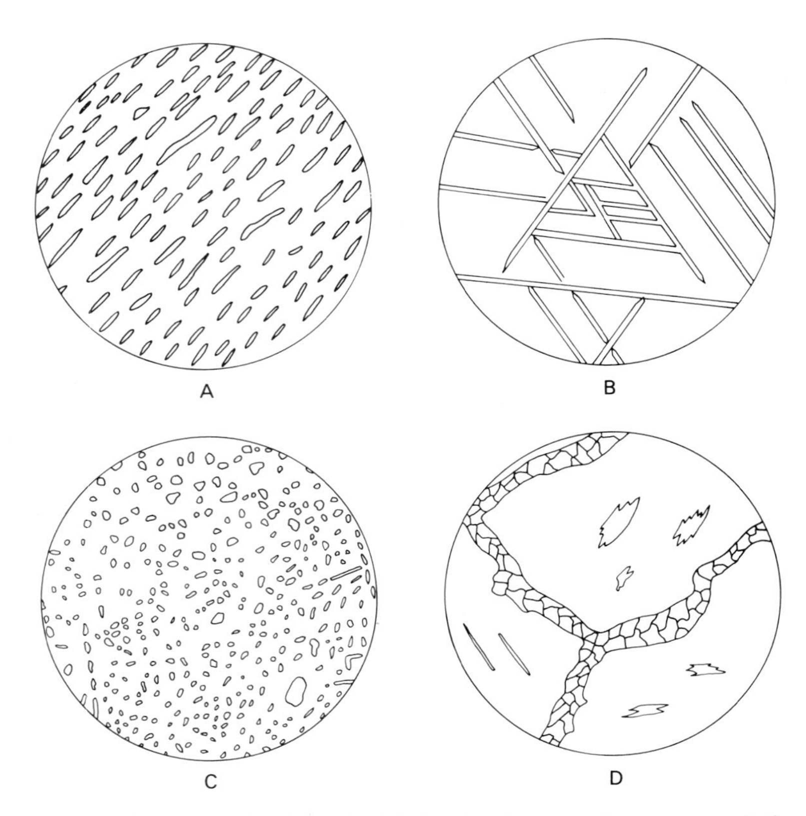

Fig. 5.1. Exsolution textures in oxide and sulphide mineral systems: A = seriate exsolution bodies of hematite-rich material in an ilmenite-rich base, × 300, Tellnes, Norway, B = exsolution lamellae of ilmenite in magnetite, × 134, Sudbury, Ontario, C = emulsoid exsolution bodies of chalcopyrite in sphalerite,* × 80, Geevor mine, Cornwall, D = rim or net exsolution texture formed by the exsolution of pentlandite from pyrrhotite, × 536, Sudbury, Ontario.

*Most workers would now agree that this density of small chalcopyrite bodies is too high to have resulted entirely from exsolution.

chalcopyrite and stannite in sphalerite	— 550°C,
sphalerite in chalcopyrite	— 400°C,
stannite in chalcopyrite	— 475°C,
sphalerite in stannite	— 325°C,
chalcopyrite in stannite	— 400–475°C.

These results, they suggested, indicated a temperature of about 600°C for the formation of the original solid solutions. There is much data in the literature suggesting that with longer annealing times lower temperatures would have been obtained and the original solid solutions might have been deposited at temperatures in the range 400–500°C. This range would agree better with the homogenization temperatures of around 400°C obtained by Groves *et al.* (1970) on fluid inclusions in cassiterites from this deposit.

The initial studies of these systems were carried out to discover those which would provide data on the temperature and pressure of ore deposition. The first results, e.g. the sphalerite geothermometer, were very promising. Further study, however, revealed that the formation of mineral assemblages and variations in mineral composition depend on many factors. Thus, Kullerud (1953) suggested that the FeS content of sphalerite gave a direct measurement of its temperature of deposition. Later, Barton & Toulmin (1963) showed that the fugacity of sulphur was also an important control but unfortunately it is often an unknown quantity.

More research into this system, however, has produced some rewards. Although the work of Scott & Barnes (1971) indicated that the composition of sphalerite coexisting with pyrrhotite and pyrite over the range 525–250°C is essentially constant (Fig. 5.2) their microprobe studies did reveal a possible geothermometer. They discovered metastable iron-rich patches within sphalerites whose composition, compared with their matrix, appeared to be constant and temperature dependent. This applies only to coexisting sphalerite-pyrrhotite-pyrite assemblages. Iron-rich patches have been found in natural sphalerites from veins but they do not appear to be present in metamorphosed sphalerite because of annealing and re-equilibration.

Scott & Barnes (1971) and Scott (1973) indicated that the iron content of

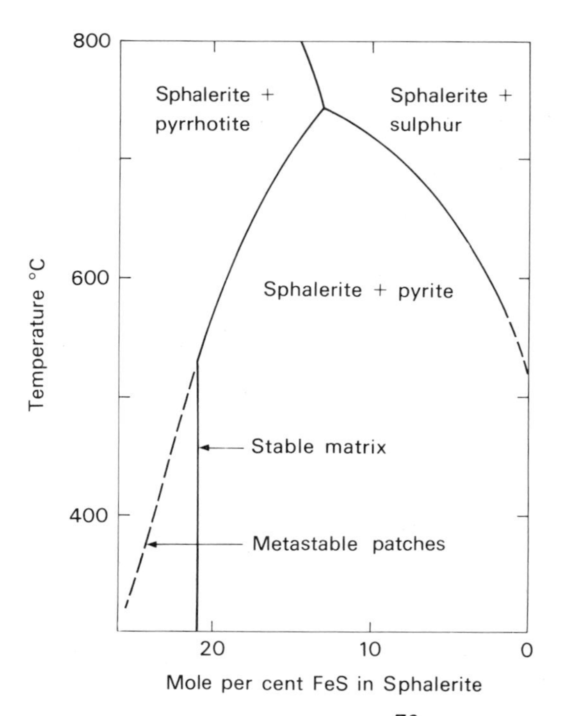

Fig. 5.2. Temperature-composition projection of a portion of the system ZnS-FeS-S showing the constant composition of sphalerite co-existing with pyrite and pyrrhotite over the range 525–300°C (vertical line), and the composition of metastable patches of iron-rich sphalerite and their variation in composition with temperature. (After Scott & Barnes 1971).

sphalerite *in equilibrium* with pyrite and pyrrhotite, although not temperature dependent below 525°C (Fig. 5.2), is strongly controlled by pressure above 300°C (Fig. 5.3). Thus, sphalerite can be used as a geobarometer provided equilibrium has been reached. This is most likely to be the case in metamorphosed ores and Lusk *et al.* (1975) have described its use in just such a case. Further refinement of this geobarometer can be found in Hutchison & Scott (1981)

This brief discussion must suffice to illustrate the use of sulphide systems in geothermometry and geobarometry. Excellent summaries of experimental techniques and problems are provided by Barton & Skinner (1979) and Scott (1974). A list of the sulphide systems studied experimentally up to 1974 is given by Craig & Scott (1974) and the use of arsenopyrite as a geothermometer is discussed by Kretschmas & Scott (1976) but Sharp *et al.* (1985) considered that its applications are very limited. There is also considerable experimental work on non-sulphide systems which is applicable to orebodies, particularly their gangue minerals (Levin *et al.* 1969). For carbonate composition variation with temperature see Goldsmith & Nowton (1969).

A very interesting recent development is the Ga/Ge-Geothermometer using sphalerite (Möller 1985). This can be used to determine temperatures in the source regions of ore solutions and to estimate the degree of mixing of hot parental ore fluids with cool, near surface waters. Möller's first applications of this method appear to be very promising and have shown that for the vein mineralization of Bad Grund and Andreasberg the ore fluid at its time of

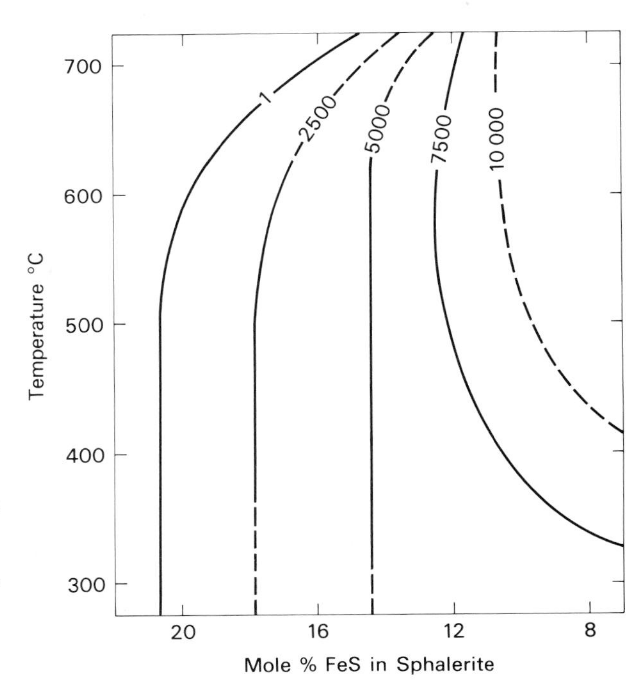

Fig. 5.3. Temperature-composition projection of the sphalerite + pyrite + hexagonal pyrrhotite solvus isobars. Pressures are in bars. Solid lines are measured isobars, pecked lines are extrapolations. (After Scott 1973).

genesis had a temperature above 220°C, but must have mixed with waters of less than 130°C at the level of mineral deposition. However, at Ramsbeck there is no difference between the Ga/Ge deduced temperature for the source region and the fluid inclusion homogenization temperatures, indicating that no significant fluid mixing occurred.

For the sediment-hosted ore deposits of various types investigated by Möller, source fluid temperatures of 180–270°C were obtained and, on the assumption that these fluids were generated by sediment dewatering in sedimentary basins, this indicates a source depth of about 6–11 km, if geothermal gradients of 25–30°C km^{-1} are assumed.

STABLE ISOTOPE STUDIES

In Chapter 4, when discussing the uses of hydrogen and oxygen isotopes to determine the origin of waters which had reacted with minerals, it was pointed out that if we know the temperature of reaction we can determine the isotopic composition of the water. Similarly, given the isotopic compositions of cogenetic minerals and water, we can determine the temperature. This is the principle of the use of oxygen isotopes on gangue minerals as a geothermometer. A much more extensively used method for ore deposits has been the sulphur isotope geothermometer.

Sulphur has four isotopes, but the major geological interest is in the variation $^{34}S/^{32}S$ because this varies significantly in the minerals of hydrothermal deposits for a number of reasons:

(1) the fractionation of the isotopes between individual sulphide and sulphate minerals in the ore varies with the temperature of deposition;

(2) the initial isotopic ratio is controlled by the source of the sulphur, e.g. mantle, crust, sea water, etc;

(3) the variable proportions of oxidized and reduced sulphur species in solution — in its simplest form the H_2S/SO_4^{2-} ratio. As this ratio depends on temperature, pH and fO_2 we can also evaluate the variation of sulphur isotopic values in terms of T, pH and fO_2, Rye & Ohmoto (1974).

Variations in $^{34}S/^{32}S$ are expressed in delta notation ($\delta^{34}S$) where:

$$\delta^{34}S = \left(\frac{R_{34}\ \text{sample}}{R_{34}\ \text{standard}} - 1\right) \times 10^3$$

Point (1) above implies that sulphides and sulphates in hydrothermal deposits will show a variation (fractionation) in $\delta^{34}S$ values from mineral to mineral which is dependent on their temperature of deposition, provided the minerals crystallized in equilibrium with each other. This fractionation is found in nature and its extent is illustrated in Fig. 5.4. The curves in this figure are based on experimental and theoretical data. They show that sulphur isotopic fractionation at 200°C between pyrite and galena is about 4.6% and between sphalerite and galena is a little over 3%. Thus, these and other mineral pairs can be used in sulphur isotope geothermometry.

Depositional temperatures obtained by fluid inclusion methods on transparent minerals, sulphur isotope methods on sulphides intergrown with them, and oxygen isotopic compositions of coexisting oxides, generally give results in close agreement, as in the work on the Echo Bay uranium-nickel-cobalt-silver-copper deposits by Robinson & Ohmoto (1973). This paper is a good illustration of the use of these methods in obtaining temperatures of mineral deposition.

Paragenetic sequence and zoning

The time sequence of deposition of minerals in a rock or mineral deposit is known as its paragenetic sequence. If the minerals show a spatial distribution then this is known as zoning. The paragenetic sequence is determined from studying such structures in deposits as crustiform banding and from the microscopic observation of textures in polished sections.

PARAGENETIC SEQUENCE

Abundant evidence has been accumulated from world-wide studies of epigenetic-hydrothermal deposits indicating that there is a general order of deposition of minerals in these deposits. Exceptions and reversals are known but not in sufficient number to suggest that anything other than a common order of deposition is generally the case. A simplified, general paragenetic sequence is as follows:

(1) silicates;
(2) magnetite, ilmenite, hematite;
(3) cassiterite, wolframite, molybdenite;
(4) pyrrhotite, löllingite, arsenopyrite, pyrite, cobalt and nickel arsenides;
(5) chalcopyrite, bornite, sphalerite;
(6) galena, tetrahedrite, lead sulphosalts, tellurides, cinnabar.

Of course, not all these minerals are necessarily present in any one deposit and the above list has been drawn up from evidence from a great number of orebodies.

ZONING

Zones may be defined by changes in the mineralogy of ore or gangue minerals or both, by changes in the percentage of metals present, or by more subtle changes from place to place in an orebody or mineralized district of the ratios between certain elements or even the isotopic ratios within one element. Zoning was first described from epigenetic vein deposits but it is also present in other types of deposit. For example, syngenetic deposits may show zoning parallel to a former shore line as is the case with the iron ores of the Mesabi Range, Minnesota; alluvial deposits may show zoning along the course of a river leading from the source area; some exhalative syngenetic sulphide deposits

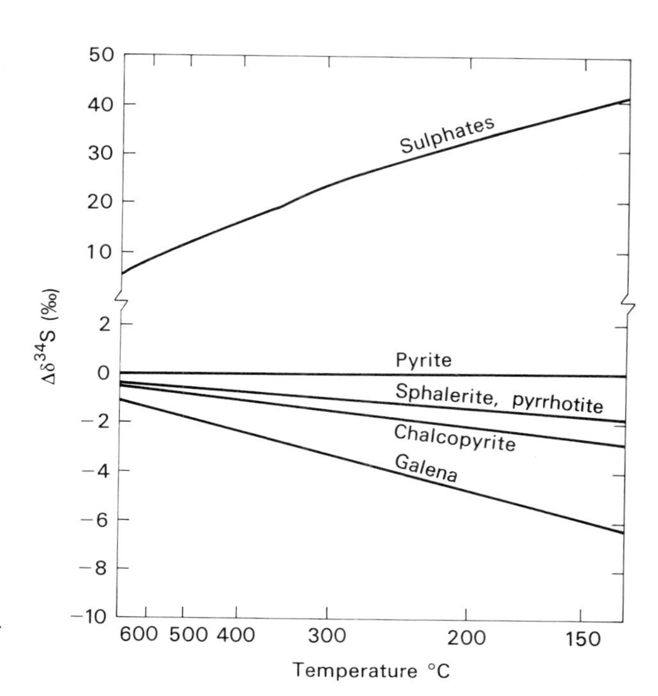

Fig. 5.4. Sulphur isotope fractionation between some hydrothermal minerals plotted with respect to pyrite. (Modified from Rye & Ohmoto 1974).

show a marked zonation of their metals and pyrometasomatic (skarn) deposits often show a zoning running parallel to the igneous-sedimentary contact. In this discussion, attention will be focused on the zoning of epigenetic-hydrothermal, exhalative syngenetic and sedimentary syngenetic sulphide deposits.

(a) *Epigenetic hydrothermal zoning*. Zoning of this type can be divided into three intergradational classes; regional, district and orebody zoning (Park & MacDiarmid 1975). Regional zoning occurs on a very large scale, often corresponding to large sections of orogenic belts and their foreland (Fig. 5.5). A number of examples of zoning on this scale are described from the circum-Pacific orogenic belts by Radkevich (1972). Some regional zoning of this type, e.g. the Andes, appears to be related to the depth of the underlying Benioff Zone which suggests a deep level origin for the metals as well as the associated magmas (Chapter 22). District zoning is the zoning seen in individual orefields such as Cornwall, England (Fig. 5.6 and Table 5.1) and Flat River, Missouri (Fig. 5.7). Zoning of this type is most clearly displayed where the mineralization is of considerable vertical extent and was formed at depth where changes in the pressure and temperature gradients were very gradual. If deposition took place near to the surface, then steep temperature gradients may have caused superimposition of what would, at deeper levels, be distinct zones, thus giving rise to the effect known as telescoping. Some geothermal systems also show metal zoning e.g. Broadlands, N.Z. where there is Sb, Au and Tl enrichment near the surface and Pb, Zn, Ag, Cu, etc. at depth (Ewers & Keays 1977). Orebody zoning takes the form of changes in the mineralization within a single orebody. A good example occurs in the Emperor

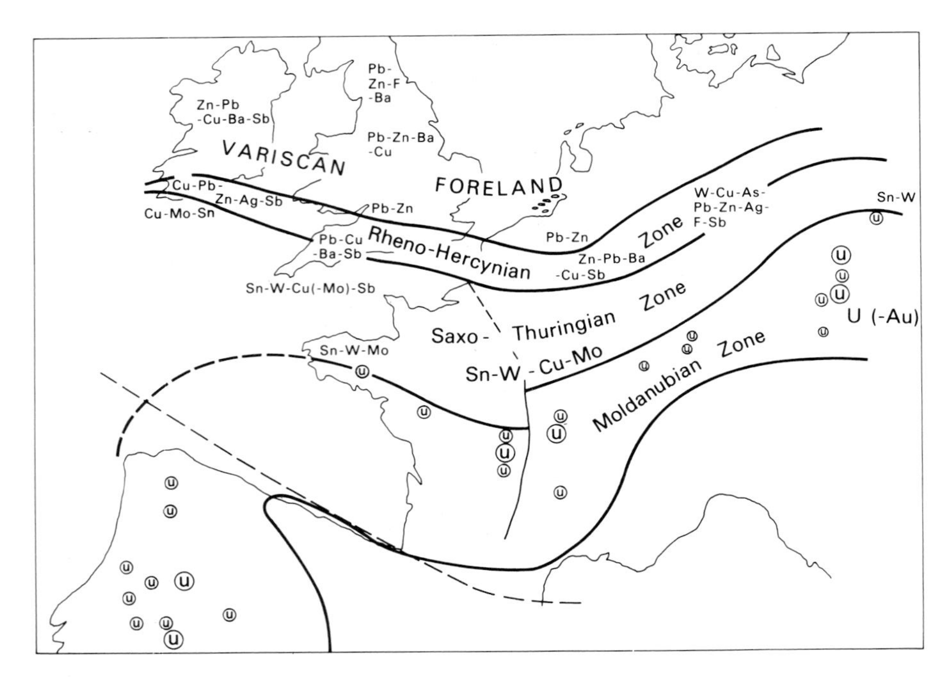

Fig. 5.5. Regional metal zonation of epigenetic deposits in the Variscan Metallogenic Province of north-west Europe. (After Evans 1967a and Cuney 1978). Sizes of symbols in Moldanubian Zone and Spain indicate relative sizes of uranium deposits.

Gold Mine, Fiji, where vertical zoning of gold-silver tellurides in one of the main ore shoots gives rise to an increase in the Ag:Au ratio with depth (Forsythe 1971). A comprehensive discussion of zoning with many examples is given in Routhier (1963).

(b) *Syngenetic hydrothermal zoning*. This is the zoning found in stratiform sulphide bodies principally of volcanic affiliation (Chapter 2). These deposits are frequently underlain by stockwork deposits and many of them appear to have been formed from hydrothermal solutions which reached the sea floor. As Barnes (1975) has pointed out, the zonal sequence is not always clearly seen in this type of ore deposit, but where it has been established, it is identical to that of epigenetic hydrothermal ores. This zonal sequence is well exemplified by the Precambrian deposits of the Canadian shield, by the Devonian Rammelsberg and Meggen deposits of Germany and the Kuroko deposits of Japan. In these deposits there is a general sequence upwards and outwards through the orebodies of Fe→(Sn)→Cu→Zn→Pb→Ag→Ba which should be compared with the similar zoning shown in Table 5.1. As in vein deposits, the zonal boundaries are normally gradational with frequent overlap of zones.

(c) *Sedimentary syngenetic sulphide zoning*. This zoning is found in stratiform sulphide deposits of sedimentary affiliation usually of wide regional development such as the Permian Kupferschiefer of Germany and Poland, and the Zambian Copperbelt deposits. Underlying stockwork feeder channelways

Table 5.1. A generalized mineral paragenesis of the mineral deposits of south-west England. (After El Shazly *et al.* 1957).

Gangue	Zone	Ore minerals		Economically important elements
		LATEST MINERALS		
	7	Barren (pyrite) Hematite Stibnite. Jamesonite		Fe Sb
Calcite	6	Tetrahedrite. Bournonite. Pyragrite? Siderite Pyrite (marcasite)	*Mesothermal and Epithermal Lodes* Generally at right-angles to granite ridges	
Chalcedony, Dolomite	5b	Argenite. Galena. Sphalerite		Ag Pb Zn
Barite, Fluorite	5a	Pitchblende. Niccolite. Smaltite Cobaltite (Native bismuth — bismuthinite?)		U Ni Co Bi
Chlorite, Hematite, Quartz	4	Chalcopyrite Sphalerite Wolframite (scheelite) Arsenopyrite Pyrite		Cu
	3	Chalcopyrite (stannite) Wolframite (scheelite) Arsenopyrite Cassiterite (wood tin)	*Hypothermal Lodes* Generally parallel to granite ridges and dykes	
	2	Wolframite (scheelite) Arsenopyrite (molybdenite?) Cassiterite		Sn W As
Tourmaline	1	Cassiterite. Specularite		
Felspar, Mica	*	Arsenopyrite. Stannite Wolframite, Cassiterite Molybdenite	Veins often in granite cusps	
	†	Arsenopyrite. Wolframite Cassiterite Molybdenite		
		EARLIEST MINERALS		

* = Greisen bordered veins
† = Pegmatites

are not known beneath these deposits and they have usually formed in euxinic environments. The zoning appears to show a relationship to the palaeography, and proceeding basinwards through a deposit it takes the form of Cu+Ag→Pb→Zn. In the Zambian Copperbelt, however, the zoning is principally one of copper minerals and pyrite, as lead, zinc and silver are virtually absent.

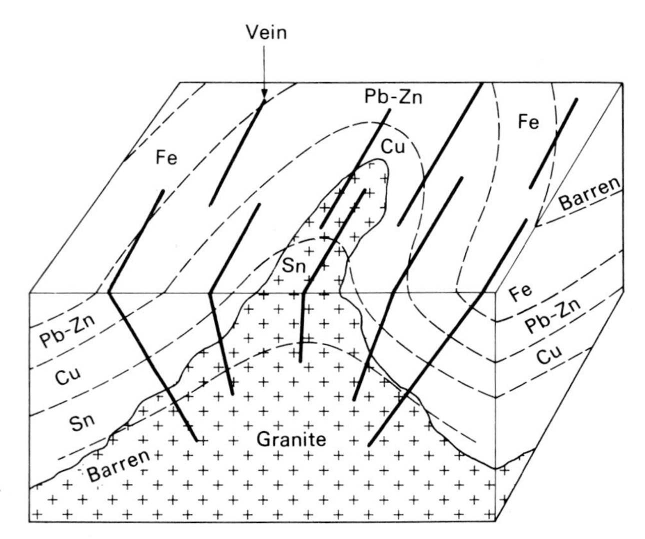

Fig. 5.6. Diagrammatic illustration of district zoning in the orefield of south-west England showing the relationship of the zonal boundaries to the granite-metasediment contact. (After Hosking 1951).

SULPHIDE PRECIPITATION, PARAGENETIC SEQUENCE AND ZONING

In Chapter 4 it was pointed out that there are two principal schools of thought concerning the transportation of metal ions in hydrothermal solutions. The first favours transportation as bisulphide complexes, and the second as complex chloride ions in chloride-rich, sulphide-poor brines. Barnes (1975) has shown how the first hypothesis can account for the zoning seen in epigenetic and syngenetic hydrothermal deposits. Clearly, the relative stabilities of complex metal bisulphide ions will control their relative times of precipitation and hence both the resulting paragenetic sequence and any zoning which may be developed. Barnes has calculated these stabilities which are shown in Table 5.2. The data in this table suggest that iron and tin would be precipitated early in the paragenetic sequence and would be present in the lowest zone of a zoned deposit, whilst silver and mercury would be late precipitates which would travel furthest from the source of the mineralizing solutions (cf. Table 5.1). Precipitation itself would be occasioned by the interplay of a number of factors including changes in pH due to wall rock alteration reactions, decrease in temperature with distance from source, reaction with carbonaceous materials in wall rocks, mixing with meteoric waters, and so on.

Syngenetic sedimentary sulphide deposits may have been precipitated from chloride-rich brines, such as those found near the Salton Sea (Chapter 4), if these brines encountered H_2S or HS^- supplied by a large volume of marine, euxinic water. On the other hand, they may have been introduced into a basin of deposition in ionic solution by rivers. It is, therefore, pertinent to look at the relative stabilities of metal sulphides which show a zonal distribution in such deposits in the presence and absence of chloride ions. Some relative solubilities are shown in Table 5.3 and 5.4 (drawn from data in Barnes 1975).

From these tables it can be seen that there is excellent agreement between

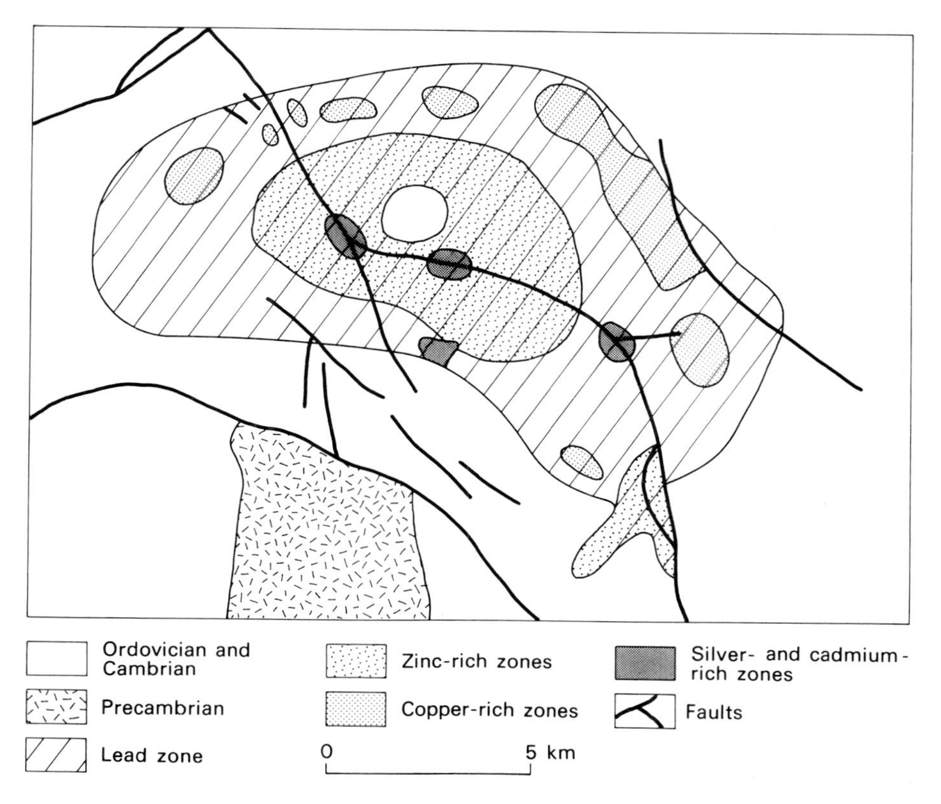

Fig. 5.7. District zoning in the Flat River area of the Old Lead Belt of south-east Missouri. (After Heyl 1969).

Table 5.2. Predicted sequence of stabilities of bisulphide complexes (in kilocalories).

← Least soluble .... Most soluble →							
Fe	Ni	Sn	Zn	Cu	Pb	Ag	Hg
79	84	126	132	135	153	157	226

Table 5.3. Relative stabilities of sulphides in chloride solutions (as expressed by the equilibrium constant for the reaction: $MeCl_2(aq) + HS^- \rightarrow MeS(s) + H^+ + Cl^-$).

	← Least soluble .... Most soluble →		
	CuS	PbS	ZnS
At 25°C	38.4	28.2	27.7

Table 5.4. Relative stabilities of sulphides in ionic solutions (as expressed by the equilibrium constant for the reaction: $Me^{2+} + HS^- \rightarrow MeS(s) + H^+$).

← Least soluble .... Most soluble →		
$CuFeS_2$	PbS	ZnS
23.7	20.7	19.2

the zonal arrangement of metals in syngenetic sedimentary deposits and the relative solubilities shown. Barnes felt that this suggests that the metals travelled in ionic solution or as chloride complexes. It must be noted, however, that studies of the transport of base metals in surface streams at the present day show that only a small proportion of the base metal contents are carried in ionic solution, which means that the above picture is a somewhat simplified version of the true situation. In addition, faunal (including bacterial) and facies factors can influence zonation significantly. Zoning, however, cannot be used to distinguish between the products of these two types of solution, because chloride complexing up to at least 100°C is not sufficiently strong to change significantly the normal order of precipitation of sulphides from that found for simple metallic ions. It might be thought from the work of Barnes that the nature of the zoning in hydrothermal deposits points to deposition from sulphide complexes, and that chloride complexes or ionic solutions may play a significant role as metal transporting agents in the formation of sedimentary sulphide deposits. The situation, however, is unfortunately not as simple as this. Different metals may be travelling as chloride and bisulphide complexes in the same solution and above 300°C chloride complexing may be dominant (Chapter 4). This leaves us with an unresolved problem — probably attributable to inadequate experimental data on ore mineral solvation reactions and the high temperature molecular properties of aqueous metal complexes (Susak & Crerar 1982).

The dating of ore deposits

When orebodies form part of a stratigraphical succession like the Mesozoic ironstones of north-western Europe their age is not in dispute. Similarly the ages of orthomagmatic deposits may be almost as certainly fixed if their parent pluton can be well dated. On the other hand, epigenetic deposits may be very difficult to date, especially as there is now abundant evidence that many of them may have resulted from polyphase mineralization, with epochs of mineralization being separated by intervals in excess of 100 Ma. There are three main lines of evidence which can be used: the field data, radiometric and palaeomagnetic age determinations. The field evidence may not be very exact. It may take the form of an unconformity cutting a vein and thus giving a minimum age for the mineralization. Similarly, if mineralization occurs in rocks reaching up to Cretaceous in age we know that some of it is at least as young as the Cretaceous, but we cannot tell when the first mineralization occurred or how young some of it may be.

RADIOMETRIC DATING

Sometimes a mineral clearly related to a mineralization episode will permit a direct dating, e.g. uraninite in a vein, but more commonly it is necessary to use wall rock alteration products such as micas, feldspar and clay minerals. In the latter case, the underlying assumption is that the wall rock alteration is coeval with some of or all the mineralization. Brownlow (1979) gives a simple summary of the theory of how ages are deduced from the radioactive decay of an unstable parent isotope to a more stable isotope of a different element (daughter) and it is assumed here that the reader is familiar with the principles of the theory. Time is but part of the story. Almost as a by-product, the isochron method also reveals the isotopic composition of the lead and strontium which were present when a rock or orebody formed; information which may point to a specific source for the element concerned. Table 5.5 summarizes the parent-daughter relationships which are of major interest for dating rocks and ores. Perusal of this table will show that in dating mineralization by these methods we are heavily dependent on the products of wall rock alteration and the assumption of its contemporaneity with the associated mineralization.

It is also important to note that our radiometric clocks can be reset. Broken Hill, New South Wales, gives us a dramatic example of this phenomenon. The lead-zinc-silver orebodies of Broken Hill are generally held to be regionally metamorphosed syngenetic deposits. Rb-Sr and K-Ar ages of biotites (450–500 Ma), however, markedly post-date the 1700 Ma of the dominant metamorphism in the country rocks (Richards & Pidgeon 1963, Shaw 1968). The heating event reflected by the mica ages may have been associated with formation of the nearby Thackeringa-type orebodies.

Porphyry coppers in many parts of the world have been dated using Rb-Sr and K-Ar methods. The results of this work have shown (a) that it is usually the youngest intrusions in any particular area that carry mineralization, and (b) that there is a close temporal relationship between porphyry copper mineralization and associated calc-alkaline magmatism. Thus at Panguna, Papua New Guinea, the mineralization dates at 3.4 ± 0.3 Ma, compared with intrusives 4–5 Ma old. At Ok Tedi, north-west Papua, alteration as young as 1.1–1.2 Ma has been found (Page & McDougall 1972). This contrasts with the oldest mineralization yet found, a banded iron formation with associated stratabound copper sulphides at Isua, West Greenland, dated at 3760 ± 70 Ma (Moorbath *et al.* 1973, Appel 1979).

Determining the age of Mississippi Valley-type deposits using radiometric

Table 5.5. Major isotopic dating methods.

Method	Generalized decay scheme	Half-life of radioactive isotope	Materials that can be dated
U-Pb	$^{238}U \rightarrow {}^{206}Pb$	$4.5 \times 10^9 a$	Uraninite, pitchblende, zircon, sphene, apatite, epidote, whole rock
Rb-Sr	$^{87}Rb \rightarrow {}^{87}Sr$	$5.0 \times 10^{10} a$	Mica, feldspar, amphibole, whole rock
K-Ar	$^{40}K \rightarrow {}^{40}Ar$	$1.31 \times 10^9 a$	Mica, feldspar, amphibole, pyroxene, whole rock

methods has proved to be about the most difficult task in ore deposit geochemistry. This knotty problem apparently arises in part from their mode of genesis; ore materials are quite probably being derived from a number of source rocks, which implies isotopic heterogeneity in the minerals of those deposits. Ore fluid mixing, reactions with wall rocks etc. lead to further difficulties. The reader will find it rewarding to read the short note by Ruiz *et al.* (1985) on this subject.

Finally, mention should be made of the model lead ages obtained on lead mineralization (normally from galenas). These, the reader should remember, record the time when the radiometric clock stopped, not the time when it was started or restarted as with the U-Pb, Rb-Sr and K-Ar methods. The Pb-Pb method is dependent on assumptions concerning the evolution of the lead isotopes in the source material, and for this mathematical models have to be deduced. The method is regarded with suspicion by many workers but if the basis of the method is borne in mind, its results can be of considerable value, not only for dating purposes but also in throwing light on the origin of the lead.

PALAEOMAGNETIC DATING

This method deserves to be more widely used than it has been. Krs & Šťovíčková (1966) applied it to veins of the Jáchymov (Joachimsthal) region of the Czechoslovakian section of the Erzgebirge, where they demonstrated the presence of Hercynian (late Carboniferous to early Permian) and Saxonian (middle Triassic to Jurassic) mineralization. Further, they showed that the Ag-Co-Ni-U-Bi mineralization was not associated temporally with the Hercynian granites but with the later Saxonian to Tertiary basaltic magmatism. Similar work by Krs in the Freiberg region of East Germany showed an excellent correlation with radiometric data on the veins he investigated (Baumann & Krs 1967).

The successful palaeomagnetic dating of ore deposits depends on a number of factors (Evans & Evans 1977). Some of the most important are:
(1) the development of magnetic minerals in a deposit or its wall rocks during one of the principal phases of mineralization;
(2) a lack of complete oxidation or alteration, which may be accompanied by overprinting with a later period of magnetization;
(3) the availability of an accurate polar wandering curve for the continent or plate in which the deposit occurs.

As is well known, magnetite and hematite are the two principal carriers of magnetization in rocks; this is also true for ore deposits. In general, magnetization carried wholly or in the main by magnetite is, with present palaeomagnetic techniques, more easily measured and interpreted. This mineral is, however, by no means common in epigenetic ore deposits and hematite-mineralized specimens often have to be used. Using specimens mineralized with hematite, Evans & Evans (1977) dated as Saxonian the primary hematite mineralization in the base metal veins of the Mendip Orefield, England, and Evans & El-Nikhely (1982) obtained a Permian age for epigenetic hematite deposits at two mines in Cumbria, England, contrary to the Triassic or post-Triassic ages proposed by previous workers.

An important development in this method of dating was the perfecting of the cryogenic magnetometer in the early seventies. This instrument may well enable us to obtain new data on the age of mineralization of many types of ore deposit including the Mississippi Valley-type, which, as mentioned above, presents many problems to the geochronologist employing radiometric dating methods. Wu & Beales (1981) have used a cryogenic magnetometer to date this type of mineralization in south-east Missouri.

6

Metallogenic Provinces and Epochs

It has long been recognized that specific regions of the world possess a notable concentration of deposits of a certain metal or metals and these regions are known as metallogenic provinces. Such provinces can be delineated by reference to a single metal (Figs 6.1, 6.2 and 6.3) or to several metals or metal associations. In the latter case, the metallogenic province may show a zonal distribution of the various metallic deposits (Fig. 5.5). The recognition of metallogenic provinces has usually been by reference to epigenetic hydrothermal deposits, but there is no reason why the concept should not be used to describe the regional development of other types of deposit provided they show a geochemical similarity. For example, the volcanic-exhalative antimony-tungsten-mercury deposits in the Lower Palaeozoic inliers of the eastern Alps form a metallogenic province stretching from eastern Switzerland through Austria to the Hungarian border.

Within a metallogenic province there may have been periods of time during which the deposition of a metal or a certain group of metals was most pronounced. These periods are called metallogenic epochs. Some epochs are close in time to orogenic maxima, others may occur later. Thus in the Variscan orogenic belt of north-western Europe and its northern foreland, which form the metallogenic province shown in Fig. 5.5, the principal epochs of epigenetic mineralization were Hercynian (end Carboniferous to early Permian) and Saxonian (middle Triassic to Jurassic). The orogenic events in this belt culminated about the end of the Carboniferous and the Saxonian mineralization and associated vulcanicity is post-orogenic.

Metallogenic provinces and epochs of tin mineralization

Tin deposits are an excellent example of an element restricted almost entirely from the economic point of view to a few metallogenic provinces. Those outside the USSR are shown in Figs 6.1, 6.2 and 6.3. Even more striking is the fact that most tin mineralization is post-Precambrian and confined to certain well marked epochs. Equally striking is the strong association of these deposits with post-tectonic granites. Among tin deposits of the whole world, 63.1% are associated with Mesozoic granites, 18.1% with Hercynian (late Palaeozoic) granites, 6.6% with Caledonian (mid Palaeozoic) granites and 3.3% with Precambrian granites. In some belts, tin mineralization of different types and ages occurs. For example, in the Erzgebirge of East Germany, Lower

Fig. 6.1. Tin belts on continents around the Atlantic Ocean. Dotted areas indicate concentrations of workable deposits. (Modified from Schuiling 1967).

Palaeozoic or Precambrian stratiform deposits of tin are developed which are probably volcanic-exhalative in origin (Baumann 1965), together with epigenetic deposits associated with Hercynian granites. It is possible that these granites represent anatectic material from the Lower Palaeozoic or Precambrian of the region and that some of the stratiform tin was remobilized when the granitic magmas were formed by partial melting.

The figures quoted above suggest that increasing amounts of tin have been added to the crust with decreasing antiquity. The puzzling feature in this connection is that there does not appear to be any concomitant increase in the level of trace element tin with decrease in geological age in granites inside or outside tin belts. As epigenetic tin deposits are developed in the cupolas and ridges on the tops of granitic batholiths, it has been suggested that the paucity of such deposits in geologically older terrains is a function of their deeper levels of erosion. If this is the case, one might ask why no Precambrian tin placers are known containing cassiterite derived from such deposits, since Precambrian gold and uranium placers are present in a number of continents. The ultimate source of tin, as for other metals, must have been the mantle — this is shown by the recent tin mineralization discovered along the Mid-Atlantic Ridge. The more immediate source, however, must have been the crust or upper mantle, as Schuiling (1967) has argued from a consideration of the south-west African-Nigerian belt, where pre-continental drift, Precambrian mineralization, syn-drift Jurassic mineralization and post-drift Eocene

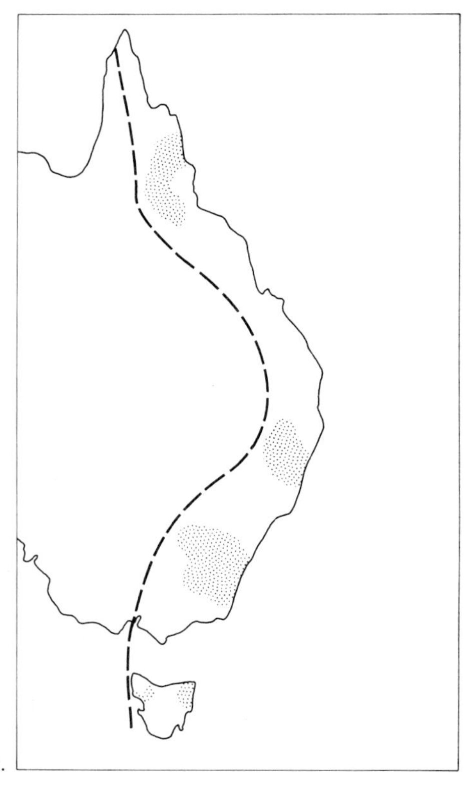

Fig. 6.2. The Palaeozoic tin belt of eastern Australia. The principal fields are shown with dotted ornamentation. (Modified from Hills 1953).

mineralization occur. If the source of the tin belonging to these different epochs had been below the African plate, then the tin deposits of different ages should occur in parallel belts. That they are not so distributed suggests derivation of the tin from within the crust or upper mantle of the African plate. The existence of tin metallogenic provinces suggests that parts of the upper mantle were (and are?) relatively richer in tin, and that this tin has been progressively added to the crust, where it has been recycled and further concentrated by magmatic and hydrothermal processes. The recognition of the development of tin provinces is of fundamental importance to the mineral exploration geologist searching for this metal. It is probable that any further discoveries of important deposits of this metal will be made within these provinces or their continuations. Continental drift reconstructions suggest that a continuation of the eastern Australian tin belt may be present in Antarctica.

Some other examples of metallogenic epochs and provinces

BANDED IRON FORMATION

These rocks, of which the commonest facies is a rock consisting of alternating quartz- and hematite-rich layers, are virtually restricted to the Precambrian.

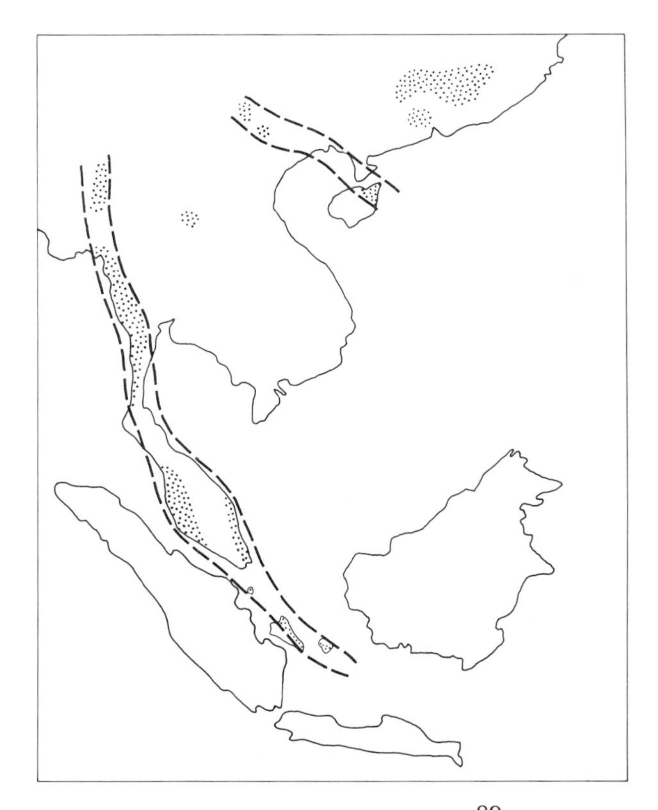

Fig. 6.3. Tin belts and fields of south-eastern Asia. These form the so-called tin girdle of south-eastern Asia, but it should be noted that if lines were drawn to join all the fields together, then the lines would cross tectonic trends at a high angle.

They occur in the oldest (>3760 Ma old) Isua sediments of western Greenland and in most Archaean greenstone belts. Their best development took place in the interval 2600–1800 Ma ago, in early Proterozoic basins or geosynclines situated near the boundaries of the Archaean cratons. There was comparatively little development of BIF after this period, and Phanerozoic ferruginous sediments are of the hematite-chamosite-siderite type. The latter also tend to be grouped into narrow stratigraphical intervals, for example the Minette ores of this class which are well developed in the Jurassic of western Europe.

NICKEL SULPHIDE DEPOSITS

These deposits are almost entirely Precambrian in age and are mainly restricted to a few Archaean greenstone belts in Ontario, Manitoba and Ungava (Canada); the Western Australian Shield; the northern part of the Baltic Shield, and in Zimbabwe. They show, therefore, a good development of metallogenic provinces, but there is considerable evidence that those with mantle-derived sulphur are confined to the Archaean and early Proterozoic (see Chapter 12).

TITANIUM OXIDE ORES OF THE ANORTHOSITE ASSOCIATION

The mid-Proterozoic mobile belts (1800–1000 Ma old) of Laurasia and Gondwanaland are characterized by post-tectonic andesine-labradorite anorthosites containing titanium-iron oxide deposits. Examples occur at Bergen, Egersund and Lofoten (Norway); St. Urbain and Allard Lake (Quebec); Iron Mountain (Wyoming), and Sanford Lake (New York). According to Windley (1984), most of these Proterozoic anorthosites crystallized in the interval 1700–1200 Ma with a peak at 1400 Ma, which indicates an important metallogenic epoch. These anorthosites are confined to two linear belts in the

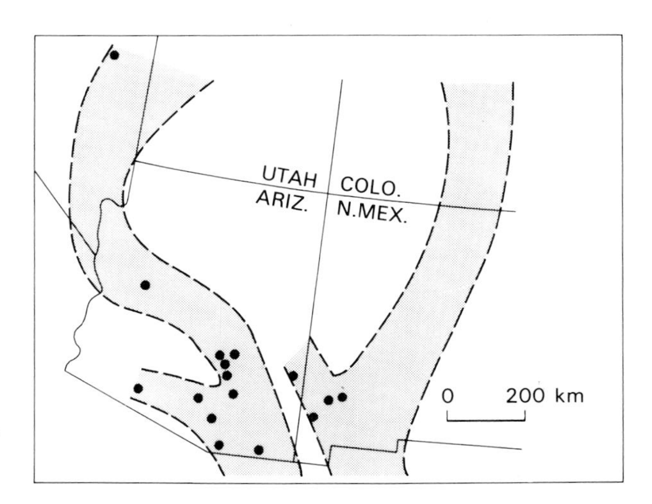

Fig. 6.4. Trace element delineated metallogenic provinces in the south-western USA. The positions of some major porphyry copper deposits are indicated. (Modified from Burnham 1959).

northern and southern hemispheres when plotted on a pre-Permian continental drift reconstruction and thus they also outline two metallogenic provinces for titanium.

TRACE ELEMENT DELINEATED PROVINCES

Burnham (1959) showed that the major ore deposits of the south-western USA lie in provinces outlined by greater than average trace element content in the crystal lattices of chalcopyrite and sphalerite. The most useful trace elements were found to be cobalt, gallium, germanium, indium, nickel, silver and tin. These exhibit well-defined geographical distributions (Fig. 6.4). Burnham suggested that the variations in trace element content are probably due to variations in the compositions of the fluids during crystal growth. The metallogenic belts appear to be independent of time, wall rock type or intrusions, and Burnham consequently suggested that they are of deep-seated origin and related to deep-seated compositional heterogeneities in the mantle.

Part II
Examples of the More Important Types of Ore Deposit

7

Classification of Ore Deposits

The classification of objects in the natural world has always been one of the prime interests of the physical and biological sciences. In geology and biology it is particularly important as it provides a shorthand method of referring to groups of objects having common properties. Without the use of classifications, the comparison of fossils from an evolutionary or palaeogeographical point of view would be practically impossible. Similarly, for the discussion of magmatic processes we must classify igneous rocks. The best classifications are generally those which have no reference to the origin of geological material. Once

Table 7.1. General characteristics of hypothermal deposits. (After Lindgren 1933).

Depth of formation	3000–15 000 m.
Temperature of formation	≈ 300–600°C
Occurrence	In or near deep-seated acid plutonic rocks in deeply eroded areas. Usually found in Precambrian terranes, rarely in young rocks. Often found in reverse faults
Nature of ore zones	Fracture-filling and replacement bodies with the latter phenomenon often more prevalent leading to irregular-shaped orebodies; nevertheless these are frequently broadly tabular. Sheeted zones common, also bedding plane deposits and short, irregular veins. Boundaries usually sharp with limited amount of ore disseminated in walls. Good persistence in depth
Ores of	Au, Sn, Mo, W, Cu, Pb, Zn, As
Ore minerals	*Magnetite*, specularite, *pyrrhotite*, *cassiterite*, *arsenopyrite*, molybdenite, bornite, chalcopyrite, Ag-poor gold, *wolframite*, *scheelite*, pyrite, galena. *Fe-rich sphalerite* (marmatite)
Gangue minerals	*Garnet*, plagioclase, *biotite*, *muscovite*, *topaz*, *tourmaline*, epidote, quartz (often originally *high quartz*), *chlorite* (*high Fe variety*), carbonates
Wall rock alteration	*Albitization*, *tourmalinization*, *rutile development*, sericitization in siliceous rocks, chloritization. Wall rocks are often crisp and sparkling
Textures and structures	Often very coarse-grained, frequently banded, fluid inclusions present in quartz
Zoning	Textural and mineralogical changes with increasing depth are very gradual over thousands of metres. Gold tellurides may give rise to spectacular bonanzas
Examples	Au of Kirkland Lake, Ontario; Kolar, Mysore; Kalgoorlie, W. Australia; Homestake, Dakota; Cu-Au of Rouyn area, Quebec; Sn of Cornwall

genetic factors are brought into a classification, difficulties arise. The student may have already noted that most igneous rock classifications in use today are essentially free of genetic parameters whereas classifications of pyroclastic (volcaniclastic) rocks have many genetic overtones. These can affect the usefulness of the classification as our ideas on genesis evolve.

In older classifications of ore deposits much emphasis was placed on the mode of origins of deposits, with the result that as ideas concerning these changed, many classifications became obsolete. A good example of this occurs with the volcanic-associated massive sulphide deposits which, thirty years ago, were generally held to be formed by replacement at considerable depths within the crust, but are now thought to be the product of deposition in open spaces at the volcanic- or sediment-sea water interface. They have, therefore, moved from the class of hydrothermal-replacement deposits to that of volcanic-exhalative deposits, and who will be so intrepid as to suggest that future generations of geologists may not postulate new theories concerning the genesis of this and other classes of deposit! If a classification is to be of any value it must be capable of including all known ore deposits so that it will provide a framework and a terminology for discussion and so be of use to the mining geologist, the prospector and the exploration geologist.

Table 7.2. General characteristics of mesothermal deposits. (After Lindgren 1933).

Depth of formation	1200–4500 m
Temperature of formation	200–300°C
Occurrence	Generally in or near intrusive igneous rocks. May be associated with regional tectonic fractures. Common in both normal and reversed faults
Nature of ore zones	Extensive replacement deposits or fracture-fillings. Boundaries of orebodies often gradational from massive to disseminated ore. Tabular bodies, sheeted zones, stockworks, pipes, saddle-reefs, bedding-surface deposits. Fissures fairly regular in dip and strike
Ores of	Au, Ag, Cu, As, Pb, Zn, Ni, Co, W, Mo, U, etc.
Ore minerals	*Native Au*, *chalcopyrite*, *bornite*, pyrite, *sphalerite*, *galena*, *enargite*, *chalcocite*, *bournonite*, argentite, *pitchblende*, *niccolite*, *cobaltite*, *tetrahedrite*, sulphosalts
Gangue minerals	Lack high temperature minerals (garnet, tourmaline, topaz, etc.), albite, *quartz*, *sericite*, *chlorite*, *carbonates*, *siderite*, epidote, montmorillonite
Wall rock alteration	Intense chloritization, carbonitization or sericitization. Walls often dull
Textures and structures	Less coarse than hypothermal ores, pyrite, when present, is often very fine-grained. Veins are often banded, large lenses usually massive
Zoning	Gradual but definite change in mineralization with depth, e.g. Butte. Good vertical range, many deposits not bottomed after 1800 m of mining
Examples	Au of Bendigo, Australia; Ag of Cobalt, Ontario; Ag-Pb of Coeur d'Alene, Idaho and Leadville, Colorado; Cu of Butte, Montana

Table 7.3. General characteristics of epithermal deposits. (After Lindgren 1933).

Depth of formation	Near surface to 1500 m
Temperature of formation	50–200°C
Occurrence	In sedimentary or igneous rocks, especially in or associated with extrusive or near surface intrusive rocks, usually in post-Precambrian rocks not deeply eroded since ore formation. Often occupy normal fault systems, joints, etc.
Nature of ore zones	Simple veins—some irregular with development of ore chambers—also commonly in pipes and stockworks. Rarely formed along bedding surfaces. Little replacement phenomena
Ores of	Pb, Zn, Au, Ag, Hg, Sb, Cu, Se, Bi, U
Ore minerals	*Native Au now often Ag-rich*, native Ag, *Cu*, Bi. Pyrite, *marcasite*, *sphalerite*, *galena*, chalcopyrite, *cinnabar*, jamesonite, *stibnite*, *realgar*, *orpiment*, *ruby silvers*, *argentite*, *selenides*, tellurides
Gangue minerals	SiO_2 as *chert*, *chalcedony* or crystalline quartz—often amethystine, (sericite), low Fe chlorite, epidote, carbonates, fluorite, baryte, *andularia*, *alunite*, *dickite*, rhodochrosite, *zeolites*
Wall rock alteration	Often lacking, otherwise chertification, kaolinization, pyritization, dolomitization, chloritization
Textures and structures	Crustification (banding) very common, often with development of fine banding, cockade ore, vugs and brecciation of veins. Grain size very variable
Zoning	Type of mineralization may vary abruptly with depth, often having only a small vertical range (telescoping) mostly bottom at 300–900 m. Grade variable with occurrence of bonanzas within low grade ore
Examples	Au of Cripple Creek, Colorado; Comstock, Nevada; Keweenawan Coppers; Sb of China

Table 7.4. General characteristics of telethermal deposits. These are believed to have formed a long way from the parent magma at low temperatures and high in the crust. Temperatures are still higher than in ground waters.

Depth of formation	Near surface
Temperature	± 100°C
Occurrence	In sedimentary rocks or lava flows, often in areas where plutonic rocks are apparently absent
Nature of ore zones	In open fractures, cavities, joints, fissures, caverns, etc. No replacement phenomena
Ores of	Pb, Zn, Cd, Ge
Ore minerals	*Galena* (poor in Ag), *sphalerite* (poor in Fe, may be rich in Cd), *marcasite* in excess over pyrite, cinnabar, etc. Viz: similar to epithermal mineralogy (Table 7.3)
Gangue minerals	Calcite, low-Fe dolomite, etc.
Wall rock alteration	Dolomitization and chertification
Textures and structures	As epithermal
Examples	Tri-State, USA, Pb-Zn ores etc.; many Hg deposits

Classifications of ore deposits have been based upon commodity (copper deposits, iron deposits, etc.) morphology, environment and origin. Commodity and morphological classifications may be of value to economists and mining engineers, but they lump too many fundamentally different deposit types together to be of much use to geologists. In the past, ore geologists have been inclined to favour genetic classifications but in recent years there has been a swing away from such ideas towards environmental-rock association classifications. Good examples of this trend are to be found in Stanton (1972) and Dixon (1979) and a comprehensive discussion of classification schemes can be found in Wolf (1981).

In 1913, Lindgren put forward the most influential classification which has ever been proposed. It is still used by many geologists, particularly when discussing epigenetic-hydrothermal deposits, and all ore geologists should be aware of the descriptive terms used by Lindgren and added to by L.C. Graton in later modifications. In this part of Lindgren's classification, epigenetic-hydrothermal deposits are classified according to their depth and temperature of formation — hypothermal deposits being deep-seated high temperature deposits; mesothermal deposits those formed at low temperatures and medium depths, and epithermal deposits being near surface. Later terms include leptothermal to cover deposits gradational from mesothermal to epithermal, and telethermal for very low temperature deposits formed at great distances from the source of the hydrothermal solutions which gave rise to them. In the field, these various deposit types have to be recognized from their mineral assemblages, type of wall rock alteration, etc. The features used for this purpose and general features of these deposits are given in Tables 7.1–7.4. One of the uses of this and other classifications is that if we can classify a particular deposit then we can compare it with others of the same class and make predictions as to its behaviour in depth. Thus, recognition that a deposit is hypothermal suggests that it will have great continuity in depth, whereas epithermal and telethermal deposits may bottom very quickly and be of limited vertical development. Again, recognition that a deposit is of massive volcanic-exhalative type should stimulate the mining geologist into searching for a possible underlying feeder stockwork which may be exploitable.

A glance at the contents pages of this book will show that an environmental-rock association classification is favoured although a whiff of genesis and morphology is included!

8

Diamond Deposits in Kimberlites and Lamproites

Long known as the hardest of naturally occurring minerals, the exciting recent work of Richardson *et al.* (1984) has shown diamonds to be among the oldest minerals in the earth, capable of being picked up as exotic fragments by magmas generated at great depths and then surviving both this ordeal and subsequent violent volcanic activity at and near the earth's surface. The endurance of this mineral is thus truly epitomized by the etymology of its name which is derived from the Greek *adamas* — unconquerable. Besides being exotic in their genesis and their properties, diamonds are still weighed in the old units known as carats. This was formerly defined as 3.17 grains (avoirdupois or troy) but the international (metric) carat is now standardized as 0.2 g. [This carat should not be confused with that used to state the number of parts of gold in 24 parts of an alloy (usual American spelling 'karat')].

The ultimate bedrock sources of diamonds are the igneous rocks kimberlite and lamproite and it is these occurrences which are dealt with in this chapter. Important amounts of diamond are recovered from beach and alluvial placer deposits which are described in Chapter 18. Not all kimberlites and lamproites contain diamond and, in those that do, it is present only in minute concentrations. For example, in the famous Kimberley Mine (RSA), 24 Mt of kimberlite yielded only 3 t of diamond, or one part in eight million. From the revenue point of view it is not necessarily the highest grade mines that provide the greatest return — what matters is the percentage of gem quality diamonds. Each mining region and even each mine has a different percentage, and some may be 90% or more while others may rely on only an occasional gem stone to boost their income. Total world production of natural diamonds is about 64 Mc p.a., of which about 45 Mc are of industrial grade. In addition, 60–80 Mc p.a. of synthetic industrial diamonds are manufactured. The leading world producers are shown in Table 8.1 and world bedrock diamond fields and the general occurrence of kimberlites in Fig. 8.1. The production figures will show a dramatic change when the newly found AK1 pipe of the Argyle Mine, Western Australia, comes on stream in 1985–6, as this mine's full production will be equivalent to a 60% addition to the total world production of natural diamonds (Anon. 1982a).

Morphology and nature of diamond pipes

Many near surface, diamond-bearing kimberlites and lamproites occur in

Table 8.1. World production of natural diamonds (Mc) in 1984. (Bedrock and placer deposits.)

Zaïre	18.5
Botswana	12.9
USSR	12.0
RSA	9.8
Namibia	0.93
Angola	0.92
Others	8.82
Total	63.87

pipelike diatremes (often just called pipes), which are small, generally less than 1 km^2 in horizontal area. They are often grouped in clusters of which the Lesotho occurrences are a good example (Nixon 1973). Lesotho probably has more kimberlites per unit area than anywhere else in the world. Seventeen pipes, twenty-one dyke enlargements or 'blows' and over two hundred dykes are listed by Nixon. In northern Lesotho over 180 kimberlites are known, i.e. about one per 25 km^2, but a high proportion are barren. Some diatremes are known to coalesce at depth with dykes of non-fragmental kimberlite (Fig. 8.2). These dykes are thin, usually less than 10 m in diameter but may be as much as 14 km long. Diatremes and dykes may be mutually cross-cutting. Their morphology and internal structures have been succinctly reviewed by Nixon (1980a) and the interested reader is strongly recommended to look at the entire contents of the book in which this paper occurs.

Some recently formed diatremes terminate at the surface in maars; these volcanic craters may be filled with lacustrine sediments to a depth of over 300 m. The sediments, which may be diamondiferous with a heavier concentration of more and bigger diamonds near the crater rim shoreline, are sometimes affected by subsidence, perhaps of a cauldron nature. The kimberlite beneath the sediments may be relatively barren. Below the flared crater area is the vertical or near vertical pipe itself which typically has walls dipping inwards at about 82° and a fairly regular outline producing the classic, carrot-shaped diatreme (Fig. 8.3) that may exceed 2 km in depth. Smaller, more eroded intrusions often have more irregular outlines (Fig. 8.4). Lower down in the 'zone of pipe generation', pipes may no longer have vertical axes and may separate into discrete root-like bodies. Upward terminating bodies ('blind pipes') are also known.

At the surface kimberlite may be weathered and oxidized to a hydrated 'yellow ground', which gives way at depth to fresher 'blue ground', and 'hardebank' (resistant kimberlite that often crops out and does not disintegrate easily upon exposure). In the upper levels of pipes the kimberlite is usually in the form of so-called agglomerate (really a tuffisitic breccia with many rounded and embayed fragments in a finer grained matrix) and tuff. The rounded fragments are often xenoliths of metamorphic rocks from deeper crust, or garnet-peridotite or eclogite from the upper mantle. Their rounded nature is attributed to a gas-fluidized origin (Dawson 1971). Magmatic kimberlite bodies

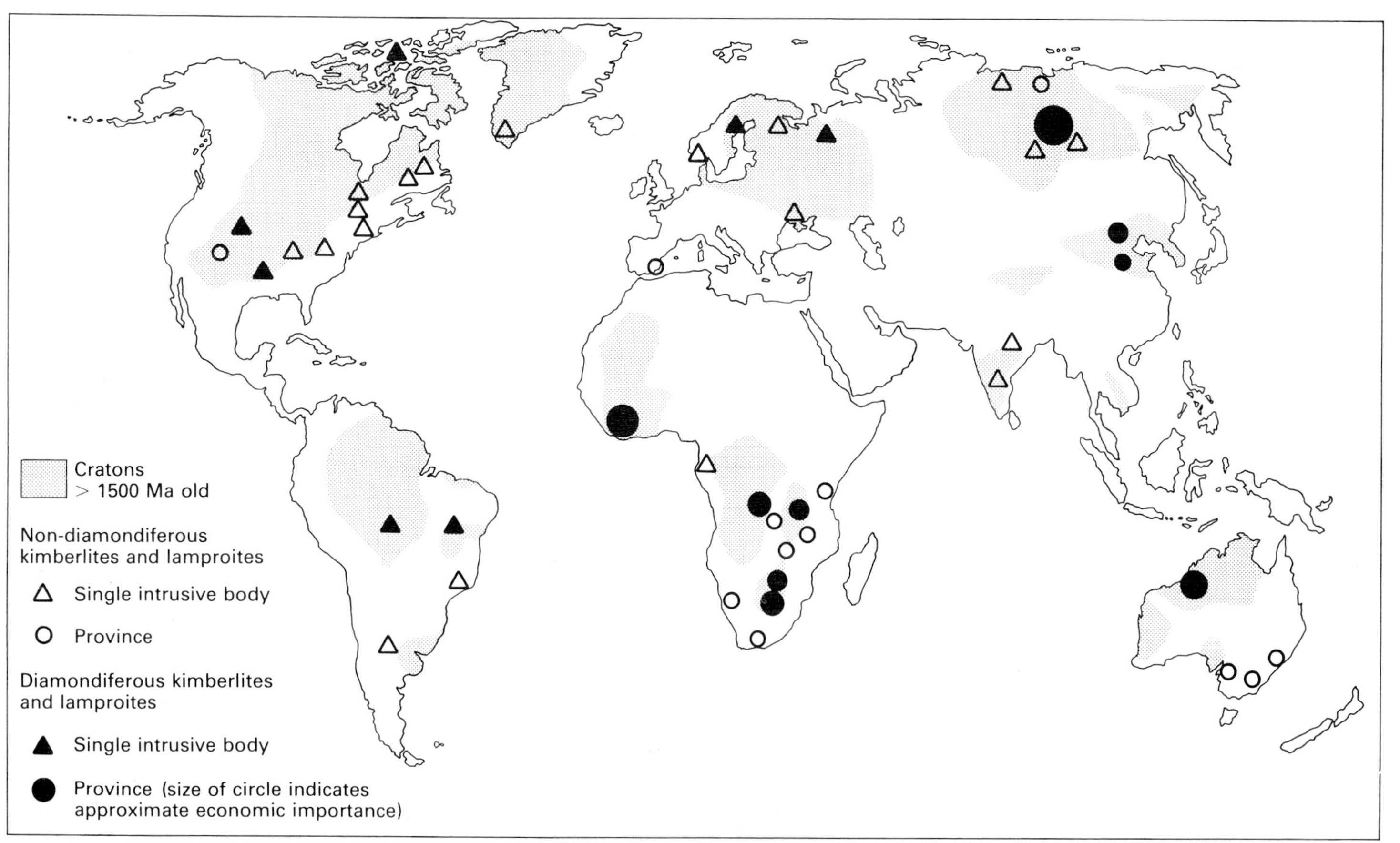

Fig. 8.1. Distribution of diamondiferous and non-diamondiferous kimberlites and lamproites. (After Ferguson 1980a and other sources).

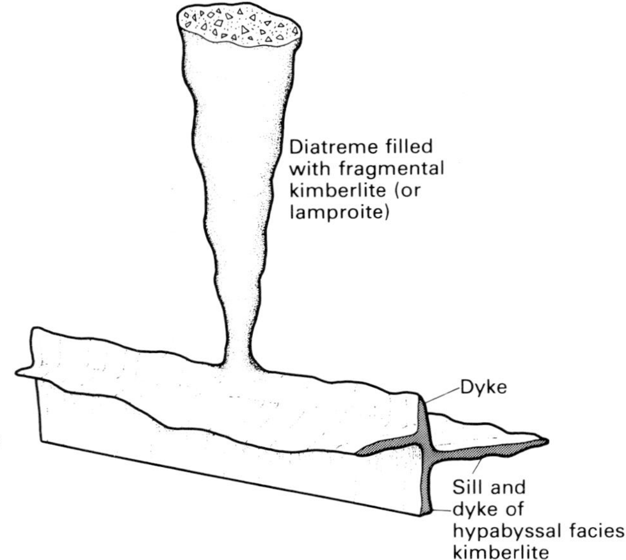

Fig. 8.2. Schematic drawing showing the relationshop between an explosive kimberlite diatreme and deep-seated sills and a feeder dyke filled with nonfragmental kimberlite magma that solidified *in situ*. (After Dawson 1971).

are mainly confined to the root zones of pipes, where they are often in the form of intrusion breccias which grade up with gradational boundaries into the agglomerate, or to sills and dykes.

Kimberlites and lamproites and their emplacement

As Scott Smith & Skinner (1984) have pointed out, kimberlite has traditionally been considered to be the only important primary source of diamond. However, both these authors and others, e.g. Jaques & Ferguson (1983) and Jaques *et al.* (1984), have pointed out that lamproites in Arkansas and Western Australia (including the AK1 pipe mentioned above) carry significant amounts of diamond, a point of such exploration importance that it is of value to look briefly at these two rock-types.

Kimberlite may be defined (Best 1982; Clement *et al.* 1984) as a potassic ultrabasic hybrid igneous rock containing large crystals (megacrysts) of olivine, enstatite, Cr-rich diopside, phlogopite, pyrope-almandine and Mg-rich ilmenite in a fine-grained matrix containing several of the following minerals as prominent constituents: olivine, phlogopite, calcite, serpentine, diopside, monticellite, apatite, spinel, perovskite and ilmenite. Some of these minerals are used as indicator minerals in stream sediment and soil samples in the search for kimberlites, e.g. red-brown pyrope, purple-red chromium pyrope, Mg-rich ilmenite, chromium diopside (Nixon 1980b) and a study of their morphology can be used to indicate the proximity of their source (Mosing 1980).

Lamproites are defined by Jaques *et al.* (1984) and Scott Smith & Skinner (1984) as potash- and magnesia-rich lamprophyric rocks of volcanic or

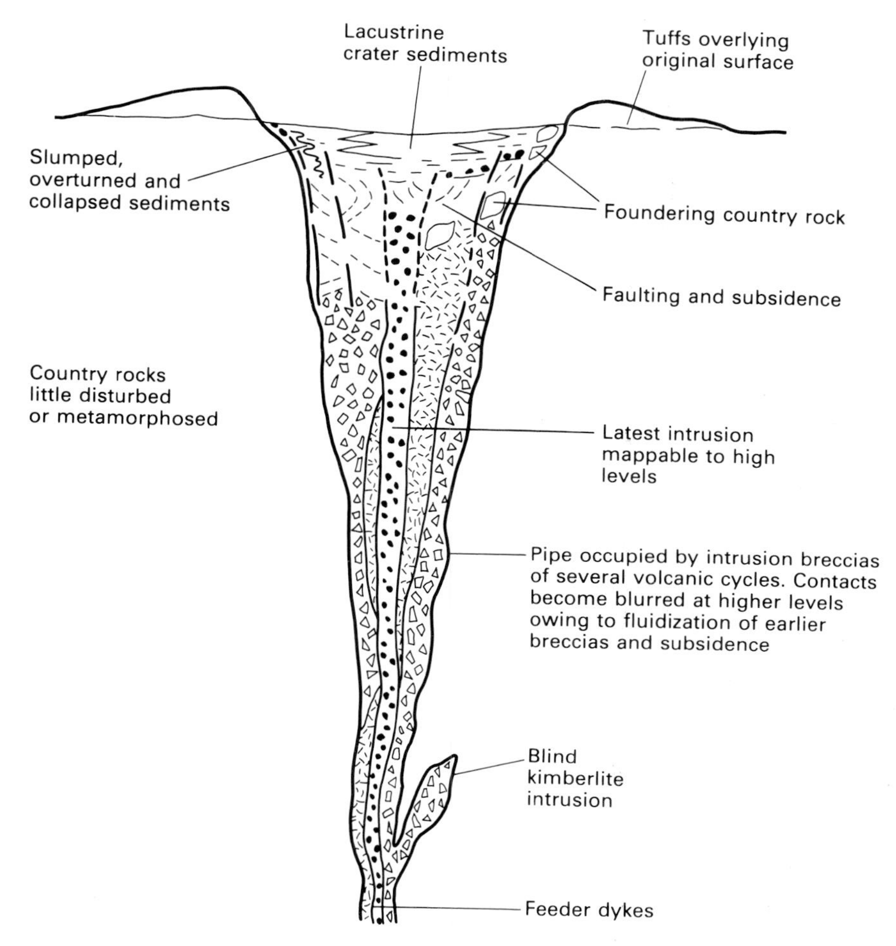

Fig. 8.3. Schematic diagram of a kimberlite diatreme (pipe) and maar (volcanic crater below ground level and surrounded by a low tuff rim). The maar can be up to 2 km across. (After Nixon 1980a).

hypabyssal origin comprising mineral assemblages containing one or more of the following primary phenocrystal and/or groundmass phases: leucite, Ti-rich phlogopite, clinopyroxene, amphibole (typically Ti-rich, potassic richterite), olivine and sanidine. Accessories may include priderite, apatite, nepheline, spinel, perovskite, wadeite and ilmenite. Glass may be an important constituent of rapidly chilled lamproites. Xenoliths and xenocrysts, including olivine, pyroxene, garnet and spinel of upper mantle origin, may be present and diamond as a rare accessory. Lamproites have greater mineralogical and textural variations than kimberlites and although they do contain some minerals characteristic of kimberlites, the chemical compositions of these two minerals are often distinctly different. Lamproitic craters are generally wider and shallower than those of kimberlites, and in the crater, magmatic

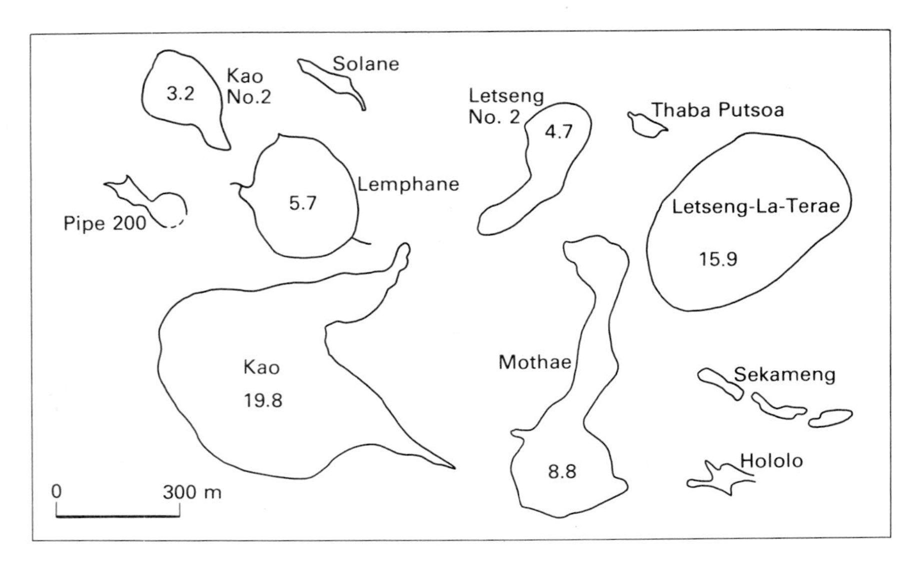

Fig. 8.4. Surface outlines and areas (in hectares) of some Lesotho kimberlite pipes from various localities. (After Nixon 1980a).

lamproite, analogous to the magmatic kimberlite found only in the deeper levels of kimberlite pipes, may be present as intrusions into the crater sediments and in the form of ponded lava lakes overlying the pyroclastic rocks and sediments. Fertile lamproites appear to be the silica-saturated orendites and madupites which carry sanidine rather than leucite (Gold 1984).

Kimberlites and lamproites are generally regarded as having been intruded upwards through a series of deep-seated tension fractures, often in areas of regional doming and rifting, in which the magmas started to consolidate as dykes. Then highly gas-charged magma broke through explosively to the surface at points of weakness, such as cross-cutting fractures, to form the explosion vent which was filled with fluidized fragmented kimberlite or lamproite and xenoliths of country rock. Quietly intruded magma was then sometimes emplaced as described above.

SOME EXAMPLES

(a) *Western Australia*. The very recent (1977) find of diamonds in Western Australia in and around the Kimberley Craton (Fig. 8.5) has led to the discovery of three diamondiferous provinces with over 100 kimberlites and lamproites. A radiometric age of 20 Ma (Jaques & Ferguson 1983) makes them the youngest known diamondiferous pipes, and the lamproitic Argyle AK1 pipe is richer in diamond than any known kimberlite. Proven reserves are 61 Mt grading 6.8 c t^{-1} and probable reserves 14 Mt at 6.1 c t^{-1}, with a gem content of 5%, cheap gem 40% and industrial diamonds 55%. It is planned to mine 3 Mt p.a. which should produce 25 Mc p.a. over a minimum twenty year

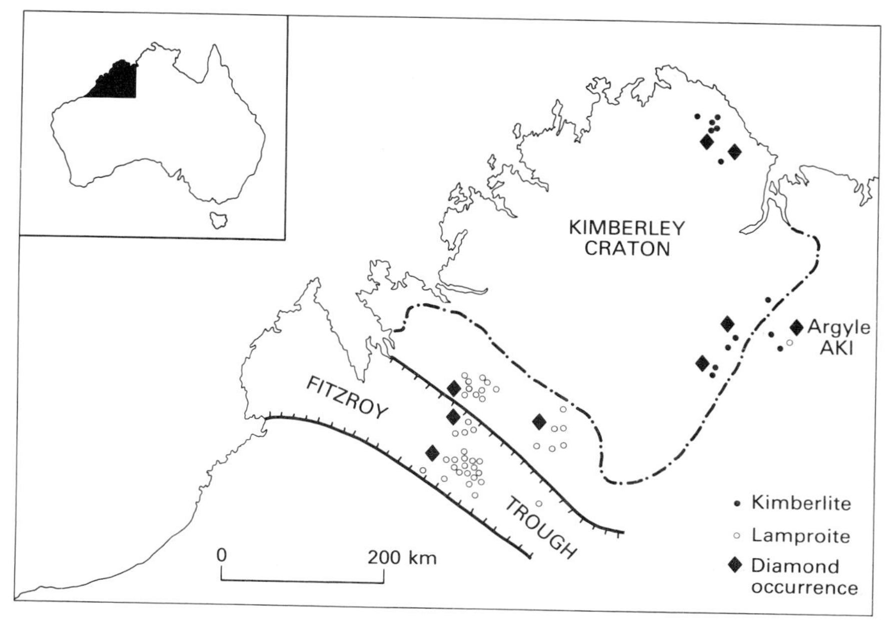

Fig. 8.5. Kimberlites, lamproites and diamond occurrences in the Kimberley region of Western Australia. (After Atkinson *et al.* 1984).

mine life. Before ore is produced, some 20 Mt of waste will be removed from the southern end of the deposit (Anon. 1983a, 1983b).

The AK1 deposit is emplaced in Proterozoic rocks aged about 1100 Ma. The crater zone is 1600 m long and 150–600 m wide (Fig. 8.6), has a surface area of nearly 50 ha and is filled with sandy lapilli-tuff (about 40% rounded quartz grains with lamproite clasts) which contains some interbedded mudstone and quartzite. Soft rock deformation structures (slumping, microfaulting) are common. In the centre of the northern end of the crater is a body of non-sandy tuff devoid of detrital quartz which is thought to be younger than the sandy tuff. Occasional late, magmatic kimberlite dykes 1–2 m thick and kimberlitic tuffisite dykes cut the sandy tuff. Diamond grades are considerably lower in the non-sandy tuff and the intrusives appear to be barren.

Jaques & Ferguson (1983) and Jaques *et al.* (1984) have suggested that the diamondiferous lamproites of the Firzroy area formed by the partial melting of metasomatized phlogopite-peridotite under H_2O- and F-rich and CO_2-poor conditions. They suggested that low degrees of melting under these conditions produced silica-saturated, olivine-lamproite melts rather than the undersaturated kimberlitic-carbonatitic melts which form under high P_{CO_2}.

(b) *Lesotho*. The deposits of this small nation, already touched on above, make an interesting contrast with those of Western Australia. Among the many non-productive kimberlite pipes the Letseng-La-Terae and neighbouring Kao pipes are notable exceptions. The main pipe is almost 776 × 366 m at the surface

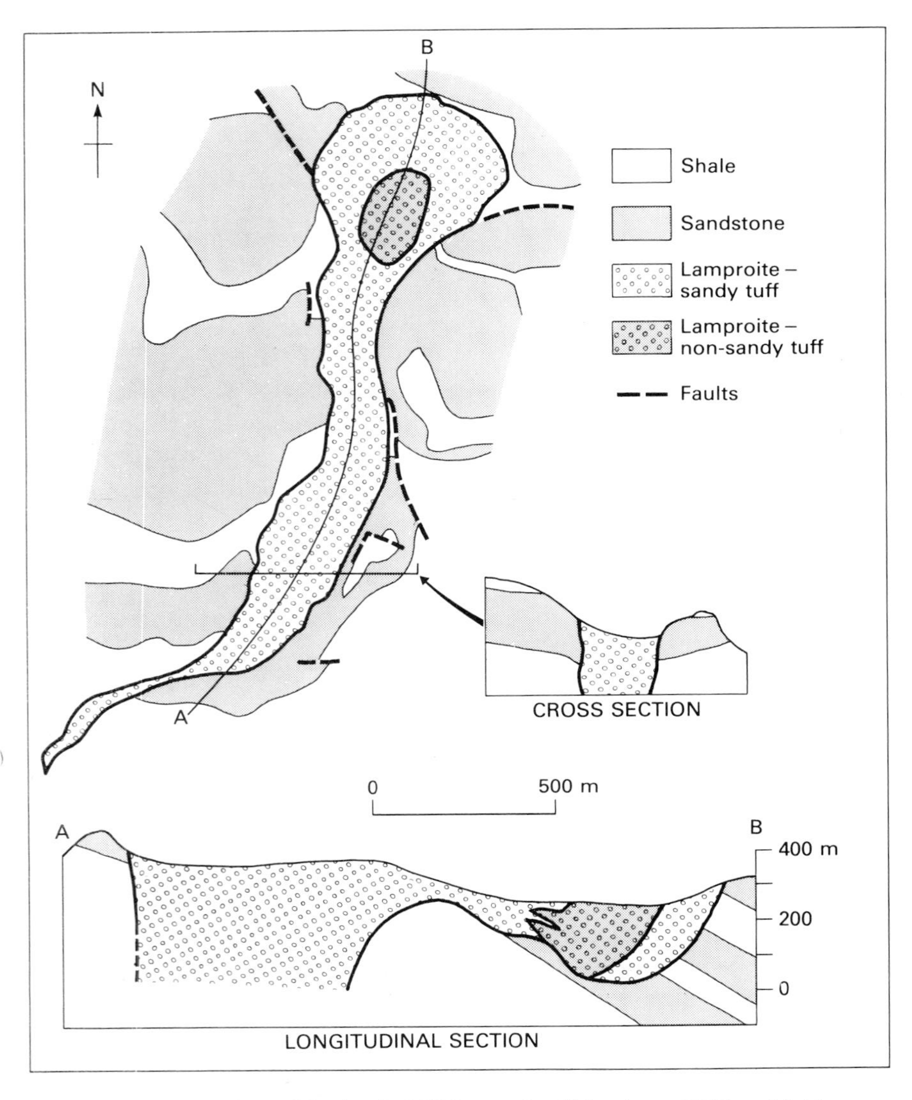

Fig. 8.6. Map and sections of the Argyle AK1 Lamproite. (After Anon. 1983b and Atkinson *et al.* 1984).

and its tadpole shaped satellite pipe is almost 426 × 131 m. The mine, owned by De Beers (75%) and the Government (25%) was opened in 1977 at a capital cost of £23 million. Although production is small, just over 50000 carat p.a. and the grade at 0.309 c t^{-1} is low, 13% of the output is of 10 c and larger stones. 'Profitability', to quote Harry Oppenheimer, De Beers' Chairman, 'depends on a small number of exceptionally beautiful stones.'

(c) *USSR*. The first diamondiferous kimberlite pipe, named Zarnitsa (Dawn), was found in Siberia in 1954. The grade of this pipe was too low for exploitation but since then exploitable pipes have been discovered, among them the rich

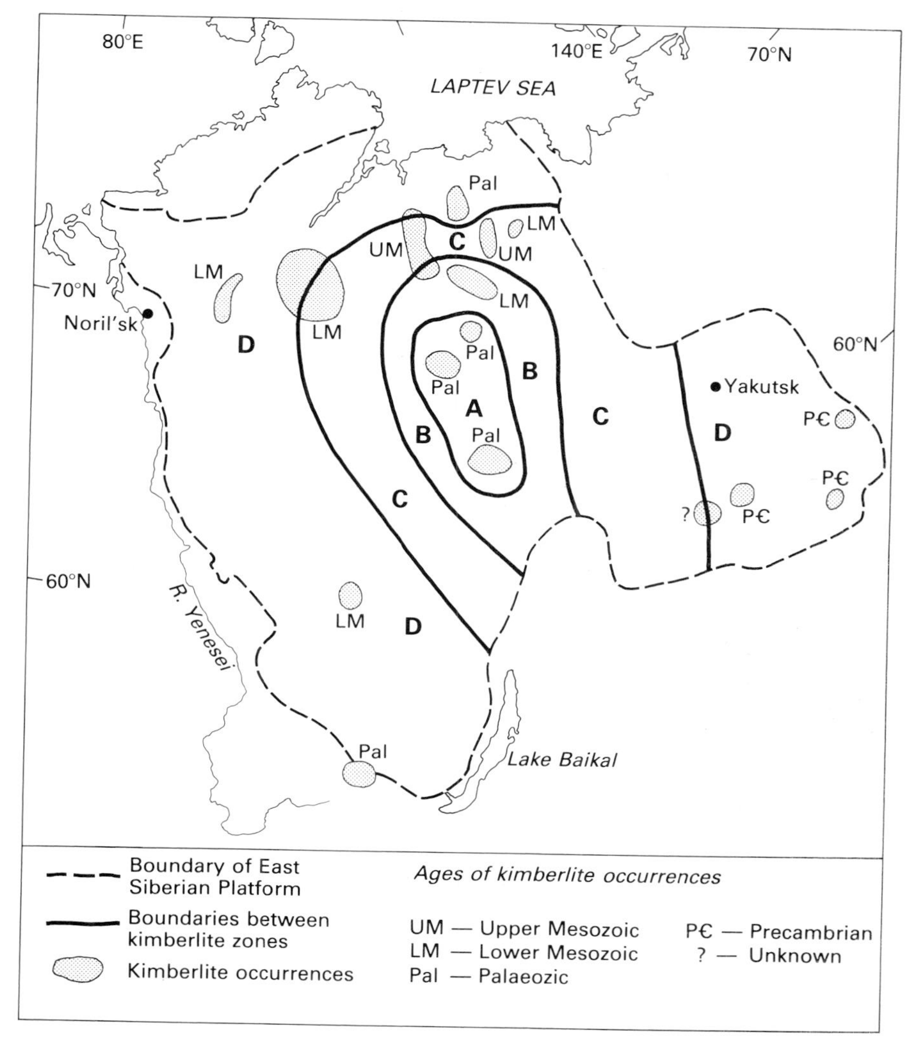

Fig. 8.7. Distribution of kimberlites in the East Siberian Platform showing the zones of different mineralogical subfacies: A = diamond, B = diamond-pyrope, C = pyrope, D = other kimberlites. (Modified from Dawson 1980).

Mir (Peace) pipe. Some 25% of USSR production is of gem quality. The Siberian kimberlites lie within the Siberian Platform (Fig. 8.7), forming groups of different ages (Dawson 1980). As in South Africa, there appears to be a regional zoning of the kimberlites. A central zone of diamondiferous kimberlites is surrounded by a zone with pyrope and low diamond values, then by a zone from which diamond is absent but pyrope present, and finally by an outer zone in which neither of these high pressure minerals is present in the kimberlites.

Origin of diamonds and some geological guides to exploration

Extremely high temperature and pressure are required to form diamond rather than graphite from pure carbon — of the order of 1000°K and 3.5 GPa, equivalent in areas of 60 km thick continental crust to a depth of about 117 km. As coesite rather than stishovite is found as inclusions in diamonds, and the inversion curve for these two silica polymorphs is equivalent to a depth of about 300 km, the approximate range for diamond genesis is 100–300 km.

For many decades there has been a very active debate as to whether diamonds crystallized from the magmas which cooled to form the igneous rocks in which they are now found (phenocrysts), or whether they were picked up by these magmas as exotic fragments derived from the diamond stability field within the upper mantle (xenocrysts). The much greater abundance of diamonds in eclogite xenoliths than in the surrounding kimberlite suggests that they have been derived from disaggregated eclogite (Robinson 1978). The former hypothesis infers that successful exploration consists of finding kimberlites and lamproites but the latter implies a more sophisticated approach (Rogers & Hawkesworth 1984). Furthermore, in southern Africa, kimberlites erupted within the confines of the Archaean craton are diamondiferous, while those in adjacent younger orogenic belts are barren (Nixon *et al.* 1983); a general point which might be deduced on a global scale from Fig. 8.1.

Most kimberlites and lamproites are younger than 200 Ma. Richardson *et al.* (1984) have now obtained convincing dates on garnet inclusions in diamonds from two southern African kimberlites (Kimberley and Finsch, 160 km apart and about 90 Ma old) of greater than 3000 Ma. It seems highly probable that the diamond hosts are of a similar great age.

This and other evidence is leading workers to the opinion that diamonds grew stably within the upper mantle, probably either in eclogite or ultramafic rocks or both. The estimated conditions of equilibration for ultramafic suite minerals co-existing with diamonds (Meyer 1985, Nickel & Green 1985) suggest that this growth takes, or took, place in a layer between about 132 and 208 km in depth beneath continents and 121–197 km beneath oceans, at temperatures of 1200–1600°K, provided that carbon is (or was) present. Thus any magma that samples a diamondiferous zone of this layer may bring diamonds to the surface if it moves swiftly enough. The speed of ascent of such magmas has been calculated to be around 70 km h^{-1} (Meyer 1985). Slow ascent could allow time for the absorption of diamonds by transporting magmas as the pressure decreased, but there is some uncertainty over this (Harris 1985).

The diamondiferous layer in the upper mantle is probably discontinuous because its formation requires the existence (and preservation) of a thick 'cold' crust above, otherwise diamond generating (and destroying) temperatures may be present at too shallow a depth. Absence of the layer may account for the absence of diamonds from many kimberlites, but another important reason is that many kimberlitic magmas were developed at too shallow a level to be able to sample it. This was probably the case in New South Wales where *most* of the kimberlites were generated at 60–70 km depth (Ferguson 1980b). A mining company exploring a particular pipe could now save itself considerable expenditure by having a study made of the equilibrium conditions of crystallization of ultramafic or eclogitic xenoliths and xenocrysts. This would determine whether they crystallized within the diamond generating layer and therefore whether the pipe was likely to be fertile or barren. (Another good example of the industrial application of what might once have been described as useless, esoteric scientific research.) This now important tool of applied science might also be used to identify other igneous rocks that had sampled diamond zones because, as Gold (1984) has pointed out, we now have reports of diamonds in alkali basalts, ophiolites and andesites, and we should keep open minds on the subject of host rocks for diamonds. Gurney (1985) has suggested the use of garnet compositions to differentiate between diamondiferous and non-diamondiferous intrusions.

Are all diamonds of great age? The answer will only come from further work. Some diamonds may have been formed from recycled 'crustal' carbon (Milledge *et al.* 1983) and these could yield post-Archaean ages. Why are diamondiferous intrusions virtually confined to old crustal areas? Is it because thicker lithosphere was stabilized during the Archaean than more recently, or because the upper mantle in Archaean times had a much higher carbon content? (Rogers & Hawkesworth 1984). Should we confine diamond exploration to continental areas underlain by old Precambrian cratons, as present experience suggests we should? Or keep a more open mind and pay some heed to the occurrence of diamondiferous breccia pipes in south Borneo, the Kamchatka Peninsula and the Solomon Islands? (Hutchison 1983).

The West Australian occurrences in cratonized Proterozoic mobile belts will surely lead diamond explorationists to investigate other environments, and the report of a diamondiferous kimberlite in British Columbia (Anon. 1985a), appears to confirm that some are already ranging far afield in the diamond hunt. Unfortunately they now have to look even more keenly over their shoulders at the back-room boys who have already taken another step forward in the manufacture of bigger and better synthetic diamonds (Anon. 1985b). 'Beware of substitutes!' — as a well known toothpaste producer used to print on its products!

Postscripts giving recent information on material in this chapter can be found in Appendix 2.

9

The Carbonatite-Alkaline Igneous Ore Environment

Carbonatites

Carbonatite complexes consist of intrusive magmatic carbonates and associated alkaline igneous rocks, ranging in age from Proterozoic to Recent. They belong to alkaline igneous provinces and are generally found in stable cratonic regions with major rift faulting such as the East African Rift Valley and the St. Lawrence River Graben. There are, however, exceptions where carbonatite complexes are not directly associated with any alkalic rocks, e.g. Sangu Complex, Tanzania, and Kaluwe, Zambia; moreover, not all alkalic rock provinces and complexes have associated carbonatites. Verwoerd (1964) pointed out that about 90% of carbonatites occur in areas of Precambrian granite or gneiss. However, Barker (1969) has shown that in North America, 66% of known alkaline rock occurrences are in Phanerozoic orogenic belts. There is no doubt that, taking a world-wide view, the majority of carbonatites occur in the marginal parts of stable cratonic regions or are closely related to large scale rift structures. An apparent exception of economic importance is the Mountain Pass carbonatite of California. This occurs within the Rocky Mountain area, but it is a Precambrian intrusion into Precambrian gneisses so that its present tectonic environment is a function of considerably later deformation (Moore 1973), and this may be the reason why it is not now in an obviously alkaline igneous province.

Carbonatites together with ijolites and other alkalic rocks commonly form the plutonic complex underlying volcanoes which erupted nephelinitic lavas and pyroclastics. Surrounding such complexes there is a zone of fenitization (mainly of a potassium nature and producing orthoclasites) which has a variable width. The emplacement of carbonatites was effected in stages (Le Bas 1977), with the dominant rock often made up of early C_1 (sövite) intrusions, usually emplaced in an envelope or explosively brecciated rocks. C_1 carbonatite is principally calcite with apatite, pyrochlore, magnetite, biotite and aegirine-augite. C_2 carbonatite (alvikite) usually shows marked flow banding and is medium- to fine-grained compared with the coarse-grained sövite. C_3 carbonatite (ferrocarbonatite) contains essential iron-bearing carbonate minerals and commonly some rare-earth and radioactive minerals. C_4 carbonatites (late stage alvikites) are usually barren. Sövites form penetrative stock-like intrusions and C_2 and C_3 carbonatites form cone-sheets

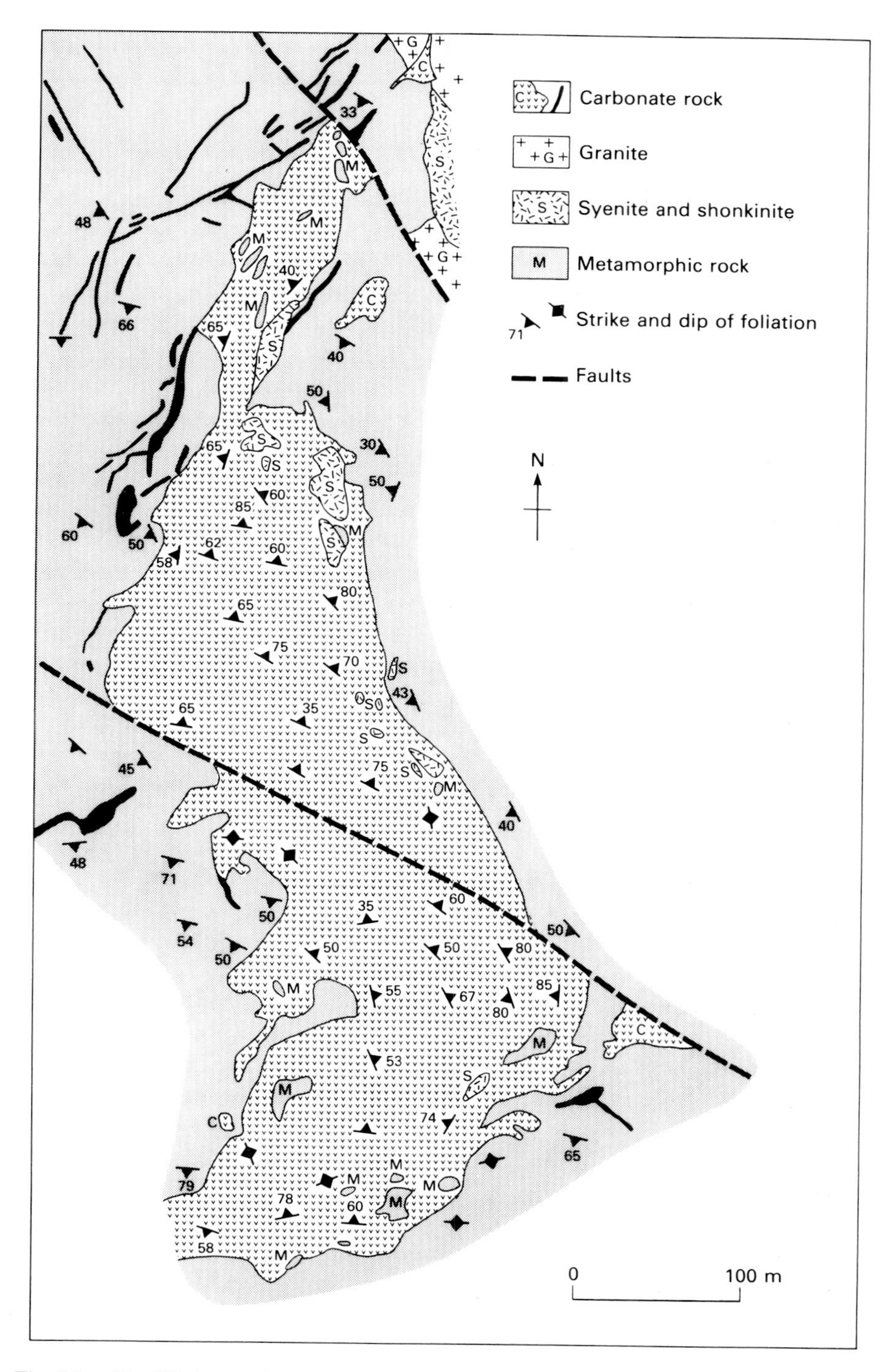

Fig. 9.1. Simplified map of the Sulphide Queen Carbonate Body, Mountain Pass District, California. (After Olson & Pray 1954).

and dykes. The intrusion of C_1 and most C_2 carbonatites is preceded by intense fenitization; other carbonatites show little or no fenitization.

Economic aspects

The most important products of this environment are phosphorus (from apatite), magnetite, niobium (pyrochlore), zirconia and rare-earth elements (monazite, bastnäsite). To date, only one carbonatite complex is a major producer of copper (Palabora, South Africa), but others are known to contain traces of copper mineralization, e.g. Glenover, South Africa; Callandar Bay, Canada and Sulphide Queen, Mountain Pass, USA (Jacobsen 1975). Other economic minerals in carbonatites include fluorite, baryte and strontianite and the carbonatites themselves are useful as a source of lime in areas devoid of good limestones. Carbonatites are highly susceptible to weathering and therefore may have residual, eluvial and alluvial placers associated with them.

THE MOUNTAIN PASS OCCURRENCES, CALIFORNIA

These lie in a belt about 10 km long and 2.5 km wide. One of the deposits, the Sulphide Queen carbonate body, carries the world's greatest concentration of rare-earth minerals (Fig. 9.1). The metamorphic Precambrian country rocks have been intruded by potash-rich igneous rocks and the rare-earth-bearing carbonate rocks are spatially and probably genetically related to these granites, syenites and shonkinites (Olson *et al.* 1954). The rare-earth elements are carried by bastnäsite and parisite, these minerals being in veins that are most abundant in and near the largest shonkinite-syenite body. Most of the 200 veins that have been mapped are less than 2 m thick. One mass of carbonate rock, however, is about 200 m in maximum width and about 730 m long and is claimed to be the largest known orebody of rare-earth minerals in the world. It is called the Sulphide Queen Mine, not because of a high sulphide content, but because it is situated in Sulphide Queen Hill.

Carbonate minerals make up about 60% of the veins and the large carbonate body; they are chiefly calcite, dolomite, ankerite and siderite. The other constituents are baryte, bastnäsite, parisite, quartz and variable small quantities of 23 other minerals. The rare-earth content of much of the orebody is 5–15%.

THE PALABORA IGNEOUS COMPLEX

This Proterozoic complex lies in the Archaean of the north-eastern Transvaal. It resulted from an alkaline intrusive activity in which there were emplaced in successive stages pyroxenite, syenite and ultrabasic pegmatoids (Palabora Mining Co. Staff 1976). The first intrusion was that of a micaceous pyroxenite, kidney-shaped in outcrop (but forming a pipe in depth) and about 6 $\times$ 2.5 km. Ultrabasic pegmatoids were then developed at three centres within the

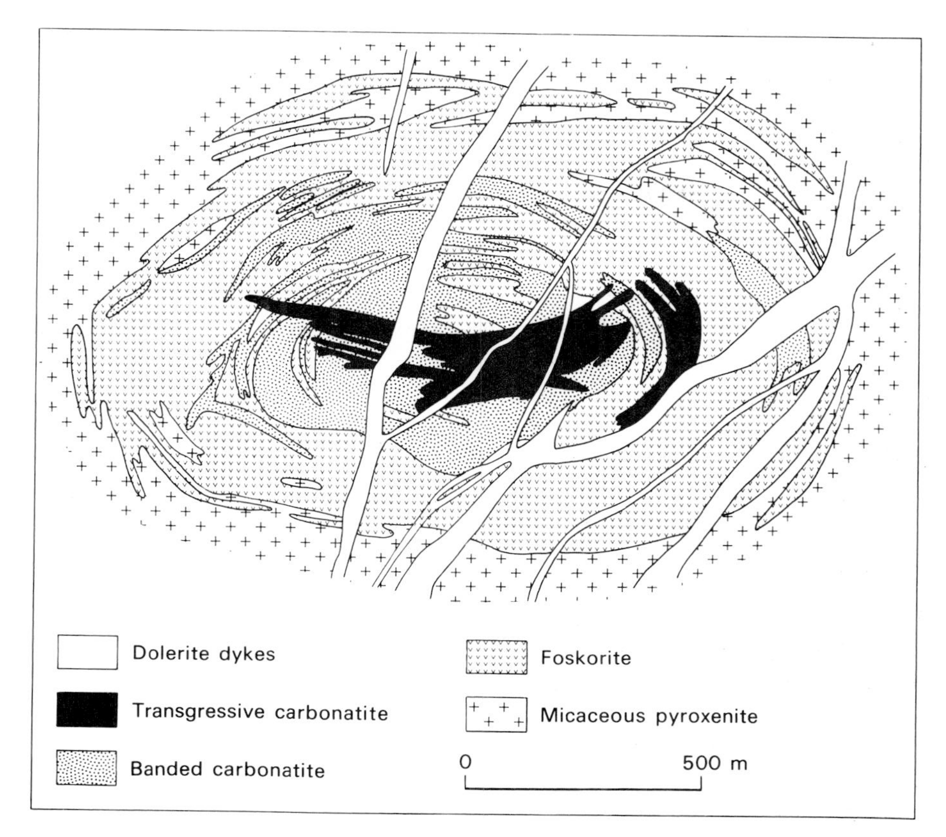

Fig. 9.2. Geology of the 122 m level, Loolekop Carbonatite Complex, Palabora, South Africa. (Modified from Jacobsen 1975). Note: this is only a small part of the whole alkaline igneous complex.

pyroxenite pipe. In the central one, foskorite (magnetite-olivine-apatite rock) and banded carbonatite were emplaced to form the Loolekop carbonate-foskorite pipe which is about 1.4 × 0.8 km (Fig. 9.2). Fracturing of this pipe led to the intrusion of a dykelike body of transgressive carbonatite and the development of a stockwork of carbonatite veinlets. The zone along which the main body of transgressive carbonate was emplaced suffered repeated fracturing, and mineralizing fluids migrated along it depositing copper sulphides which healed the fine discontinuous fractures. These near vertical veinlets occur in parallel-trending zones up to 10 m wide, although individually the veinlets are usually less than 1 cm wide and do not continue for more than 1 m. Diamond drilling has shown that the orebody continues beyond 1000 m below the surface. Ore reserves are about 300 Mt grading 0.69% Cu.

There are three separate open pits at Palabora. The carbonatite and foscorite are worked for copper with by-product magnetite, apatite, gold, silver, PGM, baddeleyite (ZrO_2), uranium, nickel sulphate and sulphuric acid. About 2 km away in the same alkaline complex is the Foskor Open Pit in an apatite-rich pyroxenite that forms the world's largest igneous phosphate deposit. Reserves in the pit area alone, which only covers part of the pyroxenite, are 3 Gt of apatite *concentrates* (36.5% P_2O_5). In a neighbouring pit vermiculite, a weathering product of phlogopite is worked.

THE KOLA PENINSULA – NORTHERN KARELIA ALKALINE PROVINCE

Alkaline igneous complexes and their associated carbonatites show a broad spatial relationship to areas of hot spot activity which may be accompanied by doming and fracturing, the Kola Peninsula (Fig. 9.3) being an excellent example. The Upper Palaeozoic alkaline igneous and carbonatite complexes of this region host a number of extremely large orebodies, of which the most important are Khibina, Kovdor and Sokli.

The USSR is the world's second largest producer of phosphate rock with much of its production coming from Khibina (Notholt 1979). This is a ring complex about 40 km across with inward dipping layered intrusions of various alkaline rock types. One apatite-nepheline orebody forms an arcuate, irregular lens-shaped mass with a strike length of 11 km and a proved dip extension of 2 km. The thickness ranges from 10 to 200 m and averages 100 m and at least 2.7 Gt of ore averaging 18% P_2O_5 are present. The apatite concentrates contain significant SrO and Re_2O_3 values. Annual production is 18 Mt from four mines. Nepheline concentrates are also produced for the manufacture of alumina.

At Kovdor (Fig. 9.4) there is a late stage development of apatite-forsterite rocks and magnetite ores. The orebody is mined by open pit methods. Ore reserves are about 708 Mt grading 36% Fe and 6.6% P_2O_5. Baddeleyite forms a by-product and vermiculite is produced from a separate mining operation. In its mineralization and rock-types this complex has affinities with Palabora. A drawback at Kovdor is the high magnesium content of the apatite concentrate, which presents processing problems at fertilizer plants.

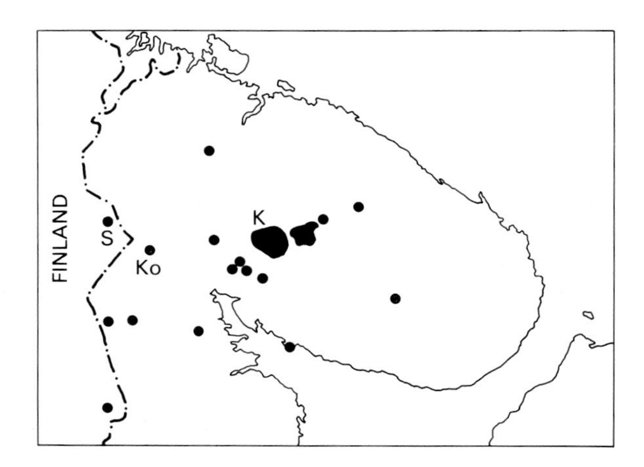

Fig. 9.3. Alkaline igneous and carbonatite complexes in the Kola Peninsula – Northern Karelia Alkaline Province. K = Khibina, Ko = Kovdor, S = Sokli. (After Vartiainen & Paarma 1979).

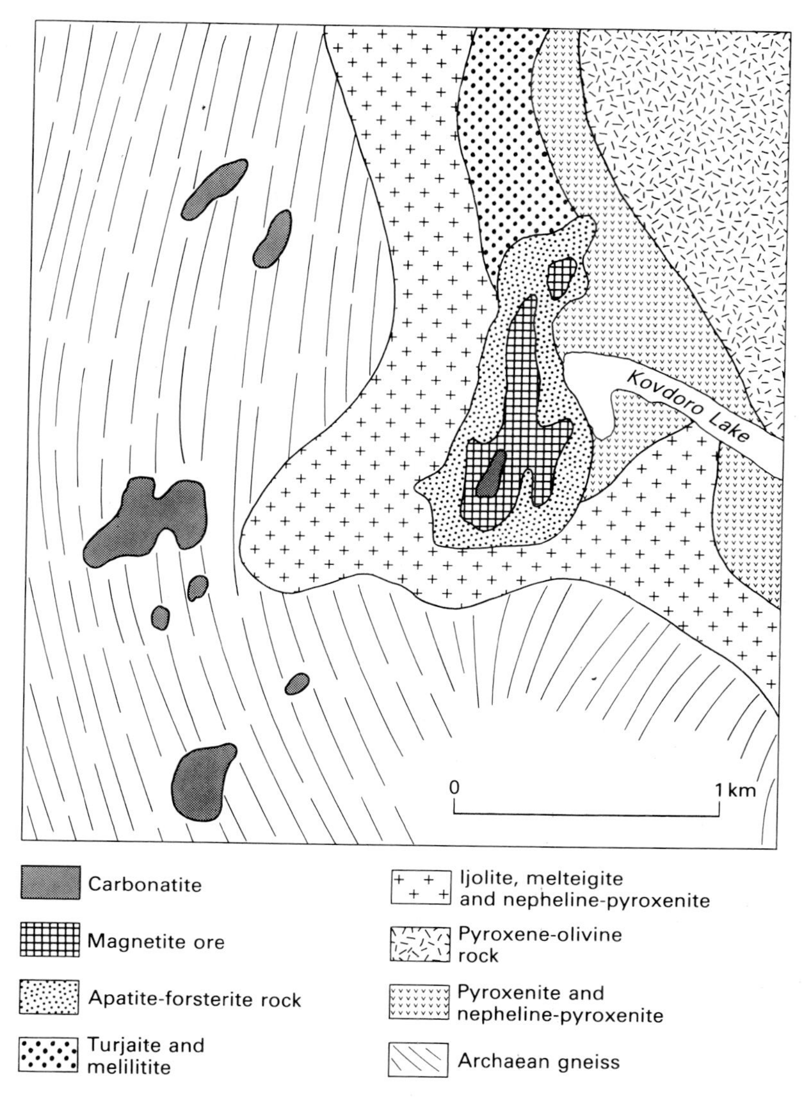

Fig. 9.4. Sketch map of part of the western section of the Kovdor Complex showing the position of the magnetite orebody. (After Rimskaya-Korsakova 1964).

Across the border in Finland is the Sokli Complex. With an areal extent of about 18 km^2 it is one of the world's largest carbonatites and it is remarkable for its partial cover of residual ferruginous phosphate rock, forming an apatite-francolite regolith of a type hitherto regarded as being of tropical origin. Proved reserves are over 50 Mt averaging 19% P_2O_5 with possible lower grade reserves of 50–100 Mt. The deposits vary from a few metres to 70 m in thickness. Pyrochlores with both U-Ta and Th enrichment occur in the carbonatites and apatite-magnetite mineralization in metafoscorites which runs

11–22% Fe (Notholt 1979, Vartiainen & Paarma 1979). Kimberlitic dykes are also present but no diamonds have been found. Seismic evidence suggests that the intrusion tapers downwards from a diameter of about 6 km at the surface to about 1 km at a depth of 5 km. Other carbonatites, e.g. Palabora, appear to taper downwards, a point of some economic significance. This evidence also suggests that at great depths carbonatites, like kimberlites, become dyke-like bodies within deep fracture zones that have tapped the levels in the mantle from which these alkaline rocks and their rare earth and associated elements have probably been derived (Samoilov & Plyusnin 1982).

Accounts of other important carbonatite deposits that the reader may like to study are as follows: Gold *et al.* (1967) on Oka, Quebec (Nb); Melcher (1966) on Jacupiranga, Brazil; and Reedman (1984) on Sukulu, Uganda (phosphate with possible Nb, zircon, baddeleyite and magnetite by-products).

10

The Pegmatitic Environment

Some general points

Pegmatites are very coarse-grained igneous or metamorphic rocks, generally of granitic composition. Those of granulite and some amphibolite facies terranes are frequently indistinguishable mineralogically from the migmatitic leucosomes associated with them, but those developed at higher structural levels and often spatially related to intrusive, late tectonic granite plutons, are often marked by minerals with volatile components (OH, F, B) and a wide range of accessory minerals containing rare lithophile elements. These include Be, Li, Sn, W, Rb, Cs, Nb, Ta, REE and U, for which pegmatites are mined (rare element or (better) rare metal pegmatites). Chemically, the *bulk* composition of most pegmatites is close to that of granite, but components such as Li_2O, Rb_2O, B_2O_3, F and rarely Cs_2O may range up to or just over 1%.

Pegmatite bodies vary greatly in size and shape. They range from pegmatitic schlieren and patches in parent granites, through thick dykes many km long and wholly divorced in space from any possible parent intrusion, to pegmatitic granite plutons many km^2 in area. They form simple to complicated fracture-filling bodies in competent country rocks, or ellipsoidal, lenticular, turnip-shaped or amoeboid forms in incompetent hosts. Pegmatites are often classed as simple or complex. Simple pegmatites have simple mineralogy and no well developed internal zoning while complex pegmatites may have a complex mineralogy with many rare minerals such as pollucite and amblygonite, but their marked feature is the arrangement of their minerals in a zonal sequence from the contact inwards. An example of a complex, zoned pegmatite from Zimbabwe is given in Fig. 10.1 and Table 10.1. Contacts between different zones may be sharp or gradational. Inner zones may cut across or replace outer zones, but not vice versa, so that inside the wall zones at Bikita no two cross-cuts expose the same zonal sequence. The crystals in complex pegmatites can be very large and at Bikita, for example, the spodumene crystals are commonly 3 m long. The Bikita Pegmatite, which is about 2360–2650 Ma old — it is notoriously difficult to obtain concordant radiometric ages for zoned pegmatites (Clark 1982) — is emplaced in the Archaean Fort Victoria Greenstone Belt. It is one of the world's largest Li-Cs-Be deposits; the main pegmatite is about 2 km long and 45–60 m thick, and the minerals of major economic importance are petalite, lepidolite, spodumene, pollucite, beryl, eucryptite and amblygonite. Cassiterite, tantalite

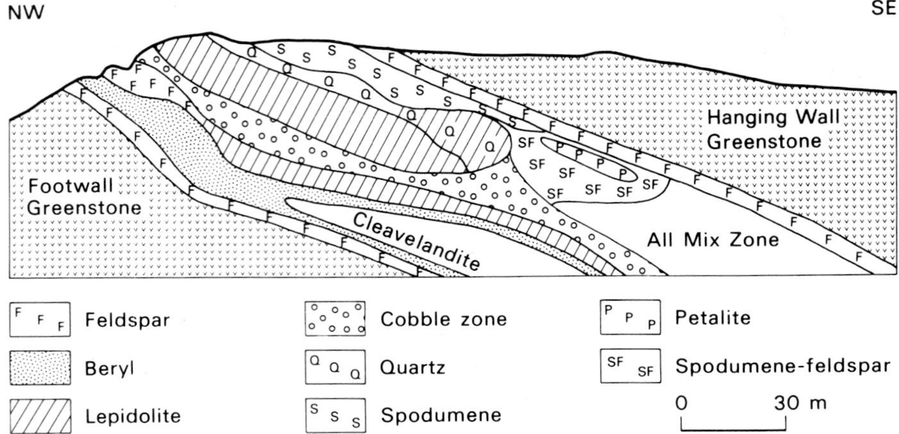

Fig. 10.1. Section through the Bikite Pegmatite showing the generalized zonal structure and the important minerals of each zone. (After Symons 1961). (Cleavelandite is a lamellar variety of white albite).

Table 10.1. Zoning in the Bikita Pegmatite, Zimbabwe. (From Symons 1961).

	Hanging wall greenstone
Border zone	Selvage of fine-grained albite, quartz, muscovite
Wall zones	Mica band. Coarse muscovite, some quartz. Hanging wall feldspar zone. Large microcline crystals
Intermediate zones	Petalite-feldspar zone Spodumene zone (a) massive (b) mixed spodumene, quartz, plagioclase and lepidolite Pollucite zone. Massive pollucite with 40% quartz Feldspar-quartz zone. Virtually devoid of lithium minerals 'All mix' zone. Microcline, lepidolite, quartz
Core zones	Massive lepidolite (a) high grade core, nearly pure lepidolite (b) lepidolite-quartz subzone Lepidolite-quartz shell
Intermediate zones	'Cobble' zone. Rounded masses of lepidolite in an albite matrix Feldspathic lepidolite zone
Wall zone	Beryl zone. Albite, lepidolite, beryl Footwall feldspar. Albite, muscovite, quartz
	Footwall greenstone

and microlite were disseminated through quartz-rich zones in lepidolite-greisen, but these have been mined out. Broadly speaking there is a world-wide similarity of zonal sequences in pegmatites, an important point which, among many others characteristic of complex pegmatites, must be explained by any proposed genetic theory. Useful discussions of internal zoning, terminology used etc., are to be found in Cameron *et al.* (1949), Černý (1982a), Jahns (1955, 1982) and Norton (1983).

Pegmatites have been classified in numerous fashions using as many as nine major criteria (Černý 1982a). A simple classification into four pegmatite formations which has exploration implications is that of Ginsburg *et al.* (1979).
(1) Pegmatite formation of shallow depths (microlitic pegmatites, 1.5–3.5 km); pegmatite pods in the upper parts of epizonal granites. Cavities may supply piezoelectric rock-quartz, optical fluorite, gemstones;
(2) Pegmatite formation of intermediate depths (rare metal pegmatites, 3.5–7 km); filling fractures in and around possible parent granites or quite isolated from intrusive granites; emplaced in low pressure (Abukumu Series) upper amphibolite facies metamorphites. Mainly or wholly magmatic differentiates;
(3) Pegmatite formation of great depths (mica-bearing pegmatites, 7–8 to 10–11 km); only minor rare metal mineralization, emplaced in intermediate pressure (Barrovian Series) upper amphibolite facies metamorphites. Largely the products of anatexis;
(4) Pegmatite formation of maximal depth (>11 km); in granulite facies terranes, usually no economic mineralization. Commonly grade into migmatites.

Reference to texts on metamorphic petrology suggests that the depths given by Ginsburg *et al.* (1979) err very much on the shallow side. For example, the pegmatites of the Bancroft field, Ontario, belonging to formation 2, are emplaced in pelitic schists carrying cordierite and almandine, which suggests a depth of emplacement of about 20 km or so (Winkler 1979).

Pegmatites may occur singly or in swarms forming pegmatite fields, and these in turn may be strung out in a linear fashion to form pegmatite belts, one of the largest being the rare metal pegmatite belt of the Mongolian Altai, about 450 km long and 20–70 km broad, which contains over 20 pegmatite fields. Pegmatites in a particular field may show a regional mineralogical-chemical zonation. When a swarm of pegmatites is associated with a parent granite pluton then the more highly fractionated pegmatites, enriched in rare metals and volatile components, are found at greater distances from the pluton. The degree of internal zoning also increases with an increase in rare metals and volatiles. The regional zoning may reflect the fact that melts enriched in Li, P, B and F have considerably lower solidi than H_2O-only saturated magma and can penetrate further from their source. This regional zoning is of course of great importance to the exploration geologist (Trueman & Černý 1982).

Most pegmatites, whether igneous or metamorphic in origin, have similar *bulk* compositions and these correspond closely to low temperature melts near the minima of Ab-An-Or-Q-H_2O systems (Černý 1982b). This is to be expected for melts developed by extreme magmatic differentiation or anatexis. Many rare metal pegmatites appear to be of magmatic origin; their unusual composition appears to be a consequence of retrograde boiling and the manner in which elements are partitioned between crystals, melt and volatile phases during the cooling of the magma. Bonding factors such as ionic size and charge largely prevent many constituents, originally present in minor or trace concentrations in the magma, from being incorporated into precipitating crystals. They thus become more highly concentrated in the residual melt which, in the case of granites, is also enriched in water, since quartz and feldspars are anhydrous minerals. The list of residual elements is long but includes Li, Be, B, C, P, F, Nb, Ta, Sn and W. Tin and tungsten are more commonly exploited in hydrothermal deposits, but there is no doubt of their orthomagmatic origin in pegmatite occurrences since there is never any evidence of hydrothermal activity (Gouanvic & Gagny 1983). A steady increase in water content, as anhydrous phases are precipitated, will also mark the crystallization of pegmatitic melts and a point may be reached where retrograde boiling produces a water-rich phase. Na, K, Si and some other residual elements listed above, will fractionate from the melt into the water-rich phase within which atoms can diffuse much more rapidly than in a condensed silicate melt. Consequently, rates of crystal growth are greatly enhanced and this might have an important bearing on the growth of giant crystals in pegmatites. London (1984) has shown how different pressures give rise to the crystallization of different phases, e.g. whether spodumene or petalite is the primary lithium aluminosilicate phase.

Broadly speaking, three hypotheses have been put forward to account for the development of internal zoning. The first is that of fractional crystallization under non-equilibrium conditions leading to a steady change in the composition of the melt with time. The second is of deposition along open channels from solutions of changing composition. The third is a two stage model: (a) crystallization of a simple pegmatite with (b) partial or complete replacement of the pegmatite as hot aqueous solutions pass through it. At the present time, most workers prefer the first or third hypothesis as the majority of the evidence, such as complete enclosure in many pegmatites of the interior zones, does not favour the existence of open channels during crystallization. As the third theory encounters difficulties in explaining the world-wide similarity of zonal sequences, the first theory finds most favour.

Some economic aspects

Pegmatitic deposits of spodumene, petalite, lepidolite and other Li minerals are exploited throughout the world for use in glass ceramics, fluxes in aluminium reduction cells and the manufacture of numerous lithium compounds. These deposits often yield by-product Be, Rb, Cs, Nb, Ta and Sn. They may be internally zoned, or unzoned as are the highly productive lithium pegmatites of King's Mountain, North Carolina. The largest known pegmatitic lithium resources are in Zaïre in two laccoliths each about 5 km long and 0.4 km wide. Reserves have been put at 300 Mt and Ta, Nb, Zr and Ti values have been reported (Harben & Bates 1984). Political insecurity, however, is a major drawback as far as overseas investors are concerned and these projects are in abeyance. Another big resource, mainly in granite, is the Echassières deposit in France which contains some 50 Mt grading 0.71% Li, 0.022% Nb, 0.13% Sn and 0.023% Ta.

The reader should note that lithium has already been produced from brines in the USA and a brine producer in Chile should come on stream soon. This new source will present an economic challenge to the hard rock producers and the market price may suffer, as by 1984 lithium production had already exceeded demand.

The traditional source of beryllium (beryl), has now been overtaken by bertrandite, which occurs on a commercial scale in hydrothermal deposits (Farr 1984). The main source of beryl is pegmatites, with the USSR (2 kt p.a.) being the world's largest producer, followed by Brazil with less than 400 t p.a.

Pegmatites are also important as a source of tantalum, but it must be noted that the largest reserves — about 7.25 Gt of Ta — are in slags (running about 12% Ta_2O_5) produced during the smelting of tin ores in Thailand. Significant reserves are present in the Greenbushes Pegmatite, Western Australia (Hatcher & Bolitho 1982), and at Bernic Lake, Manitoba (Černý 1982c) which up to 1982 supplied about 20% of the market, but is now closed down owing to the weak tantalum price. The backbone of the Greenbushes operation is tin production and most of the tin mined in Thailand is won from pegmatites, not veins (Manning 1986).

URANIFEROUS PEGMATITES

Pegmatites and pegmatitic granite have been exploited in a number of localities. Among the more important deposits are those of Bancroft, Ontario and the enormous Rössing Deposit in Namibia.

Bancroft Field, Ontario

This area is part of the south-western extremity of the Greville Province of the Canadian Shield. Granitic and related pegmatites of this province yield ages of 1100–900 Ma (Lumbers 1979), coincident with the waning stages of the high grade Grenvillian metamorphism. The Bancroft Field lies in the Central Metasedimentary Belt, which consists of a metasedimentary-metavolcanic sequence developed around a number of complex granitoid and syenitoid gneiss bodies having domal and periclinal shapes. The pegmatites occur within these gneisses, in the adjacent metasedimentary and metavolcanic sequence and in large metagabbro bodies (Ayres & Černý 1982). Most of the pegmatites are conformable with the metamorphic fabric of their host rocks and internal zoning is but poorly developed, except in the less common fracture-filling bodies that cut across the regional structures. The common presence of pyroxene, hornblende and biotite relates the pegmatitic compositions to the upper amphibolite facies grade of the enclosing rocks. Rare metal mineralization consists of a variety of U, Th, Nb–Ta, REE, Y, Ti, Zr and Be minerals. One school of thought (references in Ayres & Černý 1982) relates the origin of these pegmatites to igneous differentiation, the other to some form of ultrametamorphic activity such as anataxis (Evans 1966, Nash *et al.* 1981, and references in Ayres & Černý 1982). The large amounts of uranium and other

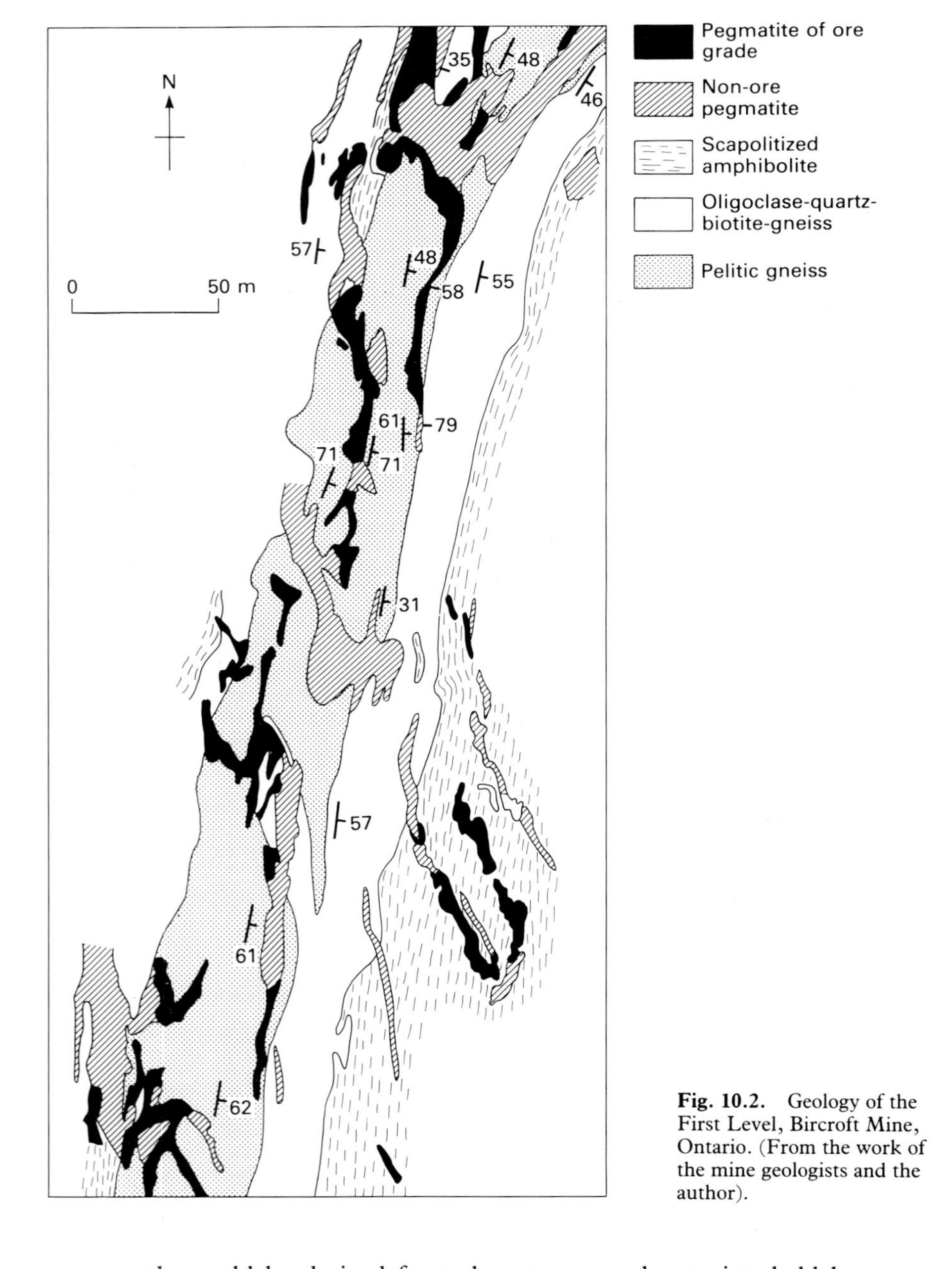

Fig. 10.2. Geology of the First Level, Bircroft Mine, Ontario. (From the work of the mine geologists and the author).

rare metals would be derived from the country rocks, a view held by many Russian workers for certain pegmatites in the USSR, e.g. Shmakin (1983). Fowler & Doig (1983), on the basis of stable isotope data have, however, suggested a mantle source for these Grenvillian pegmatites and their incompatible elements.

Average uranium grades in the Bancroft Field were a little above 0.1%

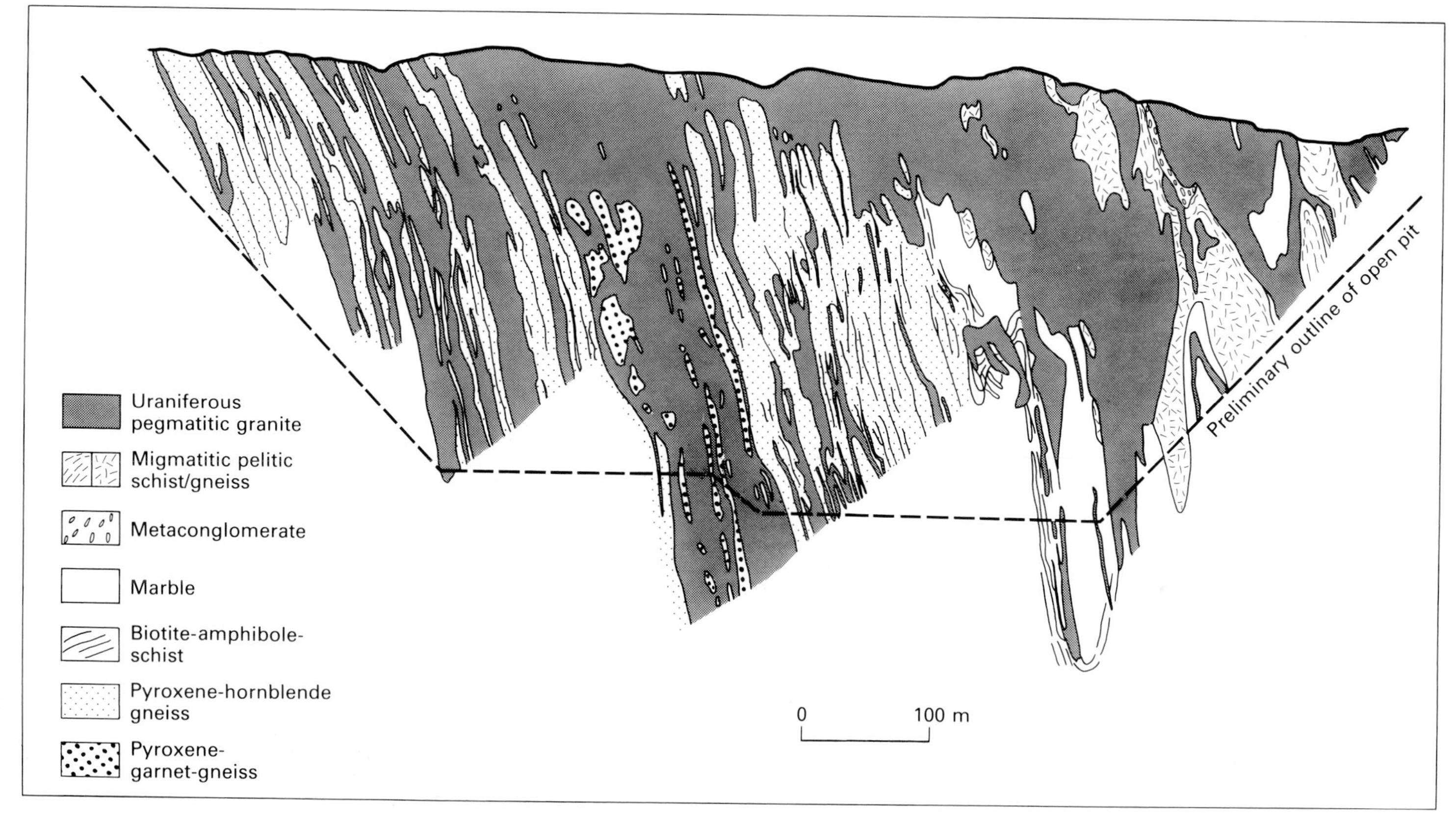

Fig. 10.3. Cross section of the Rössing uranium deposit. (After Berning *et al*. 1976).

U_3O_8 with each of the five main mines working a number of orebodies usually located in swarms of dominantly granitic pegmatite bodies. The Bicroft Mine was one of the largest operations (Fig. 10.2). The pegmatite dykes of this mine area occur in a 5 km long zone only about a quarter of which was developed (Hewitt 1967). The orebodies occurred at the footwall or hanging wall contacts but sometimes occupied the entire pegmatite. The excellent vertical continuity of the orebodies contrasted markedly with their rather tortuous planform (Bryce *et al.* 1958) and two of the largest orebodies were about 90 m long by 3 m wide but extended vertically for over 400 m. The pegmatites are very variable in their lithology, frequently carry aegirine-augite, show no signs of forcible intrusion, have a metamorphic fabric, contain non-rotated enclaves and show a tendency for their lithology to be governed by changes in the nature of their host rocks. The principal uranium minerals are uraninite and uranothorite and the mineralization is best developed where the pegmatites cross a zone containing graphitic pelitic geniss. Evans (1962) suggested that this association may indicate that the ultimate source of much of the uranium is to be found in these altered black shales.

The Rössing Uranium Deposit, Namibia

This is the world's largest uranium producer. The operation is a large tonnage low grade one — 15–16 Mt of ore p.a., grading 0.031% U_3O_8, being produced from an open pit and underground workings (Anon. 1982b). The uranium mineralization occurs within a migmatite zone (Fig. 10.3), characterized by largely concordant relationships between uraniferous, pegmatitic granites and the country rocks. These are metasediments of similar age to the Grenvillian rocks of the Bancroft area, although the pegmatites may be as young as 468 Ma (Nash *et al.* 1981). The metasediments are very similar to those of the Bancroft area and, as in that area, occupy the ground between and around granite-gneiss domes and periclines (Berning *et al.* 1976). The grade of regional metamorphism, upper amphibolite facies, is identical, with cordierite in the pelitic gneisses of both areas indicating low pressure Abukuma type metamorphism. The pegmatitic granites have a very low colour index and are termed alaskites. About 55% of the uranium is in uraninite and 40% in secondary uranium minerals. Economic uranium mineralization is concentrated where the pegmatites are emplaced in certain metasedimentary zones, including pelitic gneisses, in a manner reminiscent of the Bicroft Mine mineralization. Berning *et al.* (1976) favour an initial distribution of uranium within the metasedimentary sequence and its later concentration in anatectic melts of alaskitic composition which show only minor evidence of movement from their zone of generation.

As an example of a large tonnage, low grade, disseminated deposit, Rössing could have been included in Chapter 14 and some authors have unfortunately referred to it as a porphyry uranium deposit, but there is little justification for this description. In its geological environment, host rock type, lack of hydrothermal alteration, etc., it bears no resemblance to porphyry copper deposits.

As an appropriate ending to this chapter I refer the interested reader to the excellent volume edited by Černý (1982d) which provides a comprehensive survey of granitic pegmatites and their economic importance.

11

Orthomagmatic Deposits of Chromium, Platinum, Titanium and Iron Associated with Basic and Ultrabasic Rocks

Orthomagmatic ores of these metals are found almost exclusively in association with basic and ultrabasic plutonic igneous rocks — some platinum being found in nickel-copper deposits associated with extrusive komatiites.

Chromium

Chromium is won only from chromite [$FeCr_2O_4$]. This spinel mineral can show a considerable variation in composition with magnesium substituting for the ferrous iron and aluminium, and/or ferric iron substituting for the chromium. It may also be so intimately intergrown with silicate minerals that these too act as an ore dilutant. Because of these variations there are three grades of chromite ore. The specifications for these are somewhat variable; typical figures are given in Table 8.1.

OCCURRENCE

Three-quarters of the world's chromium reserves are in the Republic of South Africa and 23% in Zimbabwe, most deposits outside these two countries being small. The only other countries with appreciable reserves are the USSR, Albania and Turkey. Lower grade deposits were found recently in the Fiskenæsset Complex of Greenland and there are large subeconomic deposits in Canada.

All economic deposits of chromite are in ultrabasic and anorthositic plutonic rocks. There are two major types: stratiform and podiform (often referred to respectively as Bushveld- and Alpine-types). Each type yields about 50% of world chromite production, which was 9.2 Mt in 1984.

STRATIFORM DEPOSITS

This type contains over 98% of the world's chromite resources. These deposits consist of layers (see Fig. 3.2) usually formed in the lower parts of stratified igneous complexes of either funnel-shaped intrusions (Bushveld, Great Dyke) or sill-like intrusions (Stillwater, Kemi in Finland, Selukwe in Zimbabwe, Fiskenæsset); see Duke (1983).

The immediate country rocks of the complexes are ultrabasic differentiates

Table 11.1. Chromite ore grades and specifications.

	Cr_2O_3	Cr/Fe	$Cr_2O_3 + Al_2O_3$	Fe	SiO_2
Metallurgical grade	>48%	>1.5	—	—	—
Refractory grade	>30%	not critical	>57%	$\ngtr$10%	$\ngtr$5%
Chemical grade	>45%	—	—		$\ngtr$8%

of the parent gabbroic magma — dunites, peridotites and pyroxenites. The layers of these rocks have great lateral extent, uniformity and consistent positions within the complexes. The layers of massive chromite (chromitite) range from a few mm to over 1 m in thickness and can be traced laterally for as much as tens of kilometres. Orebodies may consist of a single layer or a number of closely spaced layers. Chromite in these deposits is usually iron-rich but an outstanding exception is the Great Dyke of Zimbabwe with its high chromium ores.

BUSHVELD COMPLEX

This is an enormous differentiated igneous complex in South Africa (Fig. 11.1), which is generally considered to have resulted from the repeated intrusion of two main magma types into partly overlapping conical intrusions — these eventually coalesced into three larger magma chambers corresponding to the eastern, western and northern segments (Duke 1983, van Gruenewaldt *et al.* 1985). The chromite occurs in the western and eastern outcrops of ultrabasic rocks, with layers a few cm to 2 m thick (Fig. 11.2). They make up an enormous tonnage (van Gruenewaldt 1977). Assuming a maximum vertical mining depth of only 300 m gives a figure of 2.3×10^9 t, a figure which can be multiplied by ten if lower grade deposits are included and the vertical mining depth is increased to 1200 m. Potentially the largest orebodies are the LG3 and LG4 chromite layers present only in the western Bushveld. In these layers the chromite grades 50% Cr_2O_3, Cr/Fe = 2.0, the strike length is 63 km, the thickness 50 cm and the resources (300 m vertical depth) are 156×10^6 t. The ore grades about 45% Cr_2O_3.

The Bushveld chromite zone as a whole contains as many as 29 chromite layers or groups of layers. Above this zone is the platinum-bearing Merensky Reef (Figs. 11.1, 11.2), and near the top of the basic part of the complex, vanadiferous magnetite layers occur.

GREAT DYKE OF ZIMBABWE

This consists of four layered complexes 532 km long and 5–9.5 km broad. In cross-section the layers are synclinal (Fig. 11.3). Chromite layers occur along the entire length and individual layers extend across the entire width. The

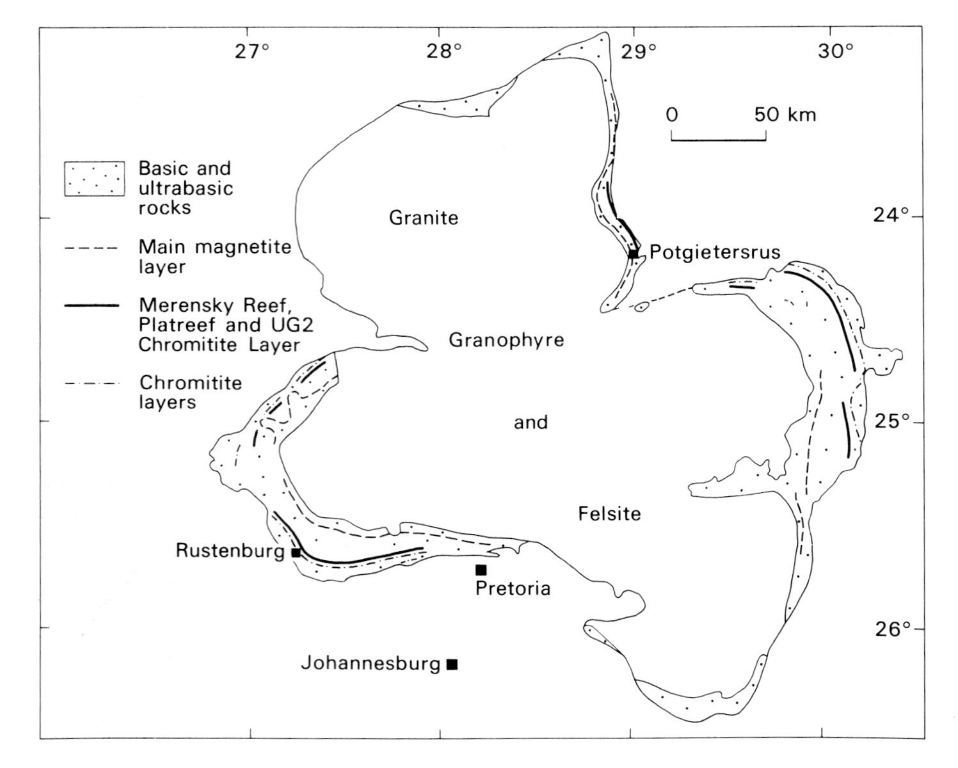

Fig. 11.1. Sketch map of the Bushveld Complex. (After van Gruenewaldt 1977).

layers are in the range 5–45 cm and nearly all the chromite is the high chromium variety. Only layers 15 cm thick or more are mined.

PODIFORM DEPOSITS

The morphology of podiform chromite orebodies is irregular and unpredictable and they vary from sheetlike to podlike in basic form, but can be very variable in their shapes — see Fig. 11.4. Their mass ranges from a few kg to several million tonnes. Throughout most of the world production comes from bodies containing 100 000 t or so and reserves greater than 1 Mt are most uncommon except in the Urals where some very big bodies occur, as in the Kempirsai Ultramafic Massif, a large ophiolite complex in the southern Urals (Fig. 11.5). Most pod deposits are of high chromium type but these deposits are also the only source of high aluminium chromite. The chromite layers are in the range 1–40 cm, or above.

Podiform deposits occur in irregular peridotite masses or peridotite-gabbro complexes of the Alpine-type which are mainly restricted to orogenic zones such as the Urals and the Philippine island arc. These are often ophiolites or parts of dismembered ophiolites. Most ophiolites are allochthonous and are

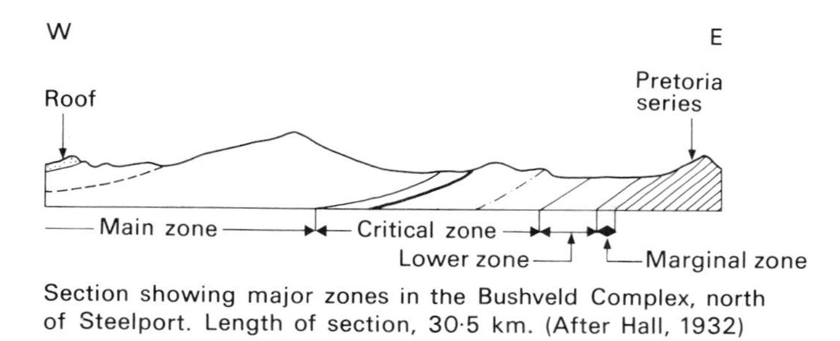

Section showing major zones in the Bushveld Complex, north of Steelport. Length of section, 30·5 km. (After Hall, 1932)

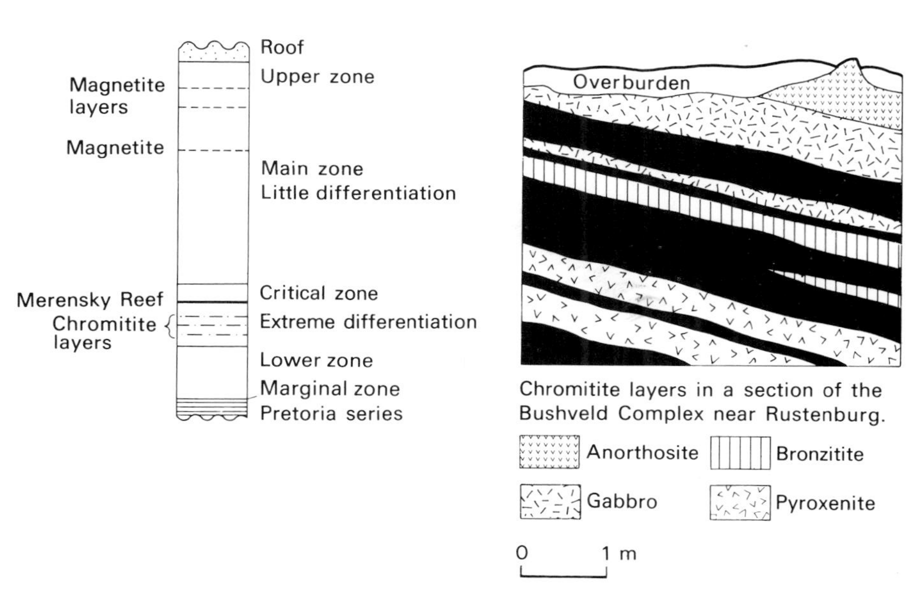

Chromitite layers in a section of the Bushveld Complex near Rustenburg.

Fig. 11.2. Sections showing the occurrence of economic minerals in the Bushveld Complex.

believed to represent transported fragments of oceanic lithosphere. Within many large complexes the chromite deposits are generally near contacts between peridotite and gabbro, but where there is no gabbro the deposits appear to be distributed at random through the peridotite. The host intrusions are usually only a few tens of square kilometres, or less, in area. They are generally strongly elongated and lenticular in shape. Large numbers of these small intrusions occur in narrow zones (serpentinite belts) running parallel to regional thrust zones and the general trend of the orogen in which they occur. The intrusions are usually layered, but the layering does not often show the perfection nor the continuity of the stratiform deposits. They are short lenses rather than extensive sheets. Compositions range from dunite to gabbro and, whilst the average composition of stratiform hosts is close to gabbro, that of Alpine intrusions is near peridotite. There are two subtypes, harzburgite and lherzolite. It is very important for exploration purposes to note that it is the harzburgite subtype which carries chromite deposits (Jackson & Thayer 1972).

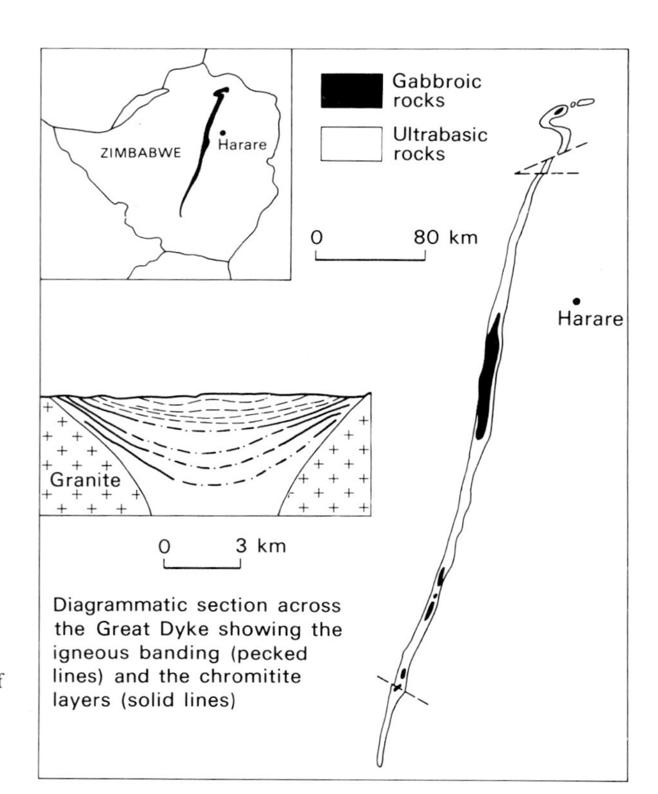

Fig. 11.3. Sketch diagrams illustrating the Great Dyke of Zimbabwe and the occurrence of chromitite layers in it. (In part after Bichan 1969).

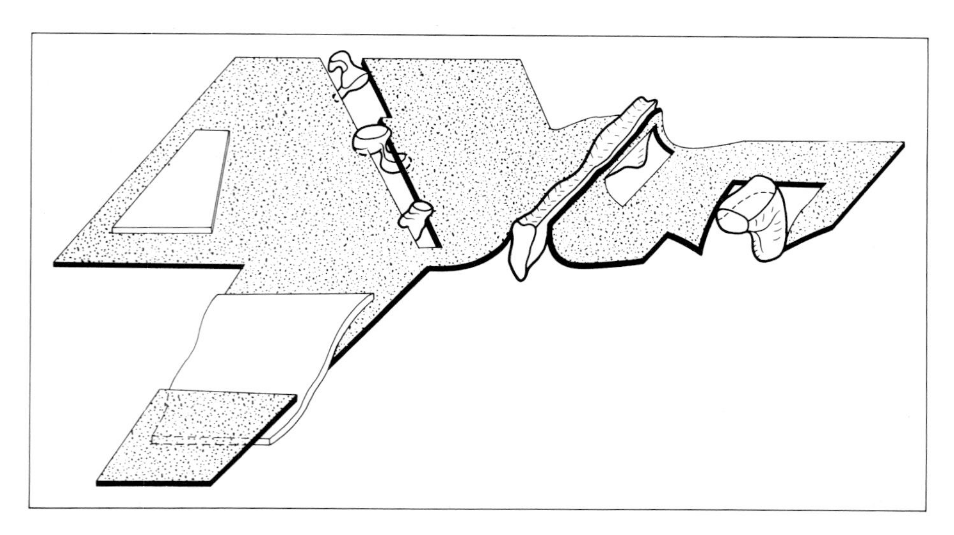

Fig. 11.4. Diagram showing the shapes of podiform chromite deposits in New Caledonia and their relationships to the plane of the foliation in the host peridotites. (The foliation is nearly always parallel to the compositional banding). Note the disturbance of the foliation around the two deposit types on the right. The tabular deposits are of the order of 50–100 kt. (After Cassard *et al.* 1981).

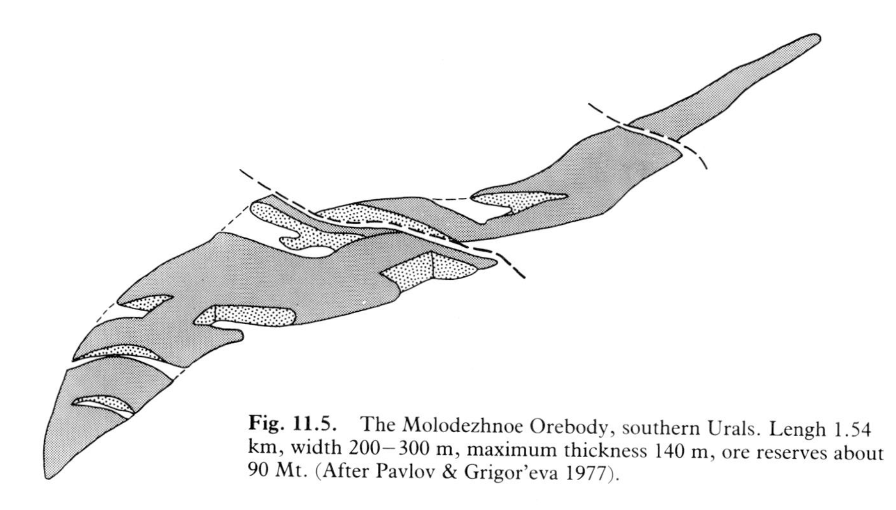

Fig. 11.5. The Molodezhnoe Orebody, southern Urals. Lengh 1.54 km, width 200–300 m, maximum thickness 140 m, ore reserves about 90 Mt. (After Pavlov & Grigor'eva 1977).

It is also important to remark that all major podiform chromite deposits are found in ophiolites formed in marginal basins and not in those formed at mid-ocean ridges (Pearce *et al.* 1984). Most podiform deposits are Palaeozoic or younger; many of them are Mesozoic or Tertiary. The largest known deposits are those of the Urals, Albania, the Philippines and Turkey. Significant deposits occur in Cuba, New Caledonia, Yugoslavia and Greece.

Whilst podiform deposits belong to mobile belts, stratiform deposits (apart perhaps from the Fiskenæsset Complex of western Greenland) were intruded into stable cratons. The Fiskenæsset Complex, being associated with what were oceanic basalts, may have been emplaced in stable oceanic crust. Thus, the two major types of chromite deposit belong to very different geological environments. Other differences are their general age — stratiform deposits with abundant chromite are Precambrian — and number of deposits. Whilst there are numerous podiform deposits, only a few stratiform deposits have produced chromite.

DISRUPTED STRATIFORM DEPOSITS

Tectonic dismemberment of stratiform deposits has led to their individual parts being identified as podiform deposits. Recognition of parts of a once continuous stratiform complex can be very important in mineral exploration as it will lead to the search for missing segments, whereas a similar programme in a podiform chromite district could be financially ruinous (Thayer 1973). The chromite deposits of Campo Formoso in Brazil were formerly thought to be isolated blocks. They have recently been shown to be parts of a highly faulted, layered complex about 18 km long.

In most, perhaps all, stratiform complexes, emplacement of the magma and crystallization took place in a stable cratonic environment and many delicate primary igneous and sedimentary features, in particular layering, are preserved. They were thus originally emplaced in the upper crust. The host rocks of podiform deposits, on the other hand, probably originally crystallized in the mantle and were then incorporated into highly unstable tectonic environments in the crust by movement up thrusts and reverse faults. They are usually part of the ophiolite suite and were probably originally developed at mid-oceanic or back-arc spreading ridges. Despite their tectonic deformation they often preserve layering and textures due to crystal settling, some identical with those in stratiform deposits, indicating a common origin. The transport of the peridotite and chromite from the upper mantle into the upper crust, where they now are, probably occurred by plastic flowage at high temperatures possibly over many kilometres; this fragmented the original layering and has produced many metamorphic features in the chromite and their host rocks. These peridotites have usually suffered a high degree of serpentinization in contrast to the stratiform deposits, in which it is relatively negligible.

There is no doubt then that the chromite deposits of both stratiform and podiform deposits are orthomagmatic. The principal genetic problem is the chromitite layers in which chromite is the only cumulus mineral. During the normal course of fractional crystallization, olivine and chromite crystallize out together and chromite comprises at most a few per cent of the solid fraction. There must be some special control which 'pushes' the magma composition into the chromite field. Possible mechanisms are discussed succinctly by Duke (1983) who, like van Gruenewaldt *et al.* (1985) and many other workers, favours a mixing of magmas for the generation of layering including the chromitite bands.

CONSTITUTION OF CHROMITE CONCENTRATIONS

As will have been gathered from the foregoing, there are some chemical differences between the chromites from stratiform and podiform deposits. These, together with the relationship of chromites from the Fiskenæsset Complex of western Greenland, are shown in Fig. 11.6. Apart from the differences brought out by the figure, it should be noted that the Fiskenæsset chromites are vanadiferous (about 1.5% V_2O_5) like the magnetites of the upper basic levels of the Bushveld Complex; and, like the Bushveld magnetite layers, the Fiskenæsset chromitites occur in the upper part of the intrusion. These differences are perhaps to be attributed to crystallization from a water-rich magma in an oceanic environment, compared with the dry magmas of the Bushveld and similar lopoliths.

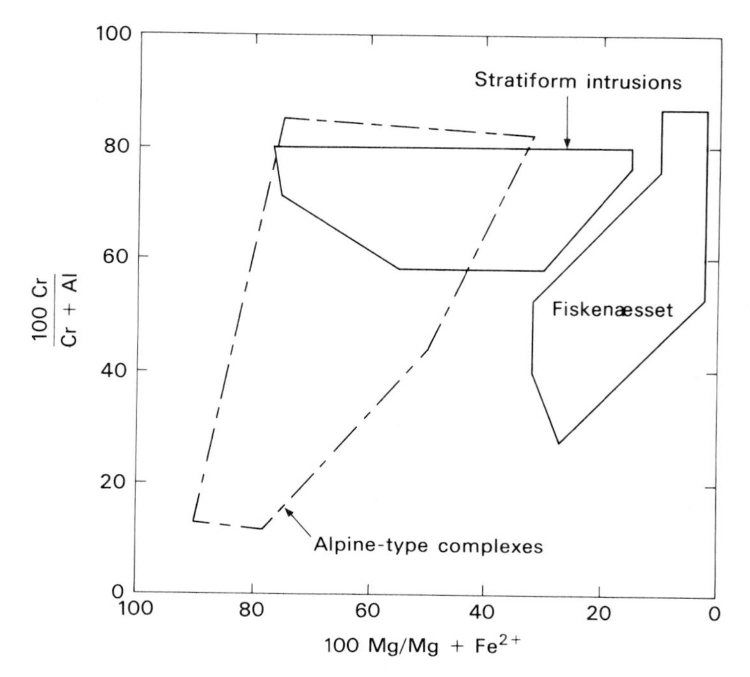

Fig. 11.6. Plot of mg against cr for spinels from stratiform and alpine-type complexes together with the Fiskenaesset Complex of western Greenland. (After Steele *et al.* 1977).

Platinum group metals

These metals used to be produced entirely from placer deposits, but now primary deposits are more important. From 1778–1823 Colombia was the only producer, then in 1822 placers were also discovered in the Urals. In 1919, the recovery of by-product platinum metals from the Sudbury copper-nickel ores of Canada started. In 1924, the South African deposits in the Bushveld Complex were found and in 1956 South African production outstripped Canadian. The principal producers are now: (i) USSR, (ii) South Africa, (iii) Canada, (iv) Colombia, (v) Alaska.

PRIMARY DEPOSITS

There are two principal types of deposit exemplified by those associated with the basic irruptive of Sudbury, Ontario, and the basic-ultrabasic intrusion of the Bushveld Complex. In the former, platinum minerals occur in liquation ores at the base of the intrusion and in the latter, platinum occurs in the Merensky Reef and similar layers some 2000 m above the base of the intrusion. Platinum-rich layers have recently been discovered in the Stillwater Complex of Montana. The platinum in these deposits occurs mainly as arsenides, sulphides and antimonides (Mertie 1969).

A third type of primary deposit is sometimes important. It consists of deposits of platinum metals in peridotites and perknites, commonly in dunite and serpentinite, less commonly in pyroxenic rocks containing no olivine. In these rocks the platinum occurs mainly as platinum alloys associated with lenticular masses of chromite or disseminated through the host rock in association with chromite. Such deposits have occasionally yielded small orebodies of phenomenally high grade, but are more important as source rocks for the formation of platinum placers.

The Sudbury copper-nickel ores will be covered in the next chapter. Attention will therefore be focused here on the Bushveld Complex.

BUSHVELD COMPLEX

In this intrusion there are three very extensive deposits: the Merensky Reef, the UG2 Chromitite Layer (both in the western and eastern Bushveld) and the Platreef in the Potgietersrus area. The first-named occurs in the Merensky Zone — an exceedingly persistent igneous zone traced for over 220 km in the western Bushveld where it has been proved to contain workable ore for over 110 km (Fig. 11.1). This zone is also well developed in two other districts. It is 0.6–11 m thick and generally consists of a dark-coloured norite.

The Merensky Reef is a thin sheet of coarsely crystalline pyroxenite with a pegmatitic habit which lies near the base of the Merensky Zone. It has a thickness of 0.3–0.6 m. Chromite bands approximately 1 cm thick mark the top and bottom of the reef and are enriched with platinum metals relative to the reef. The platinum minerals are sperrylite, $PtAs_2$; braggite (Pt, Pd, Ni)S; stibiopalladinite, Pd_3Sb and laurite, RuS_2, together with some native platinum and gold. Grades at Rustenberg are 7.5–11 g t^{-1} but grades elsewhere are lower. The average stoping width is 0.8 m and the exploitable ore contains 3.3×10^9 t. Sulphides (chalcopyrite, pyrrhotite, pentlandite, nickeliferous pyrite, cubanite, millerite and violarite) are also present, and copper (0.11%) and nickel (0.18%) are recovered.

The UG2 Chromitite Layer lies 150–300 m below the Merensky Reef. Little information about it is available. It is 90–150 cm thick, carries 3.5–19 g t^{-1} platinum together with copper and nickel values. Chromium will also be produced. The reserves are far greater than for the Merensky Reef and total about 5.42×10^9 t. The Platreef is similar to the Merensky Reef, with which it has been correlated, but it is much thicker, being up to 200 m with rich mineralization over thicknesses of 6–45 m. Grades are very irregular. Ore reserves are about 4.08×10^9 t.

The South African ores are essentially platinum ores with by-product nickel and copper. A horizon similar to the Merensky Reef has been traced for 39 km in anorthosites within the Banded Zone of the Stillwater Complex. Average grades range up to 22 g t^{-1} in parts of this zone (Conn 1979). Discontinuities and stacking of ore layers may have been created by the action of convection cells (Raedeke & Vian 1985). The huge Dufek Complex in

Antarctica may also have a platinum-rich horizon (de Wit 1985). At Sudbury, Ontario, the platinum metals are by-products of copper-nickel ores and the grades of the platinum metals are much lower, being about 0.6–0.8 g t^{-1}. Similar ores to those at Sudbury occur at Noril'sk in Siberia and the grade is said to average 2.2 g t^{-1}.

GENESIS OF PLATINUM DEPOSITS

The genesis of the platinum-rich layers in the Bushveld Complex is still very much an open question. The mystery centres on why the platinum and associated sulphides are concentrated into just a few thin layers. Many ingenious hypotheses for platinum concentration have been put forward, e.g. Vermaak (1976) — see the first edition of this book for a brief summary of his model. At the present time various complicated combinations of multiple magma injection and double-diffusive convection are the basis of most models put forward to explain the Merensky Reef and the UG2 Chromitite Layer (Gain 1985, Kruger & Marsh 1985, Naldrett *et al.* 1985, van Gruenewaldt *et al.* 1985). Both the nature and origin of the platinum mineralization in the Platreef differ completely from those of the Merensky Reef and the UG2 Chromitite Layer. The mineralization in the Platreef is thought to have resulted from contamination of the magma by country rocks (Buchanan & Rouse 1984, Cawthorn *et al.* 1985).

Titanium

The principal source of titanium is placer rutile. A few primary ilmenite deposits are, however, exploited (see pp. 90–91). These are always associated with anorthosite or anorthosite-gabbro complexes and have generally been interpreted as magmatic segregation deposits.

At Allard Lake, Quebec, the ores grade 32–35% TiO_2 and occur in anorthosites. The Lac Tio deposit contains about 125 Mt ore. The ore is coarse-grained, with ilmenite grains containing exsolved hematite blebs up to 10 mm in diameter. Gangue minerals are chiefly plagioclase, pyroxene, biotite, pyrite, pyrrhotite and chalcopyrite. The exsolution hematite is too fine-grained to be separated by grinding from the ilmenite, and dilutes the ilmenite concentrate. The orebodies form irregular lenses, narrow dykes, large sill-like masses and various combinations of these forms. Some of these clearly cut the anorthosite and appear to be later in age. It has been suggested that the titanium ores and the anorthosite are differentiates of the same parent magma.

The world's largest ilmenite orebody is at Tellnes in the anorthosite belt of southern Norway about 120 km south of Stavanger. The deposit is boat-shaped, elongated north-west, 2.3 km long, 400 m wide and about 350 m deep (Fig. 11.7). It occurs in the base of a noritic anorthosite. Proven reserves are 300 Mt of 18% TiO_2, 2% magnetite and 0.25% sulphides (pyrite and Cu-Ni sulphides). The ilmenite carries up to 12% hematite as exsolution lamellae.

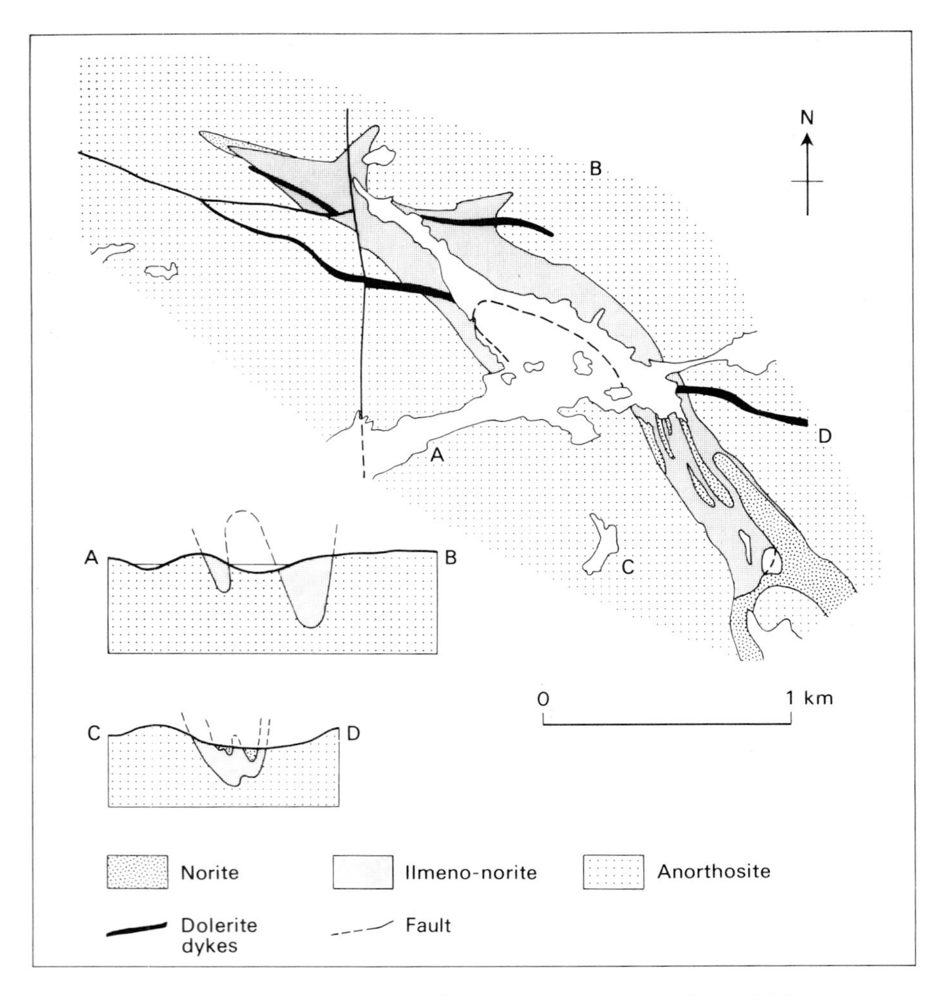

Fig. 11.7. Map and sections of the Tellnes titanium orebody. (After Dybdahl 1960).

The annual production is 2.76 Mt of ore, and extraction of this by opencast methods means that 1.36 Mt of waste have to be removed.

Iron

Many small to medium sized magnetite deposits occur in gabbroic intrusions but the really big tonnages occur in the stratiform lopoliths. Some of these deposits are worked for vanadium as well as for iron. This is the case with the vanadiferous magnetite in the Upper Zone of the Bushveld Complex (Fig. 11.1). This carries 0.3–2% V_2O_5. Values above 1.6% V_2O_5 are usually found in the Main Magnetite Layer and several thinner layers below it, but only the main layer (some 1.8 m thick) can be considered as ore. Based on a vertical limit of 30 m for opencast mining, the reserves are about 1030 Mt. Much more iron ore than this is present, but it is spoilt by a titanium content of up to 19%.

The exact mechanism by which those oxide-rich layers form is still uncertain. It is generally considered that the precipitation of copious quantities of titaniferous magnetite is triggered by episodic increases in f_{O_2}, but the process giving rise to such increases in still uncertain. The problem, as far as the Bushveld Complex is concerned, has been reviewed by Reynolds (1985) who considers that a complex interplay of factors resulted in copious Ti-V-magnetite precipitation. These included concentration of Fe, Ti and V in the residual magma and large scale, *in situ* bottom crystallization of plagioclase with development of a layer of stagnant magma above, from which the magnetite crystallized, as well as changes in f_{O_2}, temperature and f_{H_2O}/f_{H_2}.

12

Orthomagmatic Copper–Nickel–Iron (–Platinoid) Deposits associated with Basic and Ultrabasic Rocks

Introduction

Nickel metal production in the non-Communist world is currently running at about 500 000 t p.a. It is a metal which commands a high price; £3400 t^{-1} compared with about £1000 t^{-1} for copper in September 1985. Nickel is produced from two principal ore-types: nickeliferous laterites and nickel sulphide ores. We are concerned with the latter deposit type in this chapter. These deposits usually carry copper, often in economic amounts, and sometimes recoverable platinoids. Iron is produced in some cases from the pyrrhotite concentrates which are a by-product of the dressing of these ores. Nickel sulphide deposits are not common and so just a few countries are important for production of nickel sulphide ores; Canada is pre-eminent, while the USSR and Australia are the only other important producers. A small production comes from Zimbabwe, South Africa, Botswana and Finland.

The mineralogy of these deposits is usually simple, consisting of pyrrhotite, pentlandite $(Fe, Ni)_9S_8$, chalcopyrite and magnetite. Ore grades are somewhat variable. The lowest grade of a working deposit in western countries appears to be an Outukumpu (Finnish) mine working 0.2% Ni ore. This low grade can be compared with the very high grade sections of some Western Australian deposits that run about 12% Ni. Of course, the overall grade for Australian deposits is less than this because lower grade ore is mined with these high grades. Therefore, although such high grades occur at Kambalda in the Western Mining Corporation mines, the ore reserves at June 1984 were 26 Mt running 3.3% Ni. In this mining camp the ore treated in 1983–4 totalled 1.373 Mt having a mill grade of 3.43% and nickel recovery of 92.4%.

All nickel sulphide deposits are associated with basic or ultrabasic igneous rocks. There is both a spatial and a geochemical relationship in that deposits associated with gabbroic igneous rocks normally have a high Cu:Ni ratio (e.g. Sudbury, Ontario; Noril'sk, USSR), and those associated with ultrabasic rocks a low Cu:Ni ratio (e.g. Thompson Belt, Manitoba; Western Australian deposits).

Classification of ultrabasic and basic bodies with special reference to nickel sulphide mineralization

There are a number of different types of basic and ultrabasic rocks; not all of these have nickel sulphide deposits associated with them and therefore, in exploring for these deposits, it is important to know which classes of basic and ultrabasic rocks are likely to have associated nickel sulphide ores. A

Table 12.1. Classification of basic and ultrabasic bodies and associated nickel mineralization. (Based on Naldrett & Cabri 1976, Besson *et al.* 1979, Naldrett 1981 and Eckstrand 1984.)

Examples	Remarks	Examples of associated nickel deposits
Class A: Bodies emplaced in active orogenic areas		
1. Bodies contemporaneous with plate margin volcanism, largely restricted to Archaean greenstone belts		
(i) Tholeiitic suite: (a) Picritic subtype, Dundonald Sill, Ontario; Eastern Goldfields, Western Australia	Differentiated sills and stock-like intrusions. 1–10 Mt orebodies	Pechenga, USSR Lynn Lake, Manitoba Carr Boyd, Western Australia
(b) Anorthositic subtype Doré Lake Complex, Quebec; Bell River Complex, Ontario; Kamiscotia Complex, Ontario	Some examples of this class are conformable, others appear to be discordant	No known deposits
(ii) Komatiitic suite (a) Lava Flows, Munro Township, Ontario; Eastern Goldfields, Western Australia	Komatiites are principal hosts, also closely underlying small sills, in some cases adjacent metasedimentary or metavolcanic rocks. 1–5 Mt orebodies, commonly several orebodies per deposit	Langmuir, Ontario; Marbridge, Quebec; Kambalda, Windarra, Western Australia; Shangani, Inyati-Damba, Zimbabwe
(b) Intrusive dunite lenses,* Eastern Goldfields, Western Australia; Dumont, Quebec; Thompson Nickel Belt, Manitoba	Significant development in Proterozoic as well as Archaean rocks. Highly magnesian ultramafic sills larger than those in (iia). Segregated orebodies up to tens of Mt, disseminated up to 250 Mt	Agnew, Mt Keith, Western Australia; Dumont, Ungava, Que.; Pipe, Birchtree, Manibridge, Manitoba
2. Bodies emplaced during orogenesis		
(i) Synorogenic intrusions Aberdeenshire Gabbros, Scotland Rôna, Norway	Only small nickel deposits known	
(ii) Large obducted sheets New Caledonia Papua New Guinea	No significant mineralization	
(iii) Ophiolite complexes Troodos, Cyprus Bay of Islands, Newfoundland Luzon Complex, Philippines	Only small deposits known	Acoje Mine, Philippines
(iv) Alaskan-type complexes Duke Island and Union Bay, Alaska	Concentrically zoned intrusions	No economic deposits known

Class B: Bodies emplaced in cratons		
3. Largely stratiformly layered complexes Bushveld, RSA Duluth, Minnesota Stillwater, Montana Sudbury, Ontario Dufek Complex, Antarctica	Individual orebodies run up to tens of Mt. Each intrusive complex generally contains a number of orebodies. Enormous resources of disseminated mineralization present in some complexes, e.g. Duluth	Bushveld mines produce PGE with by-product nickel. Sudbury production + reserves = 700 Mt in numerous deposits
4. Intrusions related to flood basalts Noril'sk–Talnakh, USSR Insiwa–Ingeli Complex, RSA Palisades Sill, New Jersey	These intrusions generally occur in areas in which flood basalts are developed. They are chemically similar to the extruded basalts. Segregated orebodies up to tens of Mt. Duluth Complex could also be included in this class	Noril's–Talnakh, USSR
5. Medium and small-sized intrusions Skaergaard, Greenland; Rhum, Scotland		Losberg, RSA
6. Alkalic ultramafic rocks in ring complexes, kimberlite and lamproite pipes		No known examples

*Hill & Gole (1985) have questioned the intrusive nature of bodies of this group in Western Australia and have suggested that these rocks are mainly or wholly volcanic.

classification is given in Table 12.1 with an indication of the tectonic settings. Particular combinations of rock type and tectonic setting have proved to be especially productive. These are:

(1) noritic rocks intruded into an area that has suffered a catastrophic release of energy, e.g. an astrobleme (Sudbury);

(2) intrusions associated with flood basalts in intracontinental rift zones (Noril'sk-Talnakh, Duluth);

(3) komatiitic and tholeiitic flows and intrusions in greenstone belts (Kambalda, Agnew, Pechenga).

(A) BODIES IN AN OROGENIC SETTING

Two broad groups of ultrabasic and basic bodies can be seen in this setting: bodies coeval with plate margin volcanism and syntectonic intrusions.

The first group occurs in the Archaean and Proterozoic greenstone belts and can be divided into the tholeiitic and komatiitic suites. The tholeiitic suite contains the picritic and anorthositic classes. The anorthositic class is important for titanium mineralization but so far no substantial nickel mineralization has been found in rocks of this class. The picritic class is an important nickel ore carrier and ultrabasic rocks in this class occur as basal accumulations in differentiated sills and lava-flows, some having basal sulphide segregations — the Dundonald Sill of the Abitibi Greenstone Belt, Canada, is a good example. The tholeiitic activity in this and other areas was often contemporaneous with komatiitic volcanicity. The komatiitic suite is a much more important carrier of nickel mineralization. Disagreements persist regarding the definition of komatiite and the full range and authenticity of the komatiitic suite (Arndt & Nisbet 1982a, Best 1982). Many workers consider that the komatiitic suite ranges from dunite (>40 wt % MgO) through peridotite (30–40%), pyroxene-peridotite (20–30%), pyroxenite (12–20%) and magnesian basalt to basalt, thus forming a magma series with the status of the tholeiitic or calc-alkaline series. Komatiites are both extrusive and intrusive, and ultrabasic members are believed to have crystallized from liquid with up to 35 wt % MgO and carrying 20–30% of olivine phenocrysts in suspension. In some flows and near surface sills, quench textures (probably due to contact with sea water and consisting of platy and skeletal olivine and pyroxene growths) are present in the upper part. This is called spinifex texture. These flows clearly crystallized from magnesian-rich undifferentiated magma, extruded (in the case of 35 wt % MgO) at up to 1650°C. Spinifex textures resemble those of silica-poor slags, having a low viscosity and a high rate of internal diffusion — ideal conditions for the sinking of sulphide droplets to form accumulations at the flow bottom. It is very unlikely that the sulphides were in solution when rapid crystallization started or they would have been trapped before they could sink to the flow bottom. They were therefore probably already in droplet form when the magma was erupted.

The recognition of the presence of komatiites in an exploration area is of great importance and the nickel prospector should be aware of the distinguishing characteristics, both petrographic and geochemical, of these rocks. Excellent descriptions of komatiites are to be found in Arndt & Nisbet (1982b).

The synorogenic intrusions are of little importance as carriers of nickel sulphide ores, but constitute a small resource for the future. The large obducted sheets and ophiolite complexes figure large in the preceding chapter but are of little note as hosts for nickel mineralization. Alaskan-type complexes form concentrically zoned intrusions which are well developed along the Alaskan pan-handle. As a group they are distinguished from alpine-type ultrabasics or stratiform intrusions by highly calcic clinopyroxene, no orthopyroxene or plagioclase, much hornblende, more iron-rich chromite, and magnetite. The last-named occurs in concentrations that are occasionally of economic interest. Similar bodies occur in the Urals, south central British Columbia and Venezuela.

(B) BODIES EMPLACED IN A CRATONIC SETTING

There are three main groups to be noted. The first is an important metal producer because it is that of the large stratiform complexes. In the last chapter, we noted the importance of the Bushveld Complex and its by-product nickel-copper won from the platinum-rich horizons. The Sudbury intrusion also belongs to this group and it hosts the world's greatest known concentration of nickel ores. The Duluth and Stillwater Complexes may also become producing areas in the future. The overall composition of this group is basic rather than ultrabasic but a lower ultrabasic zone is usually present. Sudbury is a notable exception to this rule but it possesses a sublayer rich in ultrabasic xenoliths probably derived from a hidden layered sequence.

The second group includes intrusions related to flood basalts and usually associated with the early stages of continental rifting. Very important here are the gabbroic intrusions of the Noril'sk-Talnakh Nickel Field. The Duluth Complex and the troctolite hosting the Great Lakes nickel deposit of Ontario could be included in this group as they are both in a rift setting and are petrogenetically related to the Keweenawan flood basalts.

Thirdly there is a group of medium- and small-sized intrusions such as Skaergaard and Rhum that carry nickel mineralization of only academic interest.

Relationship of nickel sulphide mineralization to classes of ultrabasic and basic rocks

Figure 12.1 indicates the relationship between known reserves plus past production of nickel in the main deposits of the world and the rock classes given in Table 12.1. Apart from the unique position of Sudbury, production from there being responsible for almost the entire Canadian section of classes 3 and 4, komatiitic magmatism [1(ii)] is clearly the most important. Tholeiitic volcanism is much less important. Deposits near the basal contacts of the Stillwater and Duluth complexes (class 3) are low grade disseminated deposits which are unlikely to be producers in the near future.

The Noril'sk-Talnakh Field, like Sudbury, has many unusual features, which further emphasizes the importance of class 1(ii) as the best bet for further nickel exploration.

Genesis of sulphur-rich magmas

For a rich concentration of magmatic sulphides it is necessary that (1) the host magma is saturated in sulphur, and (2) a reasonably high proportion of sulphide droplets can settle rapidly to form an orebody. Slow settling may give rise to a disseminated, uneconomic ore. The production of a high proportion of immiscible sulphides is possible if the magma assimilates much sulphur from its country rocks, e.g. Duluth, Noril'sk, or if the magma carries excess amounts of mantle-derived sulphides, e.g. some komatiites. We can often differentiate between these two sources (crustal and mantle) by examining variations in $^{34}S/^{32}S$. These are reported in the delta notation (page 75). For this work the standard is troilite of the Cañon Diablo meteorite which is taken

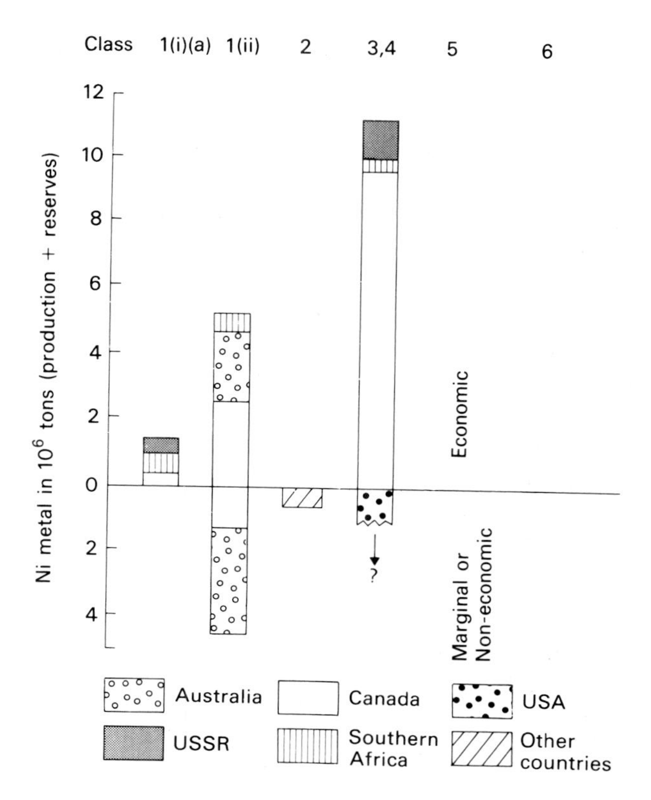

Fig. 12.1. Present reserves plus past production of sulphide nickel as a function of the class of host rock. The key to the classes is given in Table 12.1 (Modified from Naldrett 1973 and Naldrett & Cabri 1976).

to be equivalent to the earth's mantle and for which $\delta^{34}S = 0$. Originally, all the earth's sulphur was in the mantle and much of that transferred to the crust has undergone biogenic fractionation producing an enrichment in ^{34}S and giving $\delta^{34}S$ values as high as + 30. It is now held to be an oversimplification to claim that for all natural sulphur a narrow range of $\delta^{34}S$ close to zero indicates a mantle source, and a wide range of $\delta^{34}S$ a crustal source, but this statement is still probably true for the majority of magmatic systems.

Let us look first at two examples where there is no evidence to suggest that the sulphur came from anywhere but the mantle. At Sudbury, $\delta^{34}S$ varies from −0.2 to +5.9 with a mode of 1.7; this is a narrow spread of values close to $\delta^{34}S = 0$ and suggestive of a mantle origin. Similarly, ore associated with komatiites at the Alexo Mine, Ontario gives $\delta^{34}S = 4.4-6.2$; again close to a zero value but perhaps showing a little temperature fractionation. At Noril'sk, however, the values are +7 to +17. The Noril'sk Gabbro is Triassic (an unusual and perhaps unique situation as practically all economic nickel mineralization is early to middle Precambrian), and its magma is intruded through gypsum beds, where it is thought to have picked up ^{34}S-rich sulphur. A comparable example is the Water Hen intrusion of the Duluth Complex with values of +11 to +16. This is believed to have gained its sulphur by assimilating sulphur-rich sediments of the Virginia Formation in which $\delta^{34}S$ ranges from +17 to +19 (Mainwaring & Naldrett 1977). The mineralized gabbros of northern Maine have also yielded evidence of having acquired much of their sulphur by the assimilation of country rocks (Naldrett *et al.* 1984). Clearly basic intrusions in sulphur-rich country rocks should be carefully prospected for nickel mineralization!

In the sulphides of komatiitic-related deposits of Western Australia, $\delta^{34}S$ for a number of deposits ranges from −2.6 to +3.4 with a mode of nearly zero. This is clearly sulphur of direct or recent mantle origin, but it must be pointed out that the interflow sediments are rich in sulphur of almost identical $\delta^{34}S$ values (Groves & Hudson 1981). These sediments represent a mixture of volcaniclastic detritus and chemical-sedimentary material derived largely from volcanic exhalations between times of lava extrusion. The associated komatiitic magmas may therefore have brought much of their sulphur up from the mantle but could also have acquired more by assimilation of interflow sediments. Many writers at the moment favour the hypothesis that komatiites incorporate sulphur during their genesis by partial melting deep within the mantle.

We must now consider the problem of how some mantle-derived magmas could acquire a high sulphur content. This is discussed by Naldrett & Cabri (1976) with the aid of the relationships shown in Fig. 12.2 which shows the degree of melting of the mantle and the oceanic geothermal gradients for the present day and the Archaean. Consider mantle material at A; it is sufficiently above the solidus to yield a 5% partial melt. Any slight perturbation (such as a tectonic effect, accession of H_2O or magma from the descending slab of a Benioff Zone, normal convective overturn) would cause a diapir tens of kilometres across to rise adiabatically to B. Partial melting would increase to 30% and if this melt separated from the diapir it might rise to the surface along the non-adiabatic curve BF. The continued rise of the diapir would produce further partial melting but now the melt would form from depleted mantle and it would be much more magnesian since basaltic material has been removed. Partial melting might reach 30% at C (the contours hold for undepleted mantle only) to produce a much more magnesian (komatiitic) magma which might be intruded or extruded at E.

Fig. 12.2 shows that the zone of general sulphide melting intersects the Archaean geothermal gradient at about 100 km depth. Below this depth sulphides were molten and probably percolated downwards leaving a zone depleted in sulphides and producing a deeper enriched zone — this sulphur-enriched zone will lie at the level from which komatiitic magmas are ultimately derived. In this way, we can explain the relationship between komatiites and sulphide ores and the fact that they may be associated with tholeiitic picrites which themselves may carry sulphide ores as in the Abitibi region of Ontario.

It must be noted, however, that another school of thought, e.g. Keays (1982), opposes this concept. Keays contends that the low Pd:Ir ratios of komatiites implies that their magmas formed during sulphur-undersaturated

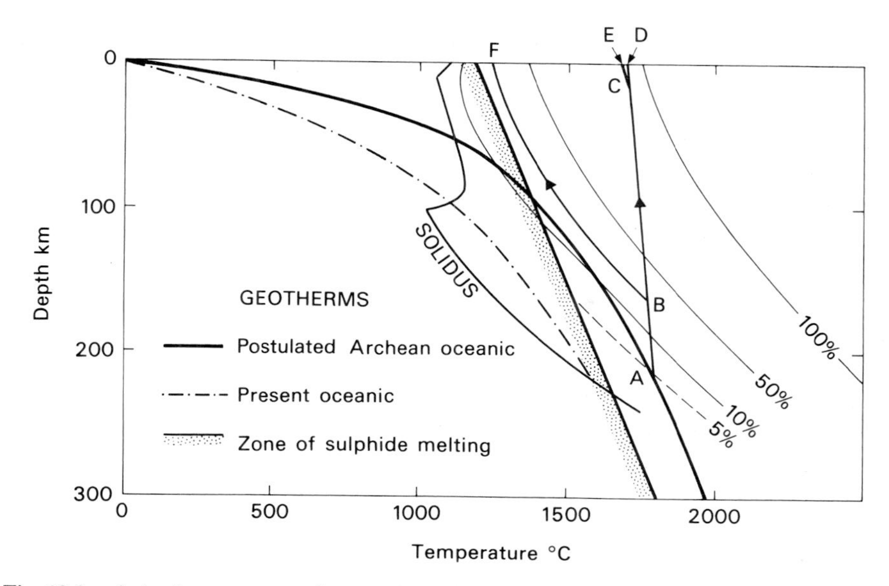

Fig. 12.2. A depth-temperature diagram showing the relationship between estimates of the modern and Archaean oceanic geotherms and melting relations of possible mantle material (pyrolite II + 0.2 wt % H_2O); drawn to illustrate the generation of komatiitic magma. (After Naldrett & Cabri 1976).

conditions and that sulphur saturation within the mantle would have led to the loss of sulphides due to pressure release during the ascent of the magma.

Nickel mineralization in time and depletion of sulphur in the mantle

The only large nickel sulphide deposits which are younger than 1700 Ma are Duluth (1115 Ma) and Noril'sk (Triassic). As we have seen, both host intrusions probably acquired most of their sulphur by assimilation of crustal rocks. This restriction of orthomagmatic nickel deposits to the Archaean and early Proterozoic may reflect sulphur depletion of the upper mantle caused by many cycles of plate tectonic or similar processes, if one accepts Naldrett's and Cabri's hypothesis. On the other hand, the concentration of nickel sulphide mineralization early in the stratigraphic column may be largely related to the concentration of komatiitic activity early in earth history, no matter whether the necessary sulphur was acquired deep within the mantle or by assimilation at or near the surface.

Origin of the metals

There is no difficulty here. Ultrabasic and basic magmas are rich in iron and in trace amounts of copper, nickel and platinoid elements. These would be scavenged by the sulphur to form sulphide droplets. The reason why other chalcophile elements such as lead and zinc are not present in significant amounts in nickel sulphide ores has been explained by Shimazaki & MacLean (1976). This process of nickel scavenging leads to the development of nickel-depleted olivines in the silicate phase, a circumstance which may have great exploration significance (Naldrett *et al.* 1984).

Examples of nickel sulphide orefields

THE VOLCANIC ASSOCIATION

Here we are mainly concerned with class 1(ii) (Table 12.1), the komatiitic suite. In some of these, sulphides occur at or near the base of the flow or sill suggesting gravitational settling of a sulphide liquid. Typical sections through two deposits are given in Fig. 12.3. These have certain features in common:
(1) massive ore at the base (the banding in the Lunnon orebody is probably the result of metamorphism);
(2) a sharp contact between the massive ore and the overlying disseminated ore which consists of net-textured sulphides in peridotite;
(3) another sharp contact between the net-textured ore and the weak mineralization above it which grades up into peridotite with a very low sulphur content.
This situation is strongly reminiscent of the billiard ball model of sulphide segregation described on page 45.

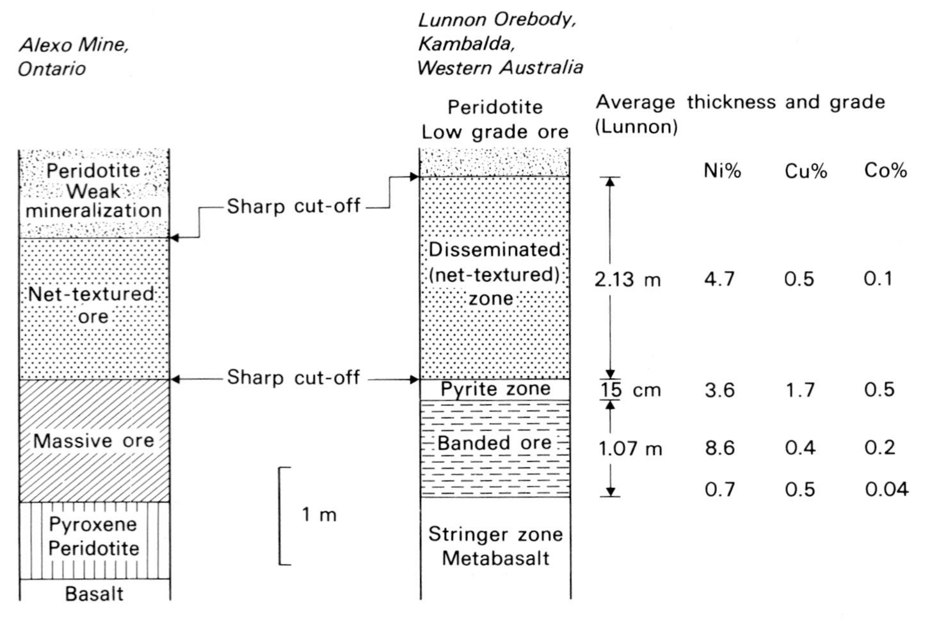

Fig. 12.3. Typical sections through two nickel sulphide orebodies associated with Archaean class 1 (ii) ultrabasic bodies. The Alexo Mine is 40 km east-north-east of Timmins, Ontario. (After Naldrett 1973).

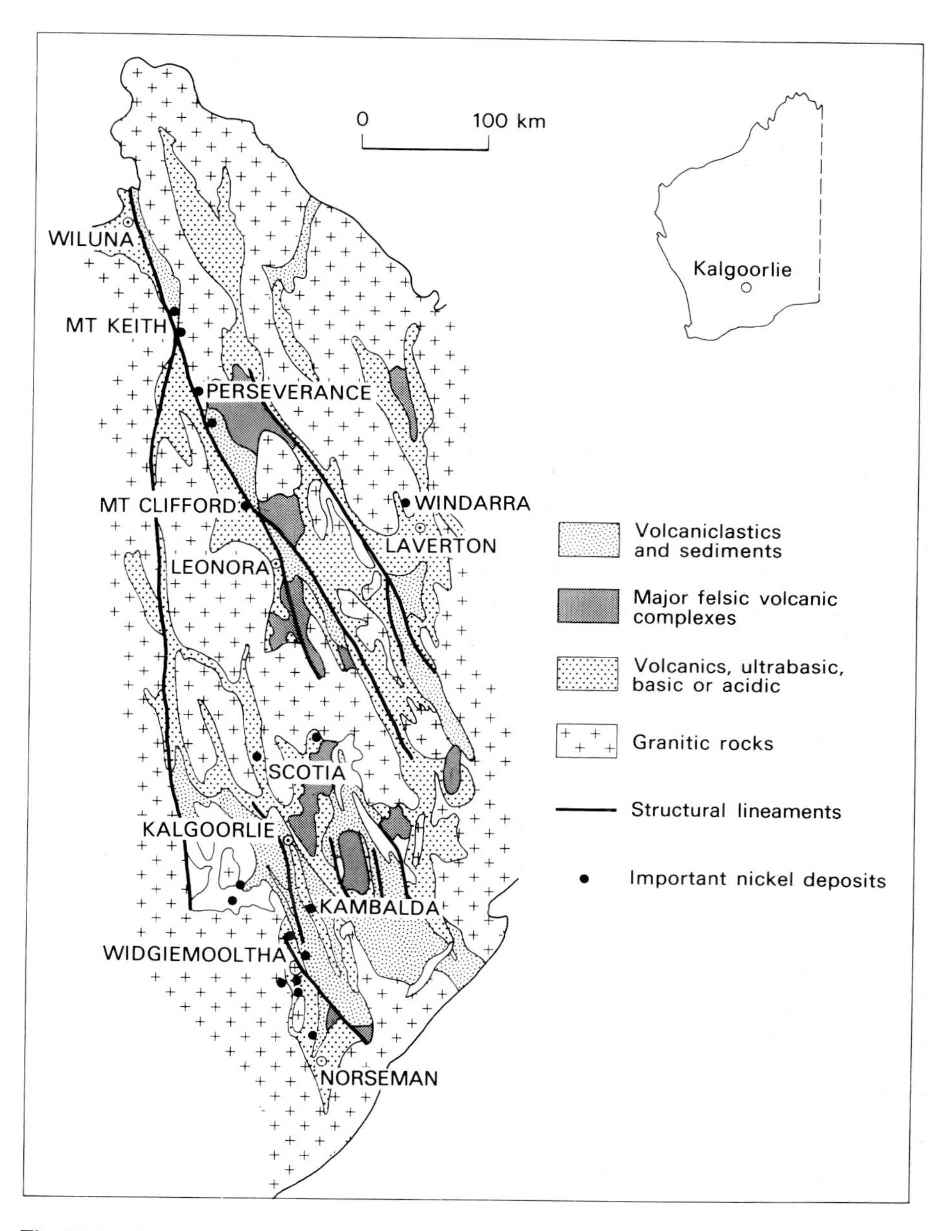

Fig. 12.4. Generalized geological map of the Eastern Goldfields Province of the Yilgarn Block showing some of the important nickel deposits. (Modified from Gee 1975).

In nearly all examples of this deposit type the orebodies are in former topographic depressions beneath the lava flow. Some of these depressions, but by no means all, appear to be fault related while others may be thermal erosion channels formed by these exceptionally hot and very fluid lavas (Huppert & Sparks 1985). Orebodies may have formed in these depressions not just by simple sinking of sulphide droplets from a static silicate liquid but also by a riffling process as the main lava stream moved for some time over these footwall embayments (Naldrett 1982).

The Eastern Goldfields Province of Western Australia is a typical Archaean region having a considerable development of greenstone belts (Fig. 12.4), with nickel sulphide deposits of two main types (Groves & Lesher 1982). The first consists of segregations of massive and disseminated ores at the base of small lenslike peridotitic to dunitic flows or subvolcanic sills at the bottom of thick sequences of komatiitic flows, e.g. Kambalda, Windarra, Scotia (Fig. 12.4), which are termed volcanic-type deposits. The second type, dykelike deposits, occurs in largely concordant but partially discordant dunitic intrusions

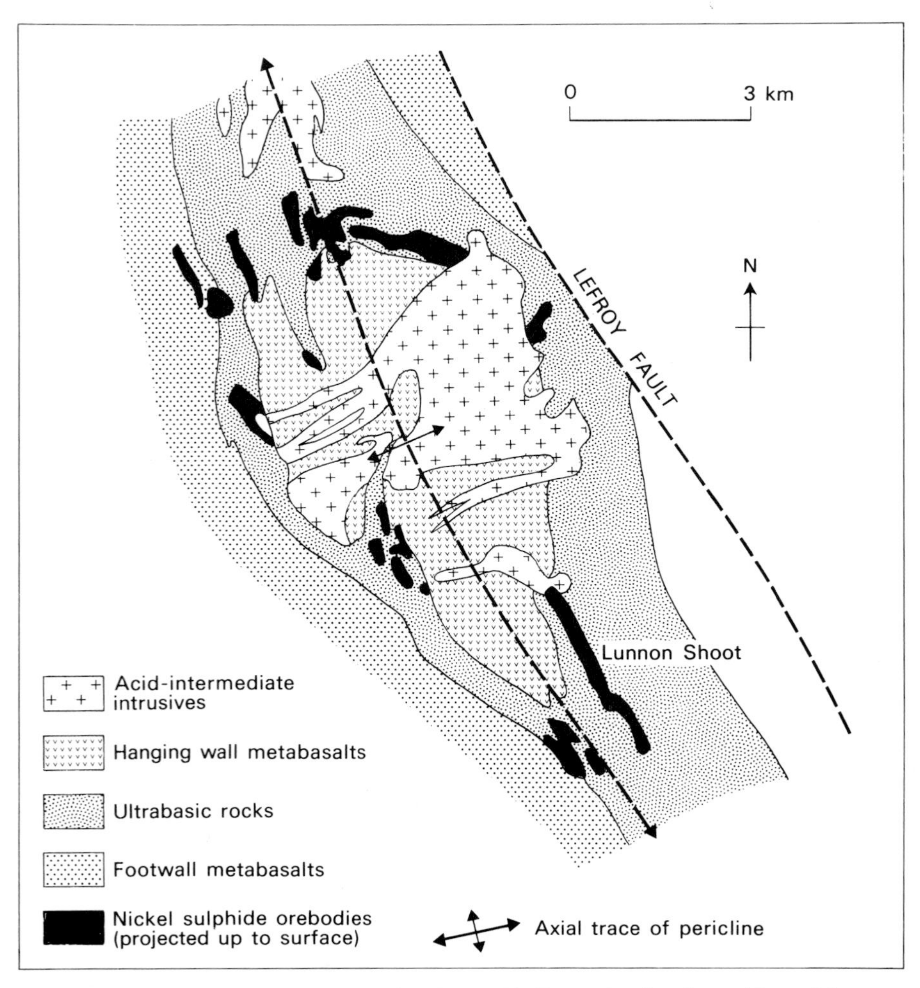

Fig. 12.5. Generalized geological map of the Kambalda Dome showing the positions of the sulphide orebodies.

emplaced in narrow zones up to several hundred kilometres in length, e.g. Perseverance, Mount Keith.

The volcanic-type deposits are commonly clustered around the periphery of granitoid periclines as at Kambalda (Fig. 12.5). These periclines trend north-north-westerly. Most ores are at, or close to, the basal contact of the mineralized ultrabasic sequence (Fig. 12.6) and they are commonly associated with, or even confined to, embayments in the footwall. The orebodies are essentially tabular with their greatest elongation subparallel to the penetrative linear fabrics in the enclosing rocks and/or the trend and plunge of both

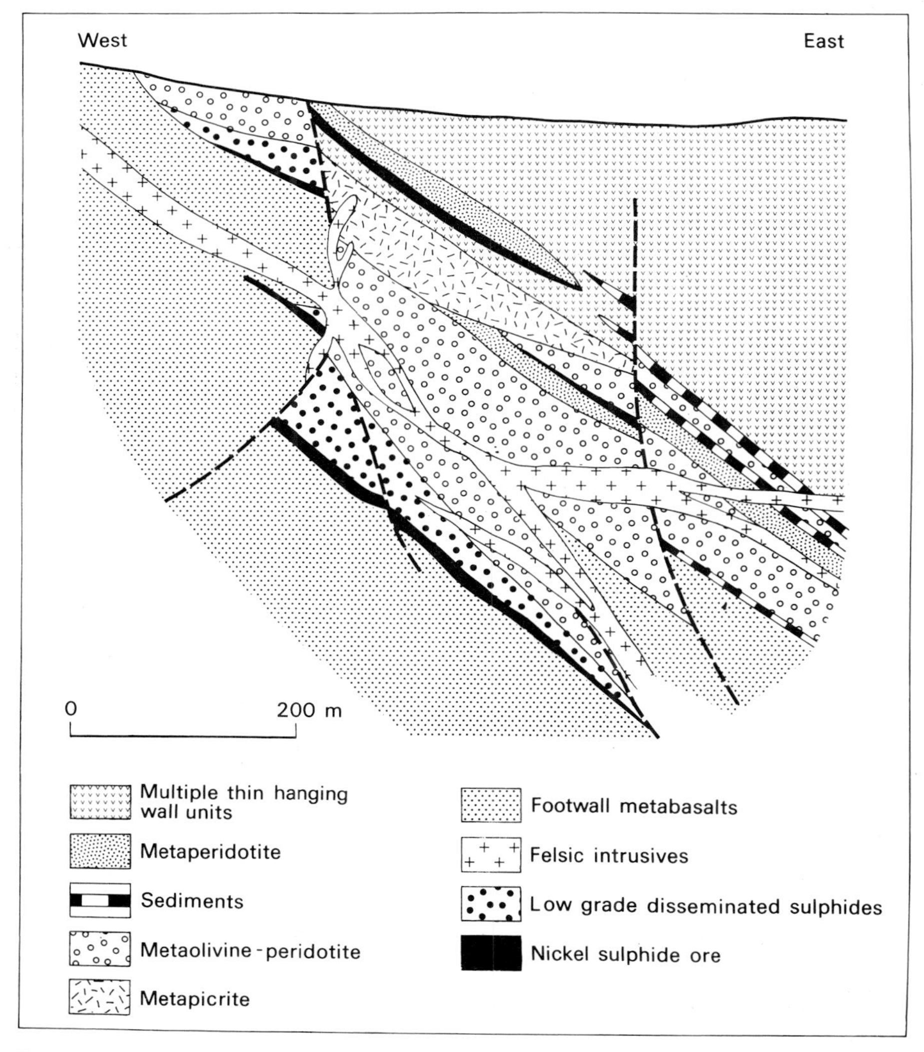

Fig. 12.6. Section through the Lunnon and neighbouring ore shoots, Kambalda, Western Australia. (Modified from Ross & Hopkins 1975).

regional and parasitic folds. The ores have been metamorphosed and complexly deformed and are only developed in amphibolite facies metamorphic domains.

The dykelike deposits are much larger than the volcanic-type. For example, at Perseverance, 33 Mt averaging 2.2% nickel with a cut-off grade of 1% have been outlined and at Mount Keith 290 Mt grading 0.6%. They occur in long dunite dykes especially where these bulge out to thicknesses of several hundred metres; e.g. at Perseverance, the host dyke thickens from a few metres to 700 (Fig. 12.7). The orebodies generally appear to be associated with areas of considerable serpentinization during which enrichment of the ores seems to have occurred. The ores are dominantly of disseminated type though some massive sulphides occur as at Perseverance.

PLUTONIC ASSOCIATION

(a) *Sudbury, Ontario*. The Sudbury Structure is a unique crustal feature lying just north of Lake Huron near the boundary of the Superior and Grenville Provinces of the Canadian Shield (Card *et al.* 1984). The structure is about 60 × 27 km (Fig. 12.8) and its most obvious feature is the Sudbury Igneous Complex (1849 Ma) which consists of a Lower Zone of augite-norite, a thin Middle Zone of quartz-gabbro and an Upper Zone of granophyre (Naldrett & Hewins 1984). The Complex is believed to have the shape of a deformed funnel. At the base of the norite there is a discontinuous zone of inclusion- and sulphide-rich norite and gabbro known as the sublayer. In the so-called offsets (steep to vertical, radial and concentric dykes which appear to penetrate downwards into the footwall from the base of the Complex) the inclusion-rich sulphide-bearing rock is a quartz-diorite. The sublayer and offsets are at

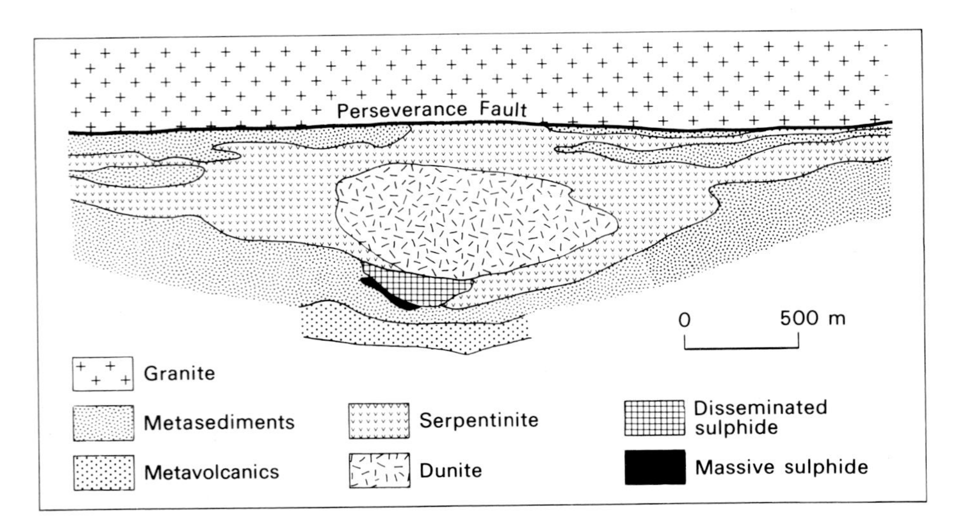

Fig. 12.7. Generalized geology of the Perserverance deposit. Many faults have been omitted. (Modified from Martin & Allchurch 1975).

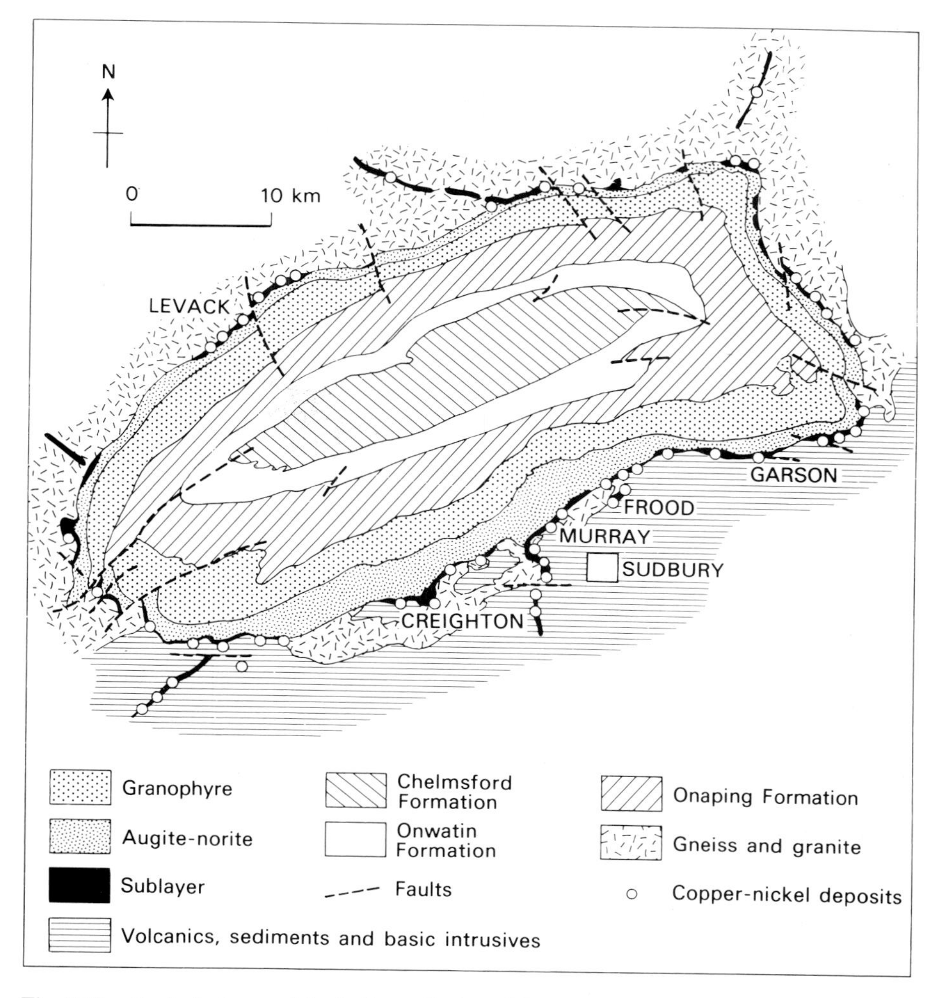

Fig. 12.8. Geological map of the Sudbury district. (After Souch *et al.* 1969 and Brocoum & Dalziel 1974).

present the world's largest source of nickel as well as an important source of copper, cobalt, iron, platinum and eleven other elements.

Inside the Complex is the Whitewater Group consisting of a volcaniclastic-like sequence (Onaping Formation), a manganese-rich slate sequence (Onwatin Formation) and the Chelmsford Formation — a carbonaceous and arenaceous proximal turbidite. These formations have only suffered slate grade regional metamorphism. No rocks which can be correlated with the Whitewater Group have been positively identified outside the basin.

To the south-east of the Complex there are metasedimentary and metavolcanic rocks of the Huronian Supergroup that were deposited unconformably upon the migmatitic Archaean basement which is exposed along the north-western side of the Complex. The Huronian has suffered a much higher grade of metamorphism than the Whitewater Group and it contains a number of basic and acid plutonic intrusions. It forms a southward facing homocline and carries a penetrative foliation caused by flattening. The rocks of the Basin are also considerably deformed (Brocoum & Dalziel 1974) and finite strain analysis suggests that the Chelmsford Formation was shortened by 30% in a north-westerly direction; this means that the Chelmsford Formation and probably the Sudbury Structure were less elliptically shaped prior to deformation.

The origin of the Sudbury Structure and the structure along which the Complex was intruded have been debated since the area was first mapped about the turn of the century. There is now an important school of thought initiated by Dietz (1964) that regards the Structure as having resulted from a meteoritic impact. Shock metamorphic features, including shatter cones, are common in the rocks around the Complex for as much as 10 km from its footwall, and in the breccias of the Onaping Formation, which is regarded by this school as being a fall-back breccia resulting from the impact. Geologists arguing against the meteorite impact hypothesis interpret the Onaping Formation as an ignimbrite deposit and have mapped a quartzite unit, believed to be a basal quartzite, lying beneath the Onaping Formation. This school generally regards the Complex as having been intruded along an unconformity at the base of the Whitewater Group. Another controversial rock-type is the 'Sudbury Breccia'. This consists of zones, a few cm to several km across, of brecciated country rocks which are sometimes hosts for the orebodies.

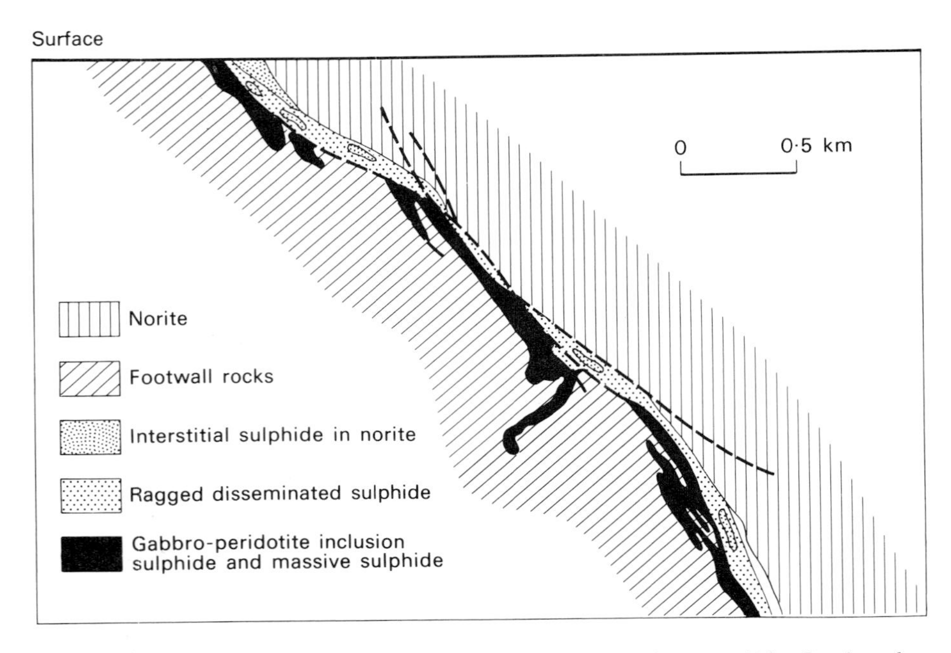

Fig. 12.9. Generalized section through the Creighton ore zone looking west. (After Souch *et al.* 1969).

The discovery of mineralization during the construction of the Canadian Pacific Railway in 1883 has led to the development of over forty mines and the total declared ore reserves of the district from the time of the original discovery to the present are of the order of 930 Mt. Of these, about 500 Mt have been exploited but new reserves are constantly being blocked out, just about keeping pace with production. According to Lang (1970), the grade of ores worked in the past was about 3.5% nickel and 2% copper. Today, with large-scale mining methods, the grade worked is around 1% for both metals.

Most of the orebodies occur in the sublayer whose magma was rich in sulphides and inclusions of peridotite, pyroxenite and gabbro. The sulphides tended to sink into synclinal embayments in the footwall giving a structural control of the mineralization. The Creighton ore zone has the greatest number of ore varieties (Souch *et al.* 1969). It plunges north-westward down a trough at the base of the Complex for at least 3 km (Fig. 12.9) and consists of a series of ore-types. The hanging wall quartz-norite above the sublayer occasionally contains enough interstitial sulphide to form low grade ore. In the upper part of the sublayer, ragged disseminated sulphide ore occurs, consisting of closely packed inclusions (several mm to 10 cm in size) in a matrix of sulphides and

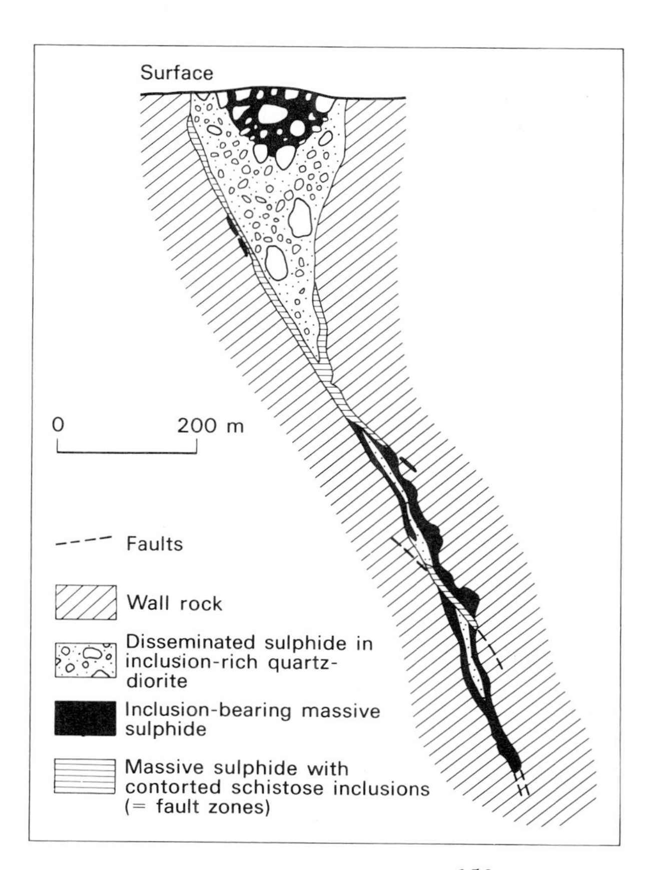

Fig. 12.10. Generalized section through the Frood orebody looking south-west. (After Souch *et al.* 1969).

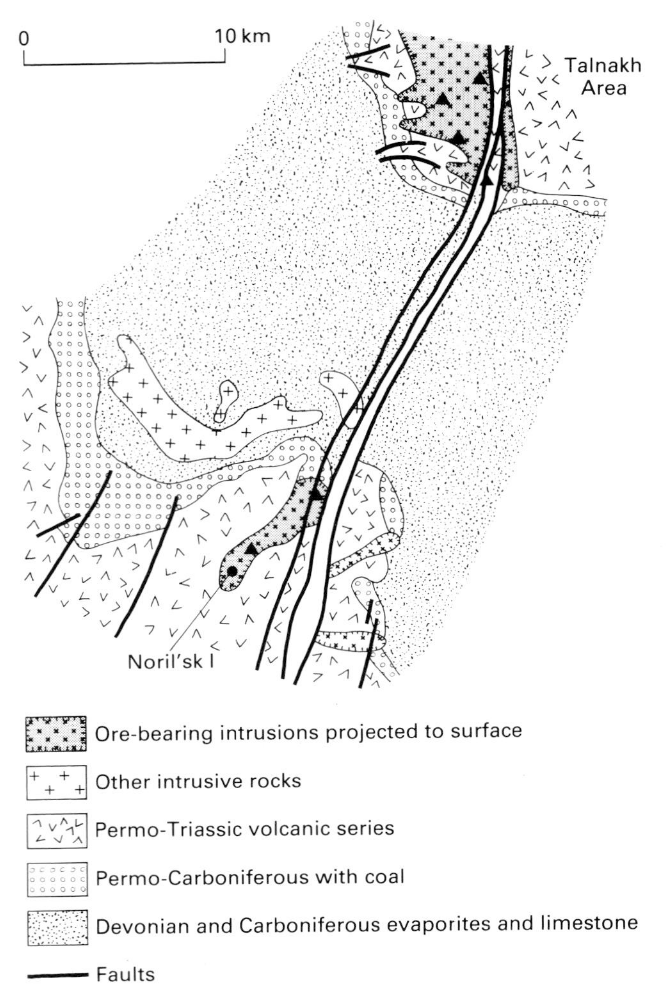

Fig. 12.11. Geology of the Noril'sk-Talnakh region after Naldrett (1981). For location see Fig. 8.7.

subordinate norite. The sulphide content increases downwards as does the ratio of matrix to inclusions, with a concomitant increase in inclusion size up to 1 m, resulting in an ore called gabbro-peridotite inclusion sulphide. This ore changes towards the footwall into massive sulphide containing fragments of footwall rocks. It is called inclusion massive sulphide and it is discontinuous and also forms stringers and pods in the footwall.

The Frood-Stobie orebody is an example of an orebody in an offset dyke. This parallels the footwall of the Complex and it has been suggested that it may once have been continuous with the Complex at a higher level. It is a huge orebody, 1.3 km long, 1 km deep and nearly 300 m across at its widest point. It consists of a wedge-shaped body of inclusion-bearing quartz-diorite (Fig.

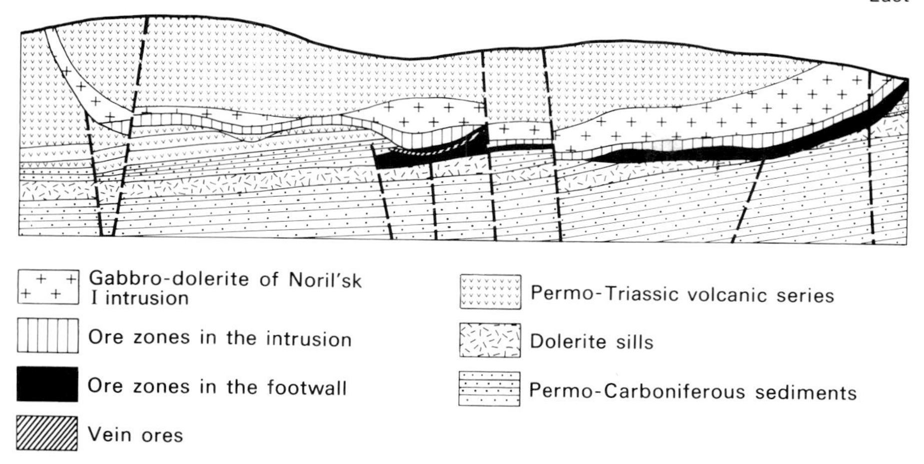

Fig. 12.12. Cross section through the Noril'sk I deposit. (After Glazkovsky *et al.* 1977).

12.10) with disseminated sulphide partially sheathed by inclusion massive sulphide. In the lower half, the ore described by Hawley (1962) as immiscible-silicate-sulphide ore occurs (Fig. 4.1). This ore-type grades into massive ore outwards and downwards. Inclusions in the quartz-diorite vary from a few centimetres to many metres in length. The largest found was one of peridotite 45 m long!

For further information on the Sudbury deposits the reader is referred to Pye *et al.* (1984), a veritable storehouse of data and ideas on the deposits and their host rocks.

(b) *Deposits in intrusions related to flood basalts*. Rifting of cratons has resulted in the development of flood basalts at various times during the earth's history, and some of the associated basic intrusions carry important nickel sulphide mineralization. These include the Jurassic Insizwa Complex, RSA, which is part of the Karoo Magmatic Event; the Lower Triassic ores of the Noril'sk-Talnakh Region in Siberia, and the 1100 Ma mineralization of the Duluth Complex, Minnesota plus the Great Lakes deposit of Ontario, which are both related to Keweenawan magmatism.

The Noril'sk-Talnakh Region (Fig. 12.11) has the largest reserves of copper-nickel ore in the Soviet Union. It lies deep in northern Siberia near the mouth of the Yenisei River (Fig. 8.7). The country rocks are carbonates and argillaceous sediments of the early and middle Palaeozoic overlain by Carboniferous rocks with coals, Permian and a Triassic basic volcanic sequence. The associated gabbroic intrusions form sheets, irregular masses and trough-shaped intrusions depending on their location in the gentle folds of the sediments lying beneath the volcanic sequence (Glazkovsky *et al.* 1977).

The Noril'sk I deposit occurs in a differentiated layered dominantly gabbroic intrusion which extends northwards for 12 km and is 30–350 m thick. In cross section it is lensoid with steep sides (Fig. 12.12). The copper-nickel sulphides form breccia, disseminated and massive ores at the base of the intrusion and vein orebodies developed in the footwall rocks and the basal portion of the intrusion. Like Sudbury there is a high Cu:Ni ratio and (Pt + Pd):Ni = 1:500.

13

Greisen Deposits

Introduction

Premoli (1985), with prophetic insight, discussed the effect on tin mining of a collapse of the International Tin Council's cartel. His prophecy was that, in the event of a severe fall in the price of tin, the small and medium sized producers operating high cost, low tonnage vein deposits, e.g. in Cornwall, would be forced to close, and that output from the old established placer deposit producers would also fall drastically as these are often undercapitalized, work deposits of declining grade and are faced with rising production costs. Premoli put forward, however, the caveat that the increasing flood of tin from Brazil, although mainly from placers, 'seems all but unstoppable'. This being the case, he forecast that the well capitalized, technically efficient miners will develop a new generation of tin mines based on comprehensive exploration and evaluation of primary tin deposits, economy of scale, and modern, cost-effective, bulk mining methods. Tin explorationists will now have to think big and their targets will be something like this: 40–60 Mt orebodies with a minimum mill grade of 0.3–0.4% Sn and a recoverability of at least 60%. By-product tungsten or other elements or minerals will of course improve the economic prospects of such deposits, whose economic viability will be somewhat variable depending on mining and mineral processing costs and the available infrastructure. Rio Algom's East Kemptville Mine in Nova Scotia, which came on stream late in 1985, will test Premoli's thesis admirably. Mineable ore reserves are estimated at 56 Mt grading 0.2% Sn, and 9000 t of ore per day will be mined, with by-product copper and zinc. The mine's economic progress will be watched with great interest by the whole industry.

The type of primary tin deposits that can come into the grade-tonnage limitations given above are:

(1) subvolcanic stockwork deposits (porphyry tins) referred to in Chapter 15;
(2) greisen deposits;
(3) large pegmatite deposits, see Chapter 10;
(4) large limestone-replacement or exhalative deposits in Dachang, China (Tanelli & Lattanzi 1985). In this chapter we will look at greisen deposits.

Tin greisen deposits

Greisens may be defined as a granoblastic aggregate of quartz and muscovite (or lepidolite) with accessory amounts of topaz, tourmaline and fluorite formed

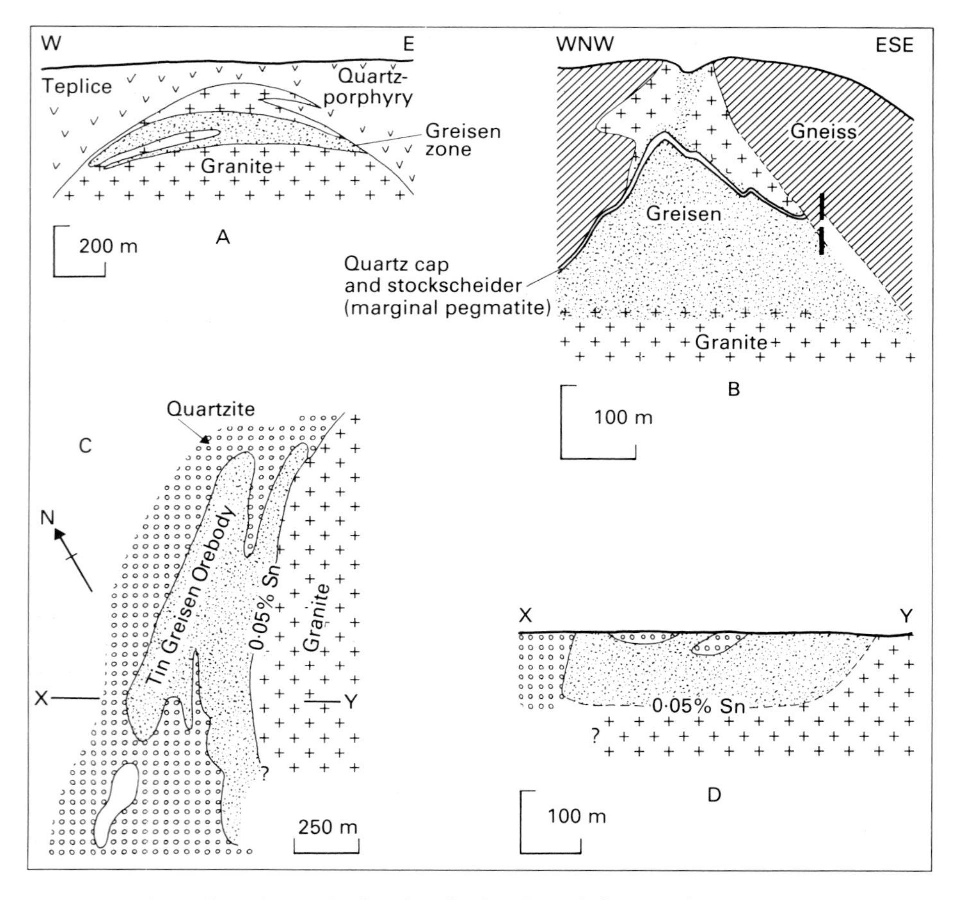

Fig. 13.1. Some tin greisen orebodies. A = Section through Cínovec, Czechoslovakia, B = Sadisdorf, DDR, C and D = Map and section of East Kemptville, Nova Scotia, Canada. A and B are after Baumann (1970) and C and D after Richardson *et al.* (1982).

by the metasomatic alteration of granite (Best 1982). They are mainly important for their production of tin and tungsten. Usually one element is predominant but there may be by-product output of the other. Greisens are usually developed at the upper contacts of granite intrusions and are sometimes accompanied by stockwork development.

Deposits of this type have been worked in the Erzgebirge on either side of the Czech-DDR border for many years, and cross sections of some typical deposits are given in Fig. 13.1 together with a plan and section of the East Kemptville orebody. The Erzgebirge deposits occur as massive greisens and greisen-bordered veins in the uppermost endocontacts and exocontacts of small lithium mica-albite granite cupolas. These coalesce at depth to form a large batholith (Štemprok 1985). The mineralization at East Kemptville is, however, developed beneath an inflection in the granite-metasedimentary contact of a discrete pluton (Davis Lake Pluton) that is part of the South Mountain

Batholith (Richardson *et al.* 1982; Chatterjee & Strong 1984). This batholith was intruded into a deformed Lower Palaeozoic metasedimentary sequence during the Acadian Orogeny and it contains a number of zoned, coalescent plutons that may average more than 10 ppm tin (Smith *et al.* 1982). They are, like the Erzgebirge granites, tin-specialized, S-type granites showing such features as high Rb/K, a marked decrease in Ba and Sr and an increase in Rb from marginal granodiorite to late stage alaskite; a pattern interpreted by Groves & McCarthy (1978) as indicative of *in situ* crystallization. These authors suggest that tin greisens and other tin deposits may develop in the upper parts of granite plutons when an impermeable roof of early crystallized cumulates has formed. Beneath this roof, water-saturated melt accumulates and eventually crystallizes. Incompatible elements, tin and other elements are concentrated in the late intercumulus liquid which eventually loses equilibrium with the cumulus minerals, and greisenization proceeds.

This model suggests some useful exploration indicators. Tin is likely to have been concentrated in zones with the lowest Ba and Sr values and confirmation of such an area, as being one where late stage crystallization occurred, can be obtained by examining the variation of concentration changes of incompatible elements such as Li and Rb with height within the granite. In addition the granites should be tin-specialized, S-type (though not necessarily in a collision tectonic setting, e.g. the tin granites of the Bushveld, RSA) with cupolas and ridge zones largely preserved from erosion, or with other traps for water-saturated melts such as the contact inflexion at East Kemptville. Further discussions of the nature of tin granites and their recognition can be found in Evans (1982) and Taylor (1979); of granophile mineral deposits in Strong (1981); and of tin and tungsten greisen deposits and other granite-related mineral deposits in Taylor (1979) and Taylor & Strong (1985), which contains more information on East Kemptville, whilst Burt (1981) has a useful discussion of the process of greisenization. A useful discussion of Premoli's ideas, with grades and tonnages of various potential deposits, can be found in Anon. (1985c).

14

The Skarn Environment

Some general points

The commonest term for these deposits used to be Lindgren's (1922) 'pyrometasomatic'; however, at the present time it seems to have been dropped by almost universal consent for the less satisfactory term 'skarn'. Skarn is an old Swedish mining term for silicate gangue, and who might want to mine gangue except perhaps for use as an aggregate?! The tyro must bear in mind that the majority of skarns are devoid of economic mineralization. The first geologists to take over this term were metamorphic petrologists who have long used it to describe rocks composed of calcium, magnesium and iron silicates that have been derived from nearly pure limestones or dolomites, into which large amounts of Si, Al, Fe and Mg were *metasomatically* introduced, frequently, but not always, at or near the contact with a plutonic igneous intrusion (Best 1982). Identical rocks formed by isochemical metamorphism of impure limestones and dolomites are termed calc-silicate hornfelses. If they contained pre-existing mineralization then this too will be recrystallized and the resultant hornfels may be mistaken for a skarn, e.g. the magnetite orebody of Dielette in Normandy, France (Cayeux 1906).

Calc-silicate hornfels can usually be distinguished from skarn by examining its field relationships, except in the difficult case of skarn development by reaction between interlayered silicate and carbonate sediments during metamorphism — such rocks are termed 'reaction skarns'. Skarnlike rocks of uncertain origin are often referred to as skarnoids.

Skarns can be classified according to the rocks they replace, and the terms exoskarn and endoskarn were originally applied to replacements of carbonate metasediments (usually marbles) and intrusive rocks, respectively, in contact zones, but some authors have extended the use of the term endoskarn to skarn formed in any aluminous rock; shales, volcanics etc. (Einaudi & Burt 1982). Both endoskarns and exoskarns may contain ore, but in an endoskarn-exoskarn couplet where the exoskarn is a converted marble then it usually contains most or all the economic mineralization. Exoskarns may be classified according to the dominant mineralogy; as magnesian if they contain an important component of Mg silicates such as forsterite, or as calcic when calcium silicates, e.g. andradite, diopside, are predominant. The majority of the world's economic skarn deposits occur in calcic exoskarns (Einaudi *et al.* 1981).

Some of the important features of skarn deposits have already been mentioned; these features and others are summarized in Table 14.1. The irregular morphology of their orebodies (Fig. 2.10) and the chief elements won from them have been discussed on pp. 18–19. On pp. 59–61 the reader will find a short description of their mineralogy followed by a discussion of their genesis, where it is pointed out that experimental work indicated that dissolution of the host marble and precipitation of sulphide could be effected by the oxidation of bisulphide solutions at 400–450°C. Such high or higher temperatures are indicated by the nature of the gangue minerals and the work of Milovskiy *et al.* (1978). Working on the scheelite-molybdenite-chalcopyrite-bearing skarns of Chorukh-Dayron, they found from a study of fluid inclusions that ore deposition began with the crystallization of scheelite from highly concentrated solutions (50–60 wt % of salts) at temperatures of 475–400°C and pressures of the order of 161 600 kPa. Metamorphic studies indicate that the previously crystallized calc-silicate gangue in skarns formed at temperatures approaching those of the associated magmatic intrusion.

Exoskarns usually show a zoning of both the silicate and ore minerals. This

Table 14.1. General characteristics of skarn deposits.

Depth of formation	Variable, one to several km
Temperature of formation	From 350–800°C
Occurrence	Adjacent to or partially inside deep-seated intrusive rocks. Often emplaced in carbonate rocks, occasionally in hornfelses, schists or gneisses
Nature of ore zones	Extremely irregular, tongues of ore may project along any available planar structure—bedding, joints, faults, etc. Distribution within the contact aureole is often apparently capricious. Structural changes may cause abrupt termination of the orebodies
Ores of	Fe, Cu, W, C, Zn, Pb, Mo, Sn, etc.
Ore minerals	*Magnetite*, specularite, *graphite*, gold, chalcopyrite, *pyrrhotite*, *scheelite*, *wolframite*, galena, sphalerite, pyrite, molybdenite, cassiterite
Gangue minerals	High temperature skarn minerals, e.g. grossularite, hedenburgite, idocrase, epidote, actinolite, wollastonite, diopside, forsterite, anorthite, etc. Quartz if present crystallized as high form. Carbonates
Wall rock alteration	Widespread development of skarn, tactite and/or marble
Textures and structures	Usually coarse-grained, may be banded due to replacement of bedded limestones
Zoning	Paragenetic sequence is usually: silicates; scheelite, magnetite, cassiterite; base metal sulphides. These groups often show a zonal arrangement—sulphides outermost. Pb and Zn usually persist to greater distances from the contact than does Cu. Barren silicate zones may be present next to the intrusive
Examples	Fe of Marmora, Ontario; Cornwall, Penn.; Cu of Mackay District, Idaho; Concepcion del Oro district, Mexico; Zn of Oslo area, Norway; W of Gold Hill, Utah; King Is. Scheelite, Tasmania; Sn of Kramat Pulai, Malaya

has been well documented for the magnetite deposits of Cornwall, Pennsylvania, by Lapham (1968), and Theodore (1977) has discussed the zoning of skarn deposits associated with porphyry copper deposits. One of his examples is reproduced in Fig. 14.1. The mineral zones in the Ely, Nevada, cupriferous skarns generally parallel the igneous contact and the bulk of the copper in the skarn was deposited in veinlets which cut the andradite-diopside rocks. The alteration envelopes along these veins contain actinolite-calcite-quartz-nontronite assemblages. These relationships indicate that a clay-sulphide stage was superposed on the earlier calc-silicate rocks.

Endoskarns also display mineral zonation resulting from a progressive addition of calcium to the igneous protolith (usually anything from granite to gabbro). A common zonation outwards towards the marble host is biotite →amphibole→pyroxene→garnet. K-feldspar disappears but plagioclase may survive and there may be such a metamorphic convergence between the mineral assemblages of an endoskarn and its adjacent exoskarn that the position of the original intrusive igneous contact is unrecognizable. According to Einaudi & Burt (1982), where skarn occurs near or over the tops of plutonic cupolas, as is the case for most skarns related to porphyry copper plutons and tin skarns, endoskarn is usually absent. It seems therefore that endoskarn development occurred in those areas where fluid flow was dominantly into the pluton or upward along its contacts with marbles, rather than where the metasomatizing fluids flowed up and out of the pluton as in the formation of porphyry copper deposits.

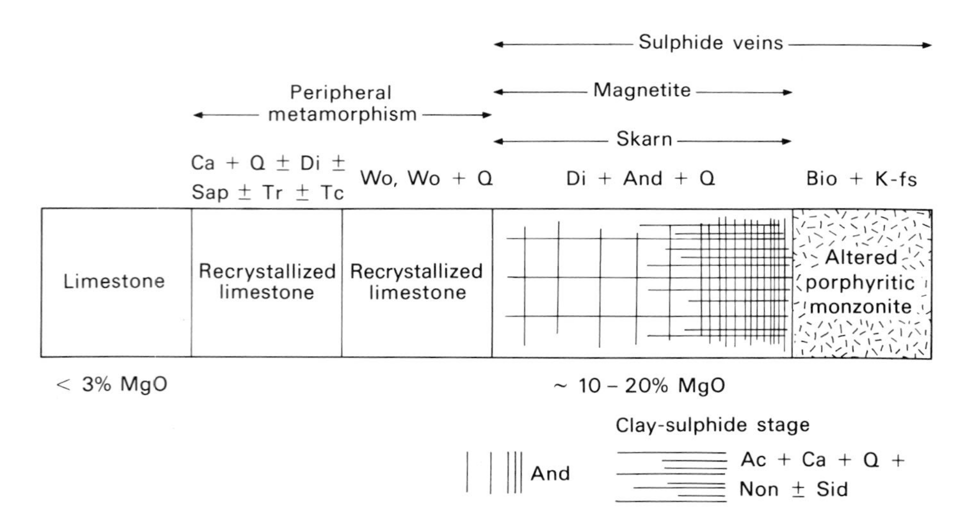

Fig. 14.1. Zonation of mineral assemblages at the Ely, Nevada, deposit. The width of the zones is schematic and the density of the patterns indicates the relative amounts of andradite (And) and the clay-sulphide stage. Abbreviations: Ac = actinolite; And = andradite; Bio = secondary biotite; Ca = calcite; Di = diopside; K-fs = secondary potash feldspar; Non = nontronite; Q = quartz; Sap = saponite; Sid = siderite; Tc = talc; Tr = tremolite, and Wo = wollastonite. (Modified from Theodore 1977).

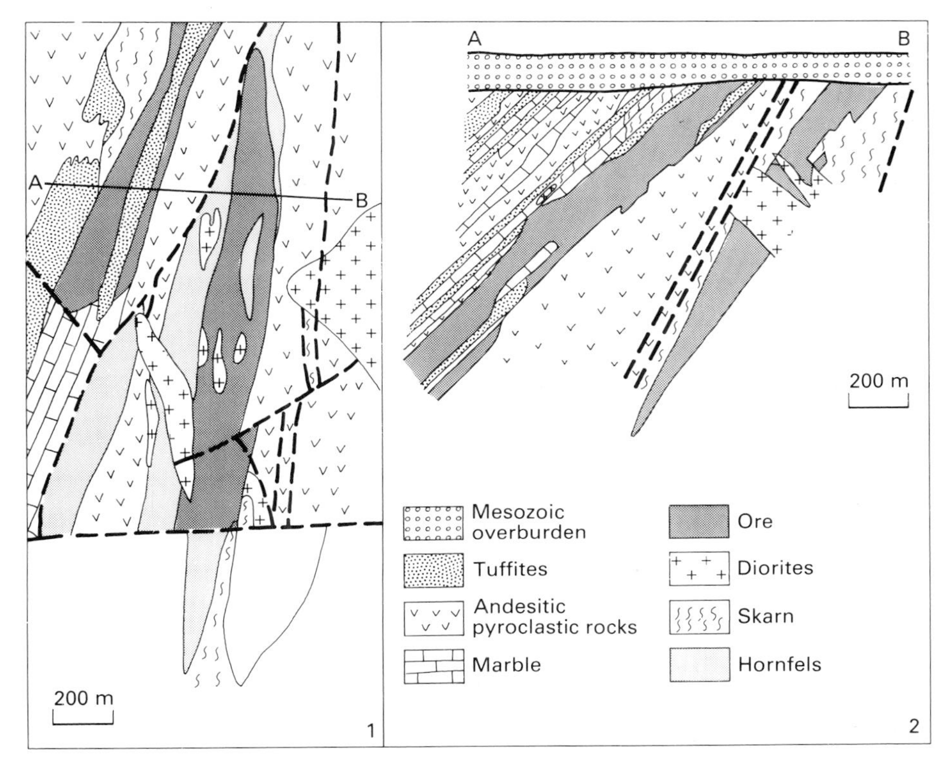

Fig. 14.2. Sarbai iron skarns, USSR. 1–Plan of 80 m level. 2–Cross section. (After Sokolov & Grigor'ev 1977).

Some examples of skarn deposits

IRON SKARN DEPOSITS

Skarns have long been important sources of iron ore and the magnetite mine at Cornwall, Pennsylvania, supplied much of the iron used during the industrial revolution in the USA. It is the oldest, continuously operated mine in North America. Mining commenced in 1737 and by 1964 93 Mt of ore had been produced with an average mill feed grade in 1964 of 39.4% Fe and 0.29% Cu, from which minor by-product cobalt, gold and silver were obtained (Lapham 1968). A pyrite concentrate was used to produce sulphuric acid, and up to 1953, when open pit operations ceased, limestone overburden was crushed and sold as aggregate. A good example of 'waste not want not!' Cornwall is a calcic iron skarn and such skarns are associated with intrusives ranging from gabbro through diorite to syenite, whilst magnesian iron skarns are normally associated with granites or granodiorites (Einaudi *et al.* 1981). The largest known deposits in either class occur in the USSR. With magnetite as the major ore mineral, these deposits are often marked by pronounced magnetic anomalies, and the detection of these by aeromagnetic surveys led to

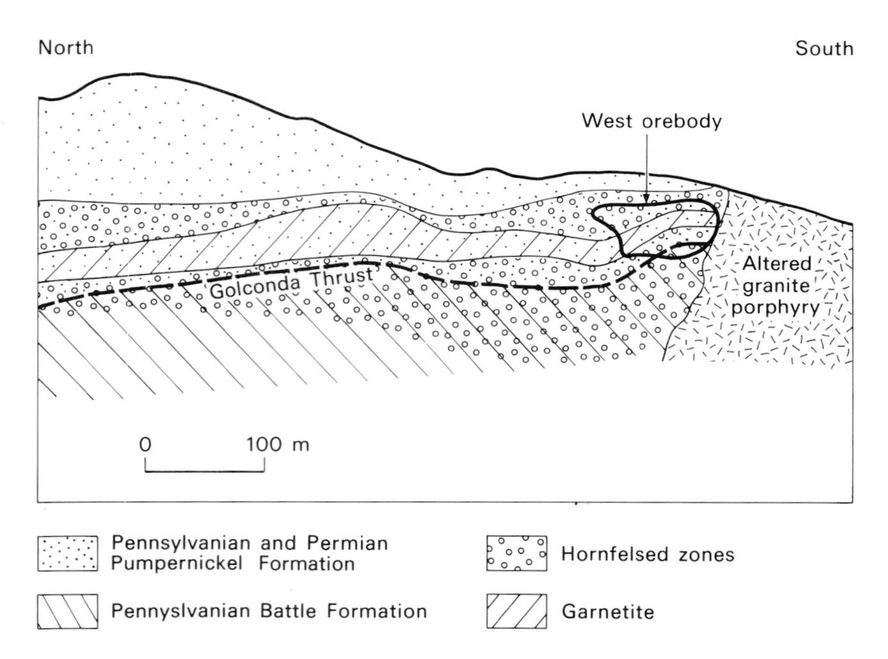

Fig. 14.3. Cross section through the copper-gold-silver skarn deposit at Copper Canyon, Nevada. Rocks peripheral to the hornfelsed zones carry secondary mica. (Modified from Theodore 1977).

the discovery *inter alia* of Sarbai in the USSR and of Marmoraton, Ontario. These deposits typically run 5–200 Mt with a grade of about 40% Fe.

Sarbai is the giant of skarn deposits with 725 Mt grading 46% Fe. It lies in the Turgai Iron Ore Province of the Kazak SSR, i.e. in the south-western part of the Siberian Platform to the east of the southern end of the Urals and about 500 km south-east of the famous and long worked iron skarns of Magnitogorsk. The orebodies (Fig 14.2) occur in a succession of metamorphosed pyroclastics, marbles and skarns developed from a Carboniferous volcaniclastic-sedimentary succession in the western limb of an anticline (Sokolov & Grigor'ev 1977). At Sarbai, as elsewhere in the Turgai Province, there is a marked development of chlorine-bearing scapolite associated with the iron skarn orebodies indicating the presence of important amounts of brine solution of during the metasomatism. The orebody dimensions are impressive: the eastern one has a strike length of 1.7 km and a thickness of up to 185 m; it has been traced down dip for more than 1 km and the western orebody for 1.8 km. There is a further possible reserve of 775 Mt!

TWO COPPER SKARN DEPOSITS

(a) *Copper Canyon, Nevada.* At this deposit, skarn has replaced calcareous shale or argillite beds just above the Golconda Thrust (Fig. 14.3) producing a

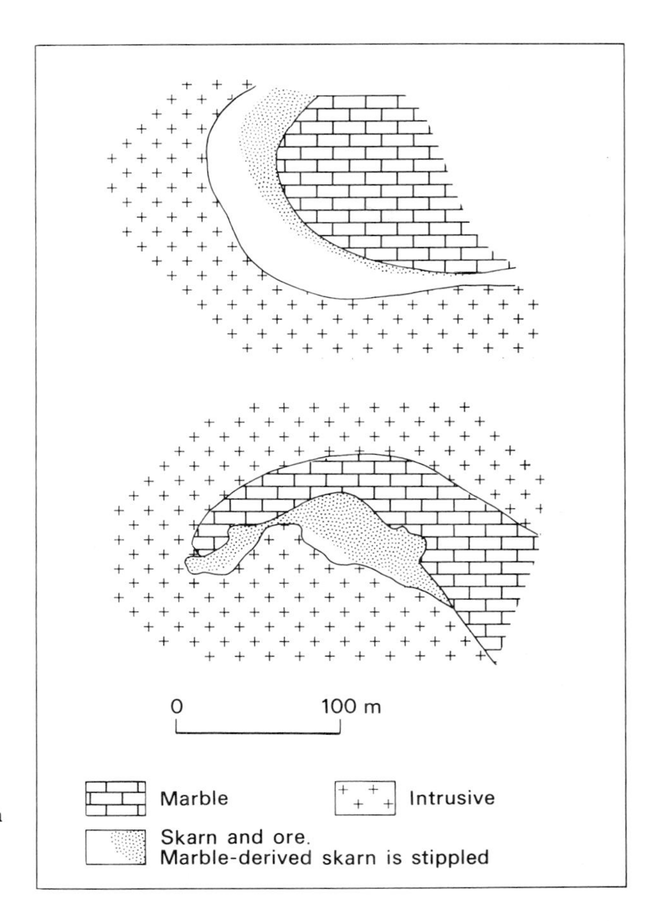

Fig. 14.4. Geological map of the 1500 ft level (above) and an east-west section (below) of the Memé Mine, Haiti. Note the concentration of skarn and ore beneath the marble. (Modified from Kesler 1968).

flat-lying tabular zone of andradite-rich rock in which most of the mineralization occurred. Although the andradite rock stretches at least 400 m from the granite porphyry stock, only that part within 180 m of the contact contains ore. Silicate zones in this deposit are symmetrical about the andradite rock, thus forming zones at right angles to the present igneous contact. For this and other reasons, Theodore (1977) suggests that the porphyry or the present site of the porphyry was not the locus from which the skarn-forming fluids emanated.

(b) *The Memé Mine, Northern Haiti*. Frequently at the contacts of skarns and intrusions there is a completely gradational contact and this is the case at the Memé copper mine where a large block of Cretaceous limestone has been surrounded by quartz-monzonite. Mineralization was preceded by extensive magmatic assimilation that formed zones of syenodiorite and granodiorite around the limestone. Following the crystallization of the magma, the limestone and neighbouring parts of the intrusion were replaced by skarn. The intrusion-derived endoskarn contains large quantities of diopside which distinguishes it

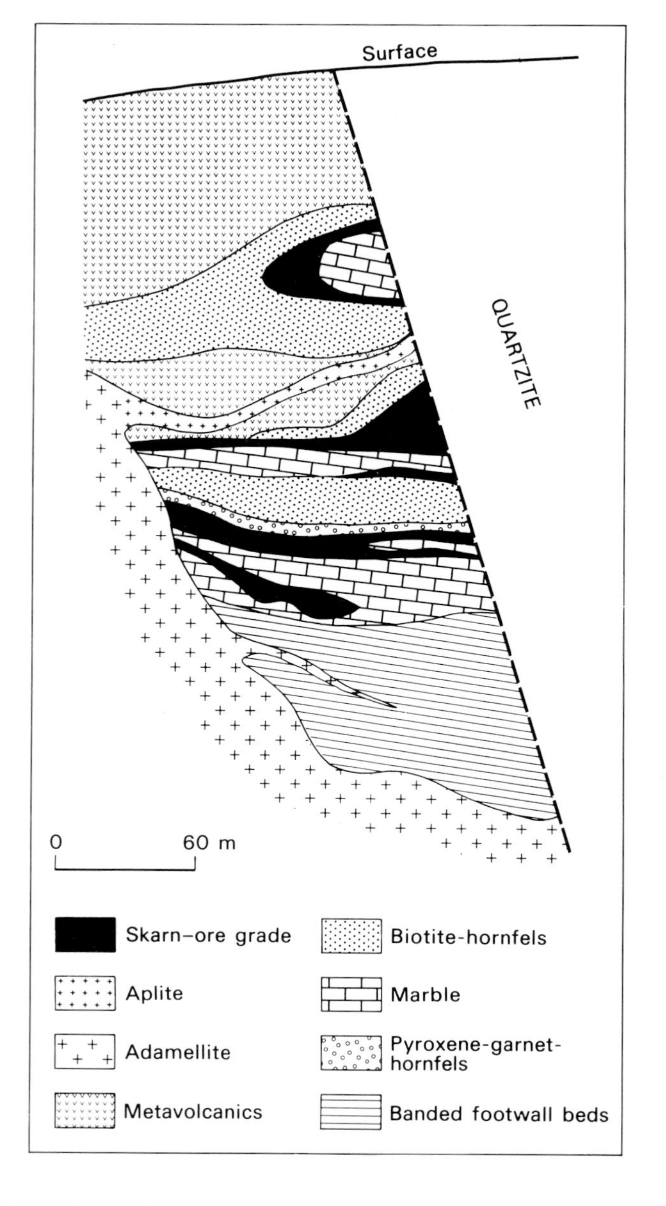

Fig. 14.5. Section through the Bold Head orebodies looking north, Tasmania. (Modified from Danielson 1975).

from the marble-derived exoskarn. There is a completely gradational contact between these skarns (Kesler 1968).

Mineralization followed skarn formation and consisted of the introduction of hematite, magnetite, pyrite, molybdenite, chalcopyrite, bornite, chalcocite and digenite, in that paragenetic order. These occur as replacement zones. The main skarn and ore development is along the lower contact with the limestone block (Fig. 14.4). Skarn formation took place at between 480 and 640°C and exsolution textures suggest that the minimum temperature of copper-iron

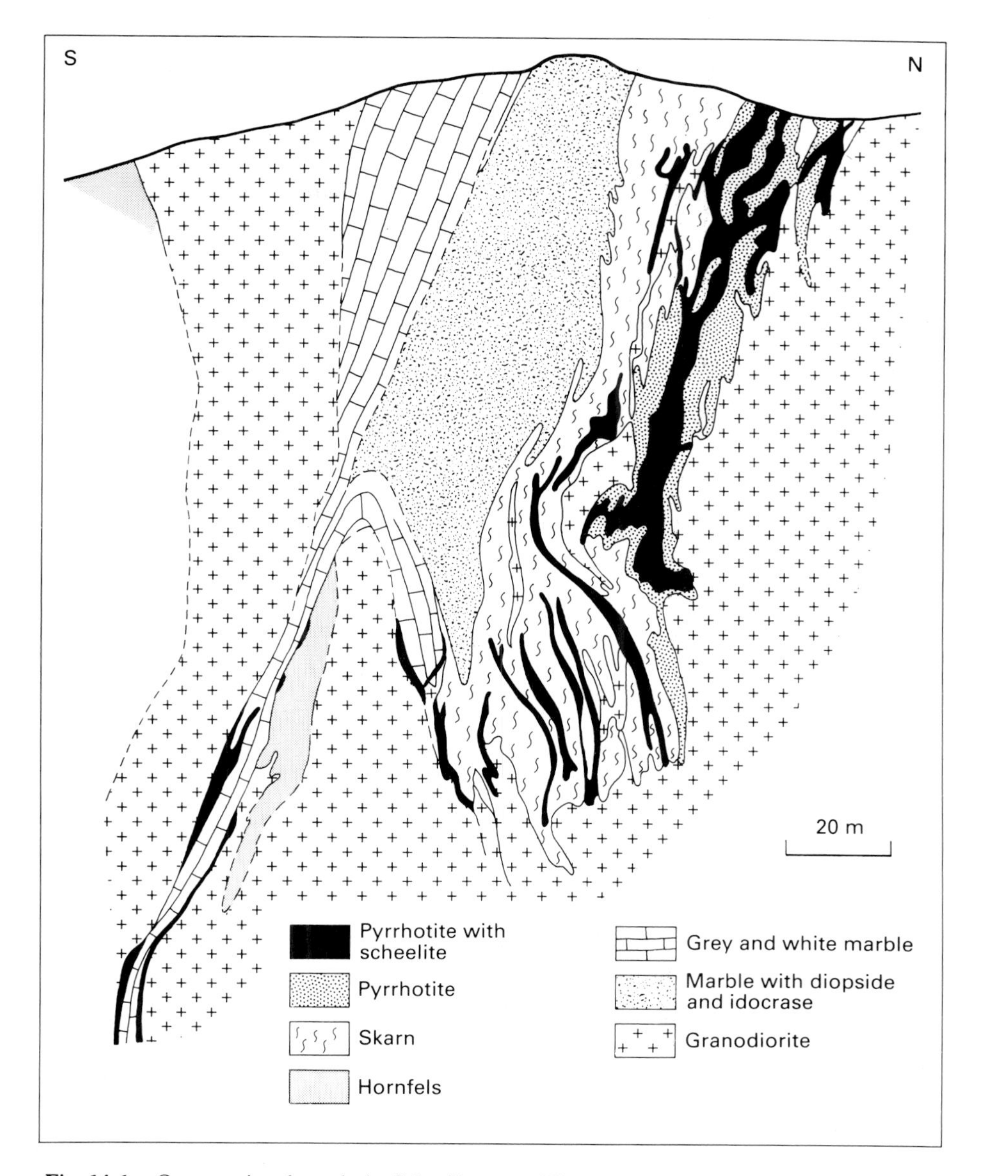

Fig. 14.6. Cross section through the Salau Tungsten Mine, France. (After Prouhet 1983).

sulphide deposition exceeded 350°C and the youngest ore minerals crystallized above 250°C. The grade is about 2.5% Cu.

TUNGSTEN SKARNS

Tungsten skarns, veins and stratiform deposits provide most of the world's annual production of tungsten, with skarns being predominant. In the western world most skarn tungsten comes from a few relatively large deposits: King Island, Tasmania; Sangdong, Korea; Canada Tungsten (NWT) and MacMillan Pass (Yukon), Canada; and Pine Creek, California, USA. To put 'relatively large' in perspective it must be remarked that western world consumption of tungsten is only about 30 000 t, i.e. the equivalent of the production of just one modestly sized copper mine which in turn is small compared to an iron mine! In this section we will look at one large and one small tungsten mine.

(a) King Island Scheelite

King Island, which lies between Australia and Tasmania at the western approach to Bass Strait, contains a number of important tungsten deposits. These are scheelite-bearing skarns developed in Upper Proterozoic to Lower Cambrian sediments which have been intruded by a granodiorite and an adamellite of lower Carboniferous age. These granites are thought to be related to the tin- and tungsten-bearing granites of the Aberfoyle and Storey's Creek district of Tasmania (Danielson 1975). Mined ore and reserves at 1980 were put at 14 Mt averaging 0.8% WO_3 (Einaudi *et al.* 1981).

Scheelite-bearing andradite skarns were formed by the selective replacement of limestone beds and as a result they form stratiform orebodies 5–40 m thick. There is a great deal of mineralized skarn below ore grade. The orebodies lie either in the exocontact of the granodiorite or the adamellite (Fig 14.5). Irregular relicts of marble occur in the skarn demonstrating its metasomatic origin and Edwards *et al.* (1956) showed that the replacement was a volume for volume process with massive addition of silica, iron and aluminium and subtraction of calcium and CO_2, cp. Lindgren's similar result p. 59. A most detailed and exhaustive geochemical and fluid inclusion investigation was performed by Kwak (1978a, 1978b), who concluded that the Edwards *et al.* mass balance calculations are in error due to poor selection of sampling sites. Kwak's work indicated that the most favourable site for ore formation was where the marbles contained numerous interlayers of hornfels or hornfelsic fragments within 400 m of the granite contact. The source of the tungsten still remains unknown.

(b) Salau Tungsten Mine, Ariège (central Pyrenees) France

Over forty significant tungsten occurrences are known along the length of the Pyrenees with two important deposits, one of which, Salau, is exploited (Prouhet 1983). Salau lies in the axial zone of the Pyrenees near the Spanish frontier where the late Hercynian Salau Granodiorite invaded Ordovician limestones and converted them at the contact into grey graphitic marbles. The orebodies occur in exoskarns in a roof pendant in the granite (Fig 14.6) and large volumes of the ore are rich in pyrrhotite with minor amounts of chalcopyrite, molybdenite, arsenopyrite and other sulphides. Grades vary from 0.1–20% WO_3 and the probable reserves before mining commenced were

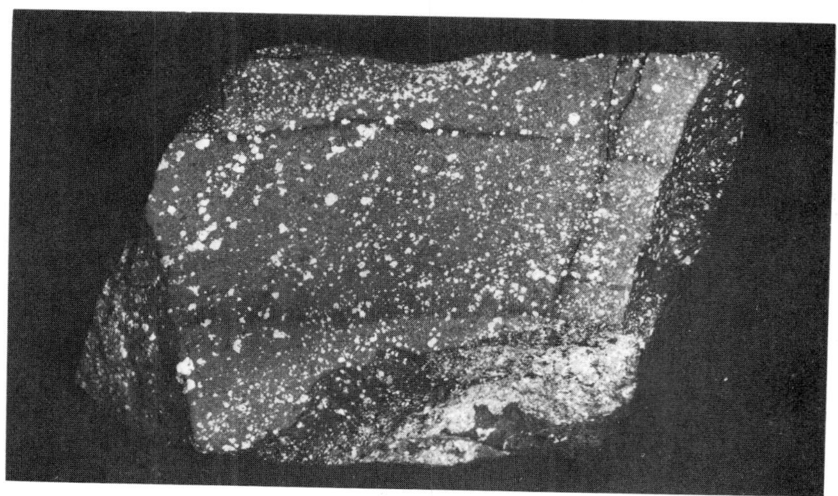

Fig. 14.7. A = Pyrrhotite-rich scheelite skarn, Salau Mine, France. Dark coloured grains are calc-silicates, hedenbergite, andradite, etc.
B = The same in ultra-violet light showing the abundant dissemination of scheelite. Scale: natural size.

1.3 Mt running 1.5% WO_3. The scheelite is richly disseminated through the pyrrhotite-bearing skarn (Fig. 14.7) and it generally runs 1–2% WO_3 but in the pyrrhotite-poor hedenbergite-garnet skarns it is only around 0.2%. Geochemical work has failed to discover any source bed for the tungsten and has also shown that the granodiorite is very poor in this element (Barbier 1982). Its ultimate origin is for the moment unknown.

OTHER SKARN DEPOSITS

Useful short summaries of skarn deposits can be found in Sawkins (1984) and Edwards & Atkinson (1986). The best exhaustive summary is Einaudi *et al.* (1981), and part 4 of *Econ. Geol.* 77 for 1982 is devoted to skarn deposits.

15

Disseminated and Stockwork Deposits Associated with Plutonic Intrusives

We are concerned in this chapter with low grade, large tonnage deposits which are principally mined for copper, molybdenum and tin. These deposits are normally intimately associated with intermediate to acid plutonic intrusives and all are characterized by intense and extensive hydrothermal alteration of the host rocks. The ore minerals in these deposits are scattered through the host rock either as what is called disseminated mineralization, which can be likened to the distribution of seeds through raspberry jam, or they are largely or wholly restricted to quartz veinlets which form a ramifying complex called a stockwork (Fig. 15.1). In many deposits or parts of deposits both forms of mineralization occur (Fig. 15.2).

The first copper deposits of this type to be mined on any scale are in some of the south-western states of the USA and it was here that the cost effectiveness of bulk mining methods was first demonstrated in the 1920s and the mining of much lower grade copper ores than had hitherto been exploited, became possible. It should however be stressed that our civilization would not have been able to produce metals so abundantly and so cheaply by these methods without the important invention, by Francis and Stanley Elmore in 1898, of that most efficient method of mineral processing, flotation. The fame and unique importance of this invention has long gone unsung and unrecognized (Wolfe 1984).

These American copper deposits are associated with porphyritic intrusives, often mapped as porphyries. The deposits soon came to be called copper porphyries, the name by which they are still generally known. On the other hand, more or less identical molybdenum deposits have been known as disseminated molybdenums although they are also called molybdenum stockworks or porphyry molybdenums. Similar tin deposits are more usually called tin stockworks, though the term porphyry tins has been used. The student will find all these names in present use. Whereas economic porphyry coppers and molybdenums are usually extremely large orebodies (50–500 Mt is the common size range), tin stockworks are much smaller, 2–20 Mt being the common size range. All three types of metal deposit may yield important by-products. Amongst these are molybdenum and gold from porphyry coppers; tin, tungsten and pyrite from the Climax molybdenum deposit (other porphyry molybdenums tend to be without useful by-products); and tungsten, molybdenum, bismuth and fluorite from tin stockworks.

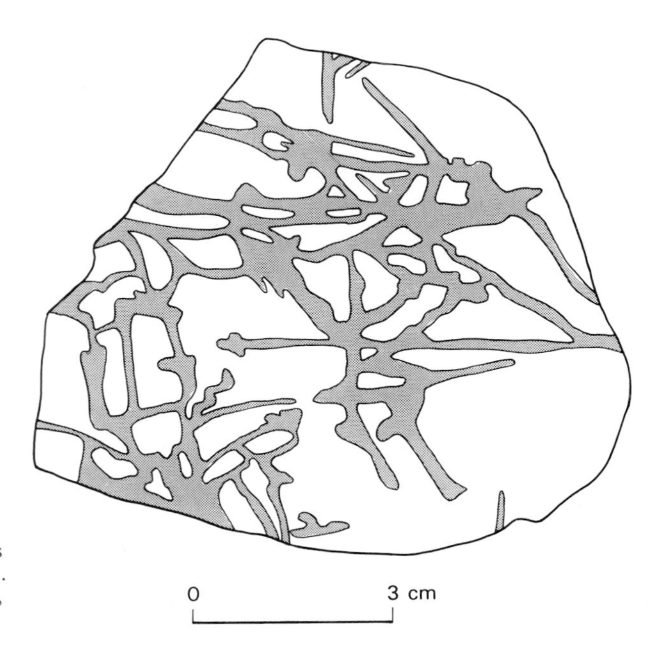

Fig. 15.1. Stockwork of molybdenite-bearing quartz veinlets in granite which has undergone phyllic alteration. Run of the mill ore, Climax, Colorado.

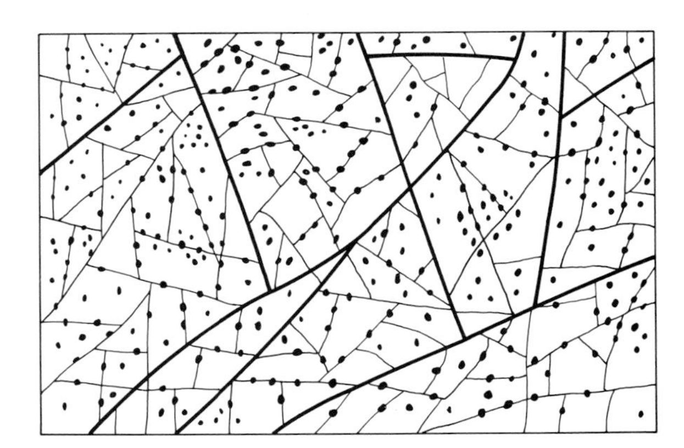

Fig. 15.2. Schematic drawing of a stockwork in a porphyry copper deposit. Sulphides occur in veinlets and disseminated through the highly altered host rock. (After Routhier 1963).

Porphyry coppers annually provide over 50% of the world's copper and over 50 deposits are in production. They are situated in orogenic belts in many parts of the world. Porphyry molybdenums in production are far fewer, about ten, and account for over 70% of world production. Tin stockworks are much less important, most tin production coming from placer and vein deposits.

Porphyry copper and molybdenum deposits are closely related and, although the next section is mainly devoted to porphyry coppers, mention will be made of some salient points concerning porphyry molybdenums.

Porphyry copper deposits

GENERAL DESCRIPTION

As has been indicated above, these are large low grade stockwork to disseminated deposits of copper which may also carry minor recoverable amounts of molybdenum gold and silver. Usually they are copper-molybdenum or copper-gold deposits. They must be amenable to bulk mining methods, that is open pit or, if underground, block caving. Most deposits have grades of 0.4–1% copper and total tonnages range up to 1000 million with a few giants being even larger than this (Fig. 15.3 and Table 15.1).

Grade and tonnage define the total amount of metal in the ground but with the continual drop in the price of copper in real terms, the emphasis in recent years has shifted to grade. At the low grade high tonnage end of the scale, development costs can turn an orebody into a wastebody, and few of the porphyry copper deposits developed in the 1970s would have covered their capital costs had today's prices prevailed when they started production. Selective mining is of course impossible and host rock, stockwork and disseminated mineralization have to be extracted *in toto*. In this way, some of the largest man-made holes in the crust have come into being.

The typical porphyry copper deposit occurs in a cylindrical stock-like, composite intrusion having an elongate or irregular outcrop about 1.5 × 2 km, often with an outer shell of equigranular medium-grained rock. The central part is porphyritic — implying a period of rapid cooling to produce the finer grained groundmass — the porphyry part of the intrusion. This raises the

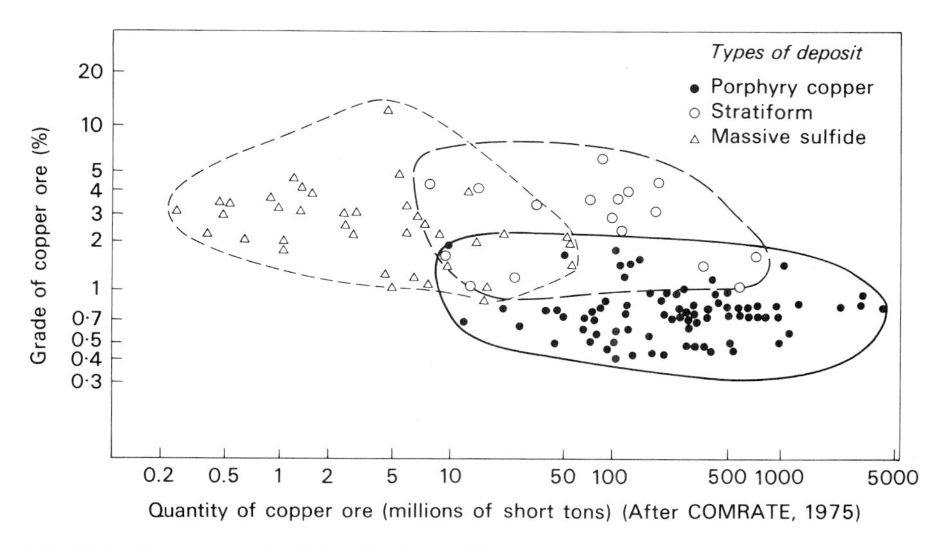

Fig. 15.3. Tonnage-grade relationships in porphyry copper, stratiform and massive sulphide deposits.

Table 15.1. Some giants and dwarfs of the porphyry copper world. (Data from Gilmour 1982 and other sources.)

Country	Deposit	Tonnage* Mt	Grade†			
			Cu%	Mo%	Au g t^{-1}	Ag g t^{-1}
USA	Bingham (Utah)	2733	0.71	0.053	—	—
	Morenci (Arizona)	848	0.88	0.007	—	—
	Ray (Arizona)	172	0.85	—	—	—
	San Manuel–Kalamazoo (Arizona)	980	0.74	0.015	—	—
	Santa Rita (New Mexico)	662	0.97	—	—	—
Canada	Lornex (British Columbia)	544	0.374	0.013	—	Tr
	Valley Copper	853	0.48	—	—	—
Mexico	Cananea	1542	0.79	—	—	—
Panama	Cerro Colorado	2000+	0.6	0.015	0.062	4.35
Chile	Chuquicamata	9423	0.56	0.06	—	—
	El Salvador	283	1.17	0.033	—	—
Papua New Guinea	OK Tedi	543	0.6	—	0.5	—
	Panguna	1085	0.48	—	0.55	1.5
Iran	Sar Cheshmeh					
	Supergene	92	1.996	Tr	Tr	Tr
	Hypogene	334	0.896	Tr	Tr	Tr
U.K.	Coed-y-Brenin	200	0.3	0.003+	0.082	
Yugoslavia	Bor	383	0.428	~0.1	0.065	~4.26

*Past production plus reserves.
†For present reserves, often higher grade worked in the past.

problem of how a late phase of rapid cooling could occur in the hot centre of the intrusion, insulated as it would be by the only just solidified and still hot outer portion. This problem will be considered in a later section.

Comprehensive reviews of porphyry copper deposits can be found in Titley and Beane (1981) and Titley (1982) and there is a useful summary in McMillan & Panteleyev (1980). One of the most comprehensive descriptions of a single deposit is that of the El Salvador deposit, in Gustafson & Hunt (1975), which is extensively utilized in the Open University case study (Open University S333 Course Team 1976) and tiros will find this to be a valuable aid to study.

PETROGRAPHY AND NATURE OF THE HOST INTRUSIONS

The most common hosts are acid plutonic rocks of the granite clan ranging from granite through granodiorite to tonalite, quartz monzodiorite and diorite. However, diorite through monzonite (especially quartz-monzonite) to syenite (sometimes alkalic) are also important host rock-types. Suggestions made in the past that diorite hosts only occur in island arcs have proved to be incorrect. Silica-poor hosts occur in both British Columbia and the central Andes.

Many authors agree that porphyry copper deposits are normally hosted by I-type granitoids, within which category it is important to distinguish the I (Cordilleran) from the I(Caledonian) intrusives, as the latter rarely carry economic mineralization (Pitcher 1983). Host intrusions in island arc settings have primitive initial strontium isotope ratios of 0.705–0.702 and are presumably derived from the upper mantle or recycled oceanic crust. The same ratios from mineralized intrusions in continental settings are generally higher indicating derivation from, or more probably, contamination by, crustal material (McMillan & Panteleyev 1980). Multiple intrusive events are common in areas with porphyry copper mineralization, with the host intrusions normally being the most differentiated and youngest of those present.

Host intrusions with associated volcanism generally formed late in the volcanic cycle and mineralization usually followed one or more pulses of magma intrusion. Thus at Ray, Arizona, early quartz-diorite was emplaced at 70 Ma, a porphyritic phase at 63 Ma and the mineralized porphyry at 61 Ma (Cornwall & Banks cited in McMillan & Panteleyev 1980). All the above points have important implications for exploration.

Attempts have been made to discriminate between barren and mineralized intrusions but without anything like universal success. Baldwin & Pearce (1982) recommended the use of a Y-MnO plot to discriminate between them and Hendry *et al.* (1985) not only showed that mineralized intrusions and their deeper level equivalents are depleted in copper compared with neighbouring barren intrusions, but also found that the latter are characterized by amphiboles of constant composition whilst fertile intrusions have amphiboles of variable composition, even within single grains.

INTRUSION GEOMETRY

The host intrusions usually appear to be passively rather than forcefully emplaced, stoping and assimilation being the principal mechanisms. They can be divided into three classes:

(1) A class in which the ore-related intrusive is simple an isolated stock. A variation on this theme could be a sill or a series of dykes or irregular bodies;

(2) In this class we no longer have a discrete isolated stock. The host is now a late stage unit of a composite, co-magmatic intrusion, often batholithic in dimensions. Examples belonging to this class occur in both continental and island arc settings;

(3) This class is not yet known to carry economic mineralization but it is clearly related to porphyry copper deposits. Occurrences belonging to this class take the form of extensive alteration zones carrying weak mineralization and occurring in the upper parts of equigranular intrusions.

HYDROTHERMAL ALTERATION

In 1970, Lowell & Guilbert described the San Manuel-Kalamazoo orebody

(Arizona) and compared their findings with 27 other porphyry copper deposits. From this study they drew up what is now known as the Lowell–Guilbert model. In this invaluable and fundamental work they demonstrated that the best reference framework to which we can relate all the other features of these deposits is the nature and distribution of the zones of hydrothermal wall rock alteration. They claimed that generally four alteration zones are present as shown in Fig. 15.4. These are normally centred on the porphyry stock in coaxial zones which form concentric but often incomplete shells and they are frequently used as a guide to ore in exploring porphyry copper deposits. In the Lowell–Guilbert model they are as follows:

(a) *Potassic zone*. This zone is not always present. When present it is characterized by the development of secondary orthoclase and biotite or by orthoclase-chlorite and sometimes orthoclase-biotite-chlorite. Sericite may also be present. These secondary minerals replace the primary orthoclase, plagioclase and mafic minerals of the intrusion. Anhydrite may be prominent in this zone. The secondary potash feldspar is generally more sodic than the primary potash feldspar. It may also be present in the quartz veinlets forming the stockwork. There is often a low grade core to this zone in which chlorite and sericite are prominent.

(b) *Phyllic zone*. This is alteration of the type known in other deposits as sericitization and advanced argillic alteration. It is characterized by the assemblage quartz-sericite-pyrite and usually carries minor chlorite, illite and rutile. Pyrophyllite may also be present. Carbonates and anhydrite are rare. The inner part of the zone is dominated by sericite while further out, clay minerals become more important. The sericitization affects the feldspars and primary biotite, alteration of the latter mineral producing the minor rutile. These are silica-generating reactions, so much secondary quartz is produced (silicification). The contact with the potassic zone is gradational over tens of

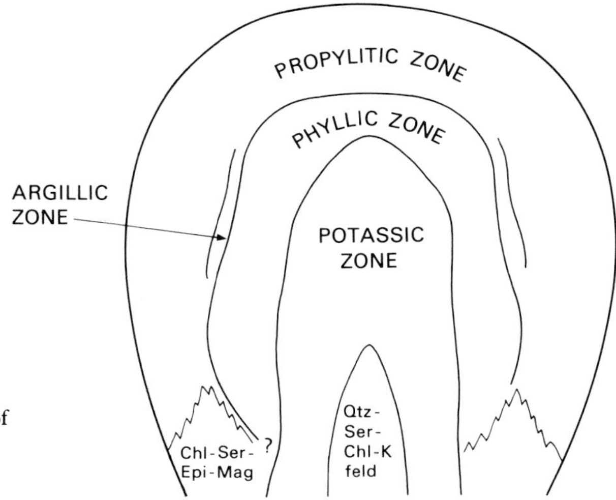

Fig. 15.4. Hydrothermal alteration zoning pattern in the Lowell–Guilbert model of porphyry copper deposits. (After Lowell & Guilbert 1970).

metres. When the phyllic zone is present it possesses the greatest development of disseminated and veinlet pyrite.

(c) *Argillic zone*. This zone is not always present. It is the equivalent of what is called intermediate argillic alteration in other deposits. Clay minerals are prominent with kaolin being dominant nearer the orebody, and montmorillonite further away. Pyrite is common, but less abundant than in the phyllic zone. It usually occurs in veinlets rather than as disseminations. Primary biotite may be unaffected or converted to chlorite. Potash feldspar is generally not extensively affected.

(d) *Propylitic zone*. This outermost zone is never absent. Chlorite is the most common mineral. Pyrite, calcite and epidote are associated with it. Primary mafic minerals (biotite and hornblende) are altered partially or wholly to chlorite and carbonate. Plagioclase may be unaffected. This zone fades into the surrounding rocks over several hundreds of metres.

Obviously, in many deposits the behaviour of these zones in depth is poorly known and for some deposits there are no data at all. What evidence there is suggests that the zones narrow in depth and quartz-potash feldspar-sericite assemblages become more frequent, with chlorite replacing biotite.

HYPOGENE MINERALIZATION

The ore may be found in three different situations. It may be (i) totally within the host stock, (ii) partially in the stock and partially within the country rocks (Fig. 2.9) or (iii) in the country rocks only. The most common shape for the orebody in the examples analysed by Lowell & Guilbert (1970) is that of a steep walled cylinder (Fig. 15.5). Stubby cylindrical to flat conical forms and gently dipping tabular shapes are also known. The orebodies are usually surrounded by a pyrite-rich shell.

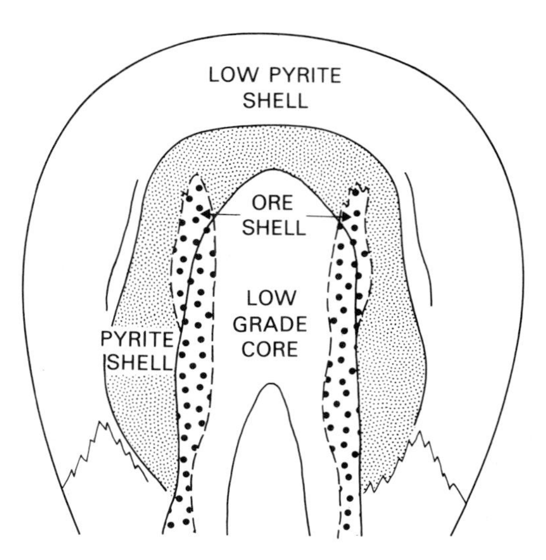

Fig. 15.5. Schematic diagram of the principal areas of sulphide mineralization in the Lowell–Guilbert model of porphyry copper deposits. Solid lines represent the boundaries of the alteration zones shown in Fig. 15.4. (After Lowell & Guilbert 1970).

Like the alteration, the mineralization also tends to occur in concentric zones (Fig. 15.5). There is a central barren or low grade zone with minor chalcopyrite and molybdenite, pyrite usually forming only a few per cent of the rock, but occasionally ranging up to 10%. The mineralization appears to be disseminated rather than fracture-controlled but Titley and Beane (1981) contend that this is simply a matter of scale and the 'disseminated' mineralization was located by microfractures and hair line cracks not noticed by some observers. Passing outwards, there is an increase first in molybdenite and then in chalcopyrite as the main ore shell is encountered. Veinlet mineralization is now more important. Pyrite mineralization also increased in intensity outwards to form a peripheral pyrite-rich halo with 10–15% pyrite but only minor chalcopyrite and molybdenite. The shells show a spatial relationship to the wall rock alteration zones (see figures) with the highest copper values often being developed at and near the boundary between the potassic and phyllic zones. Mineralization is not commonly found in the argillic zone, which usually appears to have been formed later and to be superimposed on the other alteration zones (Beaufort & Meunier 1983). Weak, non-economic mineralization continues outwards into the propylitic zone.

BRECCIA ZONES AND PIPES

Breccia zones and pipes are common in a number of deposits and are often mineralized, while some deposits consist mainly of mineralized breccia pipes. These breccias may occur within the porphyry body or in its wall rocks. Some appear to be the result of hydrothermal activity — fluidized breccias with rounded clasts and rock flour cements — whilst others appear to be angular collapse breccias. The former are frequently referred to as pebble dykes.

VERTICAL EXTEND OF PORPHYRY BODIES

Sillitoe (1973) suggested that porphyry copper deposits occur in a subvolcanic environment associated with small high level stocks and he emphasized their close association with subaerial calc-alkaline volcanism. He envisaged the host pluton as being overlain by a stratovolcano (Fig. 15.6), and contended that evidence from Chilean and Argentinian deposits shows that the propylitic alteration extends upwards into the stratovolcano which is itself likely to carry native sulphur deposits. The zones of alteration close upwards. The potassic zone dies out, sericitic and argillic alteration become important and the upper limit of economic mineralization is reached. At the same time, the porphyry stock becomes smaller in size and hydrothermal breccias appear over large areas.

Francis *et al.* (1983) proposed a modification of this picture. Whereas Sillitoe appeared to infer that the overlying stratovolcanoes were all large structures, these authors contend that the volcanic tops of porphyry systems may vary a lot in size and that many may be capped by small dacitic domes.

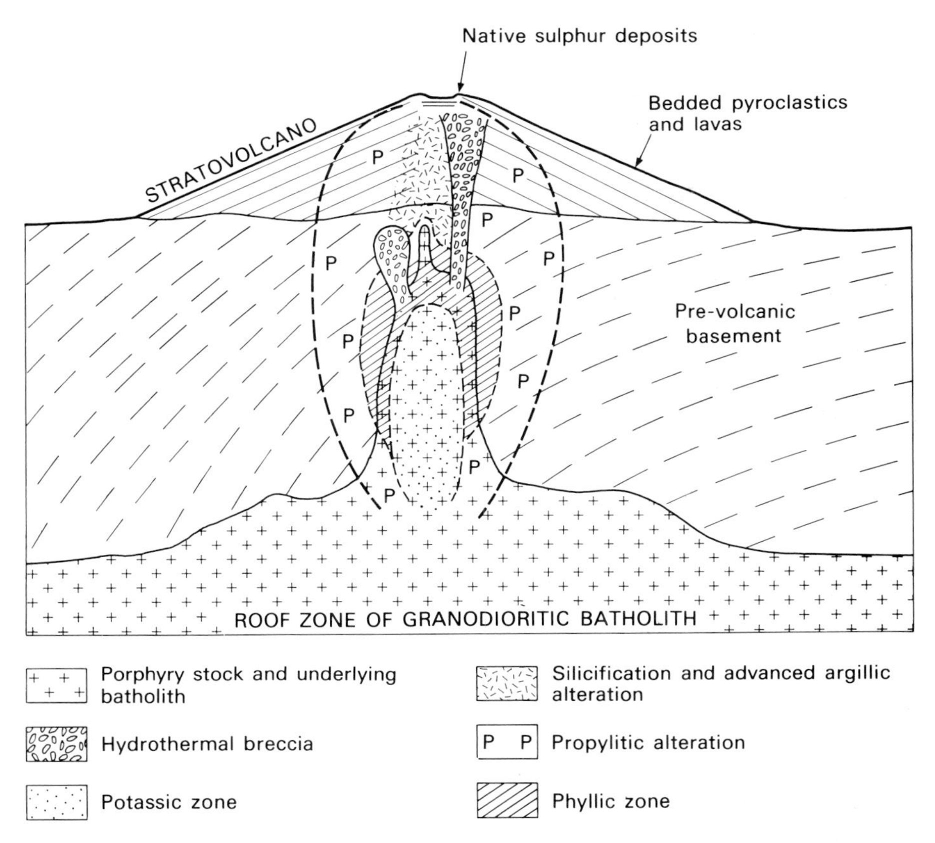

Fig. 15.6. Diagrammatic representation of a simple porphyry copper system on the boundary between the volcanic and plutonic environments. (After Sillitoe 1973).

In the lower parts of porphyry copper deposits the available evidence suggests that there is often a downward transition from porphyry into an equigranular plutonic rock of similar composition which forms part of a pluton of much larger dimensions. With this textural change the mineralization dies out in depth.

THE DIORITE MODEL

Subsequent to Lowell & Guilbert's classic work it has been recognized that some porphyry copper deposits are associated with intrusives having low silica:alkali ratios. Various names have been suggested for this type. The one which has won general recognition is 'diorite model', although the host pluton may be a syenite, monzonite, diorite or alkalic intrusion (Hollister 1975). Diorite model deposits differ in a number of ways from the Lowell–Guilbert model; one of the main reasons appears to be that sulphur concentrations were relatively low in the mineralizing fluids. As a result, not all the iron oxides in the host rocks are converted to pyrite and much iron remains in the chlorites and biotites while excess iron tends to occur as magnetite which may be present in all alteration zones.

(a) *Alteration zoning*. The phyllic and argillic alteration zones are usually absent so that the potassic zone is surrounded by the propylitic zone. This zonal pattern is present in both island arc and continental porphyry copper deposits. In the potassic zone, biotite may be the most prominent potassium mineral and when orthoclase is not well developed, plagioclase may be the principal feldspar.

(b) *Mineralization*. The main difference from the Lowell–Guilbert model is that significant amounts of gold may now occur and molybdenum/copper is usually low. The fractures containing gangue silicate minerals and copper sulphides may be devoid of quartz. On the other hand, chlorite, epidote and albite are fairly common.

(c) *Comparison of the Lowell-Guilbert and diorite models*. Table 15.2 (based on Hollister 1975) contrasts the principal features of each model and lists further features of the diorite model.

METAL ABUNDANCES IN PORPHYRY COPPER DEPOSITS

The vast majority of deposits can be divided into two classes according to whether the principal accessory metal is molybdenum or gold (Kesler 1973). Generally speaking, copper-gold deposits appear to be concentrated in island arc settings and copper-molybdenum where continental crust is present. There are, however, some notable exceptions, e.g. Cerro Colorado, one of the largest island arc deposits, has a high molybdenum/gold ratio and Sillitoe (1980) has reported the presence of porphyry molybdenum mineralization in the Philippines. A plot of these metals for porphyry deposits in western Canada shows a gradation from Cu-Mo to Cu-Au types rather than a clear-cut difference (Sinclair *et al.* 1982) and other exceptions were previously noted by Titley (1978). Although Kesler's contention is still generally true as viewed on a world-wide basis we must keep an open mind on this subject.

REGIONAL CHARACTERISTICS OF PORPHYRY DEPOSITS

The distribution of porphyry copper and molybdenum deposits is shown in Fig. 15.7. From this map it can be seen that the majority of porphyry deposits are associated with Mesozoic and Cenozoic orogenic belts in two main settings — island arcs and continental margins. The major exceptions are the majority of the USSR deposits and the Appalachian occurrences of the USA. These exceptions belong to the Palaeozoic. Only a few porphyry deposits have so far been found in the Precambrian. These facts are of great importance from the exploration point of view.

(a) *South-western USA and Mexico*. The deposits of this region form an oval cluster in the USA which contrasts with the linear belts of deposits in the

Table 15.2. Comparison of Lowell-Guilbert and diorite models of porphyry copper deposits.

Feature	Lowell-Guilbert model	Diorite model
Host pluton		
Common rock-types	Adamellite, granodiorite, tonalite	Syenite, monzonite
Rarer rock-types	Quartz-diorite	Diorite
Alteration		
Central core area	Potassic	Potassic
Peripheral to core	Phyllic Argillic Propylitic	Propylitic
Mineralization		
Quartz in fractures	Common	Erratic
Orthoclase in fractures	Common	Erratic
Albite in fractures	Trace	Common
Magnetite	Minor	Common
Pyrite in fractures	Common	Common
Molybdenite	Common	Rare
Chalcopyrite/bornite	3 or greater	3 or less
Dissemination of chalcopyrite	Present	Important
Gold	Rare	Important
Structure		
Breccia	May occur	Rare
Stockwork	Important	Important

Andes and elsewhere (Lowell 1974), but which continues south-eastwards as a linear belt into Mexico for at least 1900 km (Damon *et al.* 1983).

There are over 100 deposits, with about 30 in production. The ages range from 20 to 195 Ma with a peak around 60 Ma (Laramide Orogeny). Some deposits lie along marked lineaments with a tendency to be developed at their intersections but this is not true of all the deposits. They are spread over a considerable area and some are so far inland that, if the genesis of the host intrusives is to be linked to a subduction zone, then it is necessary to postulate very low dipping or multiple subduction zones.

(b) *Northern (Canadian) Cordillera of America*. In this complex region there is a range of copper, copper-molybdenum and molybdenum porphyries. From their study of a large area in central British Columbia, Griffiths and Godwin (1983) have related these to magma generation connected with Benioff Zones active during late Triassic-Jurassic and Createaceous to early Tertiary times. The host granitoids are I-type and the Cu/Mo ratios of the deposits correlate with the composition of the host plutons, decreasing from diorite to granite and implying a genetic link between the major element chemistry and the metal content of the magmas.

(c) *Appalachian orogen*. According to Hollister *et al.* (1974) and Hollister (1978), porphyry deposits were first developed in this region during the Cambrian and Ordovician and later in the Devonian and Carboniferous. Considering all the deposits together, copper porphyries formed first followed by the coeval development of separate copper and molybdenum porphyries with a final period when only molybdenum porphyries were formed. Accompanying

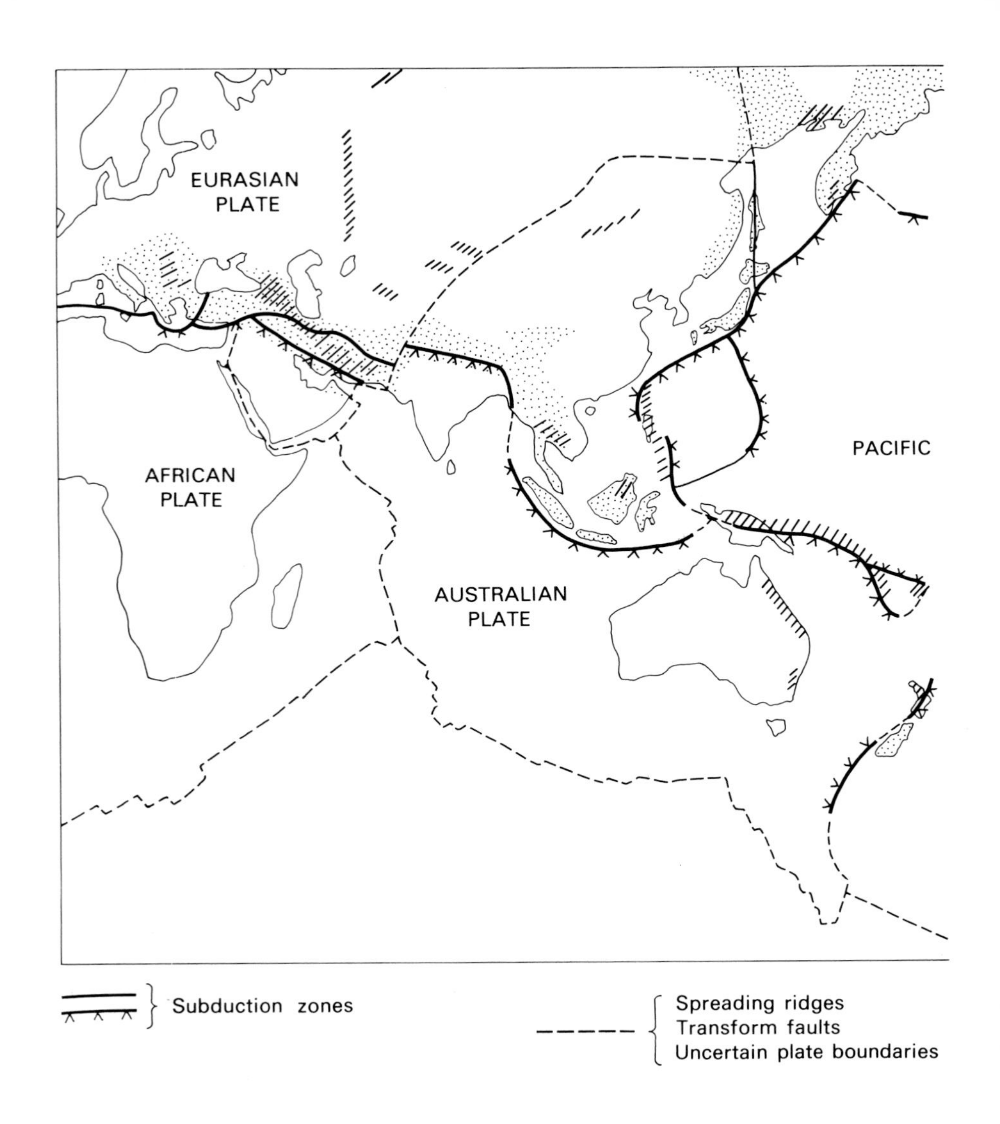

this change was a variation in the composition of the magmas associated with the mineralization from quartz-monzonitic to granitic. The copper and molybdenum porphyries either lack or have a very small development of the phyllic zone. This may be an effect of deep erosion. On the other hand, this fact and the low content of pyrite may indicate a deposit type transitional to the diorite model.

Prior to continental drift, this orogen was continuous with the Caledonian Province of the British Isles where porphyry copper mineralization has been found in North Wales (Cambro-Ordovician) and Scotland (Devonian).

(d) *The Andean province.* As Fig. 15.8 shows, this has a most marked linear distribution of deposits. These occur in a region where erosion has not cut

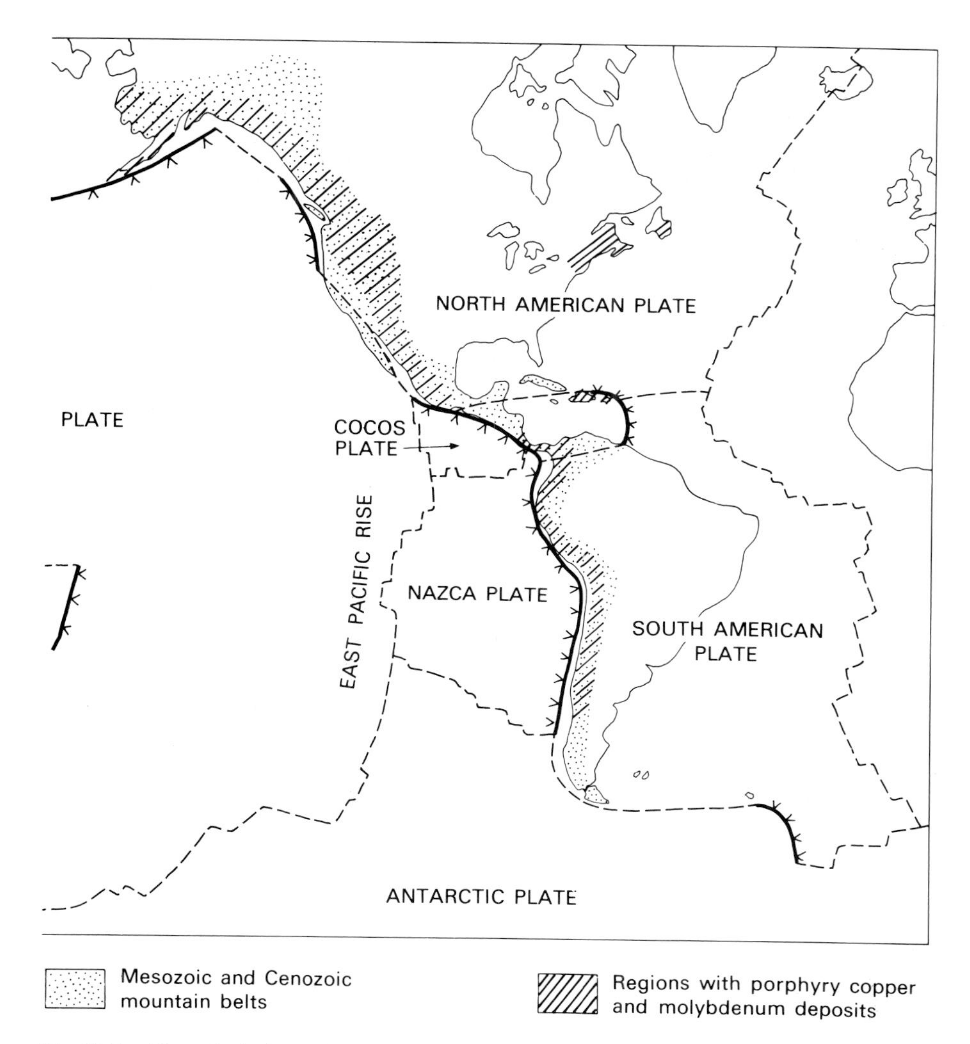

Fig. 15.7. The principal porphyry copper and molybdenum regions of the world. Also shown are present plate boundaries and Mesozoic-Cenozoic mountain belts.

down as deeply as it has in the coastal range batholiths to the west, but has cut down more deeply than in the volcanic belt to the east. The age range of most of the deposits is 59–4.3 Ma, nearly all have a significant molybdenum content and the underlying crust appears to be entirely continental.

Most deposits are related to stocks of intermediate composition and calc-alkaline affinity, although a few are associated with stocks of the shoshonitic suite. There are fifteen major porphyry copper deposits and at least fifty occurrences in Chile and Argentina. The belt continues northwards into Colombia and Central America. The exploitation of the Chilean deposits has made that country the western world's leading copper producer. A good summary of the Chilean deposits can be found in Sillitoe (1981a).

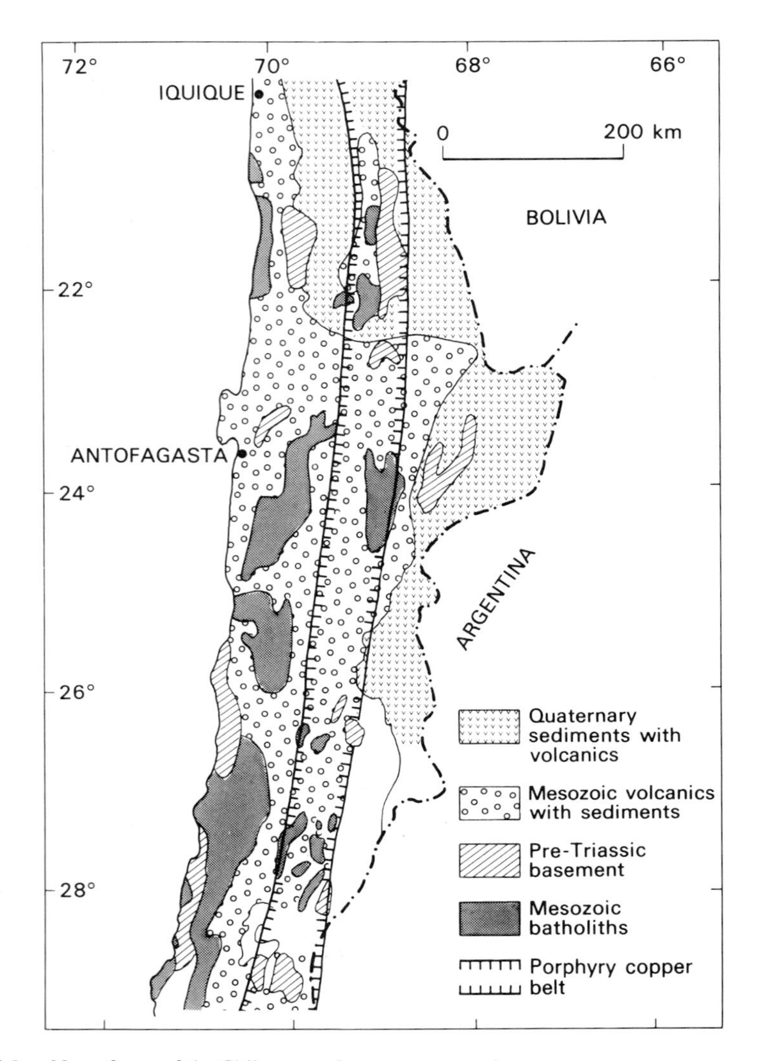

Fig. 15.8. Map of part of the Chilean porphyry copper province.

(e) *The South-western Pacific island arcs.* The island arcs of this province generally lack continental crust; gold is usually an important by-product and molybdenum uncommon. Lowell-Guilbert and diorite model deposits occur together. Most of the deposits occur in arcs containing thick sequences of pre-ore rocks. The host porphyries have penetrated to high structural levels and are all very young (less than 16 Ma). The deposits formed at a late stage of arc evolution just before volcanic activity ceased.

Useful descriptions of the deposits in Papua New Guinea with invaluable summaries of economic data can be found in Amade (1983). Readers interested in the controversy surrounding the development of one of these, Ok Tedi, should read Jackson (1982).

(f) *Porphyry copper and molybdenum deposits in the USSR*. The majority of Soviet deposits are Palaeozoic with a peak in the Carboniferous. The distribution of the major fields is shown in Fig. 15.7. The most important are in Kazakstan, the Caucasus, Uzbekistan and the Batenevski Range in Siberia. Most, if not all, deposits are related to present or former subduction zones and about a quarter are molybdenum-poor copper porphyries which have formed in island arc settings. Copper-molybdenum deposits lie along continental or microcontinental margins like those of the south-western USA and the Andes. Porphyry molybdenum deposits in any given area are always younger than associated porphyry copper deposits (Laznicka 1976). The north-eastern USSR is a good example of an area where it is important from the prospecting point of view to distingush between I(Cordilleran) and I(Caledonian) type granites for, whilst the former type in Kamchatka carries porphyry copper-type mineralization, the I_{CAL} granitoids of the Verkhoyansk Belt to the west and north of the Kolyma-Omolon Plate, like the type I_{CAL} granitoids of Scotland, do not carry any economic porphyry style mineralization (Shilo *et al.* 1983, Pitcher 1983, pers. comm.).

GENESIS OF PORPHYRY COPPER DEPOSITS

The principal arguments over recent years have been concerned with a magmatic versus a meteoric derivation for the mineralizing fluids and the origin of the metals and sulphur. In considering the formation of these deposits we must remember that the most striking characteristic of porphyry copper deposits when compared with other hydrothermal orebodies is their enormous dimensions. The size and shape of these deposits imply that the hydrothermal solutions permeated very large volumes of rock, including country rocks, as well as the parent pluton. That at least some of these solutions originated in the host pluton is suggested by the existence of crackle brecciation.

(a) *Crackle brecciation and its origin*. Crackle brecciation is the name given to the fractures which are usually healed with veinlets to form the stockwork mineralization. The zone of crackle brecciation is usually circular in outline, always larger than the orebodies and it fades out in the propylitic zone. It is often less well developed near the centre of the deposit, particularly if potassic alteration is present. This brecciation is thought to be due to the expansion resulting from the release of volatiles from the magma (Phillips 1973).

The host magmas of porphyry copper deposits appear to have reached to within 0.5–2 km of the surface before equigranular crystallization commenced in their outer portions. The intrusions would then be stationary and the confining pressure would not fluctuate. With the steady development of crystallization, however, anhydrous minerals form and the liquid magma becomes richer in volatiles, leading to an increase in the vapour pressure. If the vapour pressure rises above the confining pressure, then what is called retrograde boiling will occur and a rapidly boiling liquid will separate. If retrograde boiling occurs in a largely consolidated rock, the vapour pressure has to overcome the tensile strength of the rock as well as rising above the confining pressure. This will result in expansion and extensive and rapid brecciation (Fig. 15.9). The reason for this is that water released at a depth of about 2 km at 500°C would have a specific volume of 4 and, if 1% by weight formed a separate phase, it would produce an increase in volume of about 10%. At shallower depths the increase would be even greater and the degree of fracture intensity higher (Burnham & Ohmoto 1980). Evidence for the development of retrograde boiling in porphyry copper deposits is common in the form of the widespread occurrence of liquid-rich and gas-rich fluid inclusions in the same thin section (Chapter 3).

It is important to notice that whilst the crystallization of solid phases is an exothermic one, bubble formation is an endothermic process. Rapid nucleation and the adiabatic expansion of the vapour would absorb a great deal of heat, taking up the latent heat of crystallization and significantly lowering the temperature of the system. This would result in a second phase of rapid cooling in the central part of the intrusion which would considerably increase the number of nucleation sites producing a period of rapid crystallization, which, in turn, would be responsible for the fine-grained groundmass and hence the porphyritic nature of the intrusion.

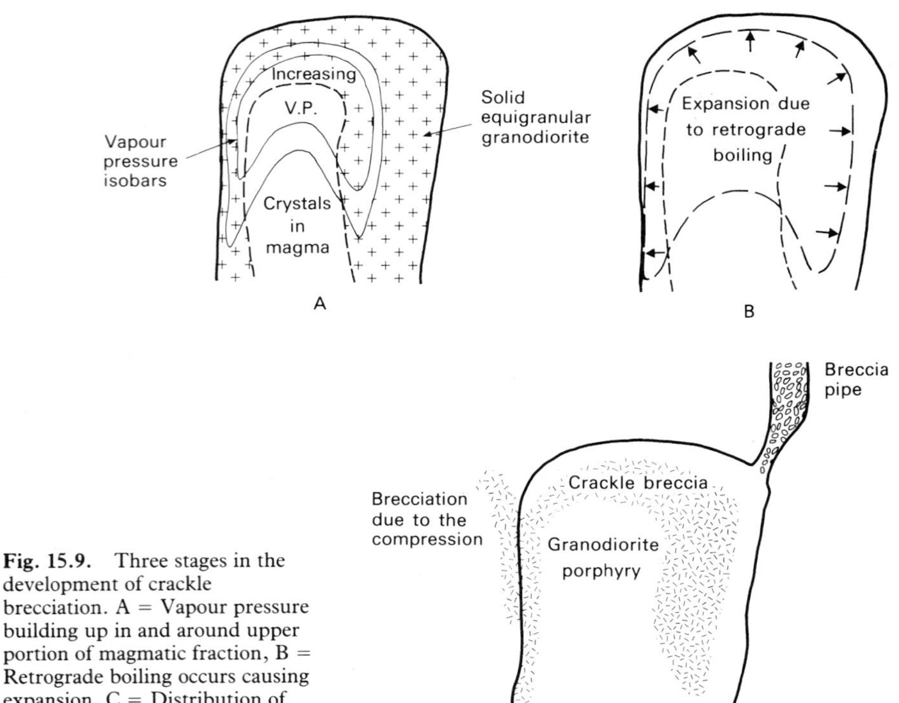

Fig. 15.9. Three stages in the development of crackle brecciation. A = Vapour pressure building up in and around upper portion of magmatic fraction, B = Retrograde boiling occurs causing expansion, C = Distribution of resulting brecciation. (After Phillips 1973).

(b) *Some chemical processes in the formation of porphyry copper deposits.* Retrograde boiling produces an aqueous phase (hydrothermal solution) in a porphyry system and chloride ion is partitioned strongly into it as is bisulphide ion, provided a sulphide mineral such as pyrrhotite is not stable (Burnham 1979). The presence of chloride ion supplies a transporting mechanism for the base metals that also fractionate strongly into the aqueous phase, and the bisulphide provides the sulphur for the eventual precipitation of sulphides. However, there are important controls on the fractionation of sulphur into the aqueous phase, as Burnham & Ohmoto (1980) have pointed out. One of these is the fugacity of oxygen (f_{O_2}) in the magma. Although sulphur is in solution in hydrous silicate melts as HS^-, it exists in the aqueous phase as both H_2S and SO_2, and the partition coefficient for total sulphur increases with an increase in the SO_2/H_2S ratio, which itself goes up with an increase in f_{O_2}. Now f_{O_2} in a magma prior to retrograde boiling is largely determined by the Fe^{3+}/Fe^{2+} ratio of the magma, which in turn is largely dependent on the source rock from which the magma was generated. Thus aqueous fluids that separate from I-type magmas with relatively high f_{O_2} tend to produce sulphur-rich porphyry copper mineralization, whereas fluids from S-type magmas may deposit the more sulphur-poor tin oxide mineralization. Sulphur isotopic studies indicate that sulphur in porphyry copper deposits is largely of upper mantle or remelted oceanic crust origin (McMillan & Panteleyev 1980), which emphasizes the importance of the above control.
(c) *Evidence from isotopic and geochemical investigations.* Further evidence of a magmatic derivation of at least some of the hydrothermal solutions comes from stable isotope investigations (Sheppard 1977). Waters in equilibrium with potassium silicate alteration assemblages and formed at 550–700°C are isotopically indistinguishable from primary magmatic waters (Fig. 15.10). On the other hand, waters associated with sericites from the phyllic zone of alteration are depleted in ^{18}O relative to the biotites of the potassic zone. Comparison with Fig. 4.10 suggests that connate waters from the country rocks were involved in the sericitization; in other words, meteoric water played a significant role in the hydrothermal fluids responsible for the phyllic alteration. The isotopic data for advanced and intermediate argillic alteration show an identical pattern to that for the phyllic alteration data. Field and microscopic evidence suggest that the phyllic and argillic alterations were later than the potassium silicate and propylitic alterations and were superimposed to varying degrees upon them. These two stages of development are depicted in Fig. 15.11.

It appears that after intrusion of the porphyry body, solidification occurs and a magmatic-hydrothermal solution evolves. This solution reacts with the porphyry and to a varying extent with the surrounding country rocks, giving rise to the development of a central zone of potassium silicate alteration. The introduction of much of the metals and sulphur probably accompanies this stage. Further out from the intrusion, thermal gradients set up a convective circulation of water in the country rocks, and this is responsible for the

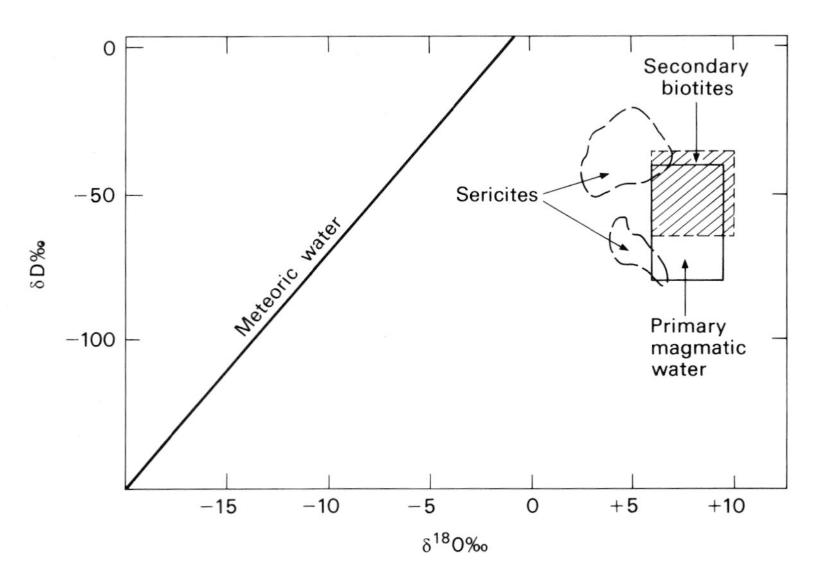

Fig. 15.10. Isotopic compositions of hydrothermal waters associated with secondary biotites from the potassic zones and sericites from the phyllic zones of five porphyry copper deposits. (Modified from Sheppard 1977).

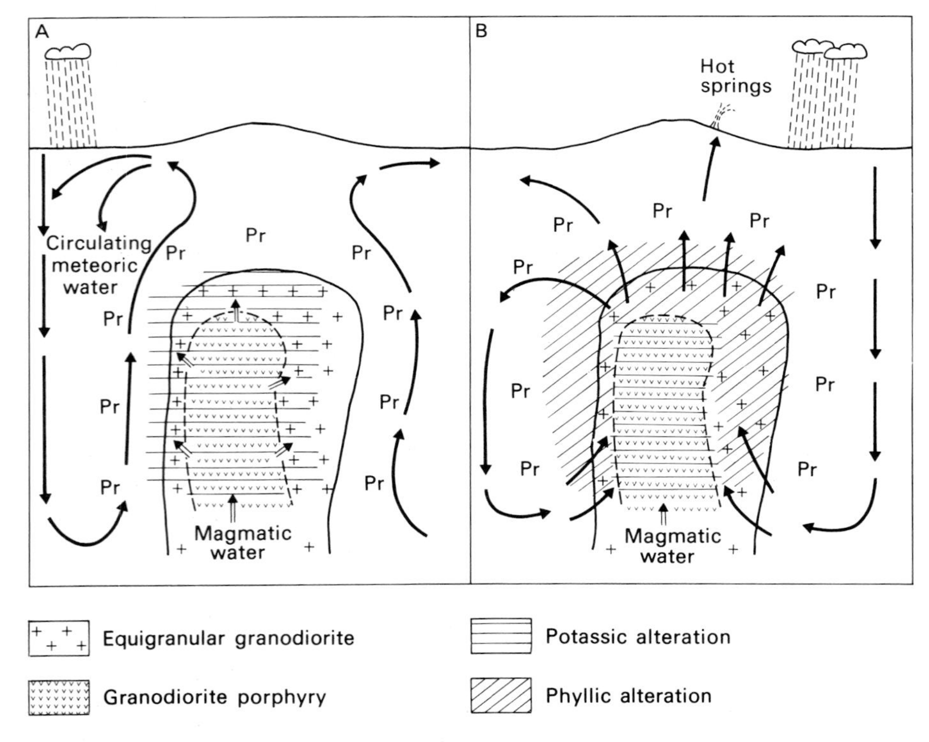

Fig. 15.11. Diagrammatic sections through a porphyry copper deposit showing two stages in the development of the hydrothermal fluids leading to the formation of a Lowell-Guilbert model deposit (B). Pr = propylitic alteration. (Modified from Sheppard 1977).

propylitic alteration (diorite model conditions) as Taylor (1981) has shown for the diorite model deposit of Bakirçay, Turkey. In this study he used Sr isotopes to demonstrate that fluids that formed the potassic alteration were magmatic-hydrothermal and those that caused the propylitic alteration were meteoric-hydrothermal.

When the intrusion cools, this meteoric-connate hydrothermal system may encroach upon and mix with the waning magmatic system leading to the development of lower temperature minerals: sericite, pyrophyllite and clay minerals. These would replace in particular the feldspar and biotite of the outer part of the original potassium silicate zone. The relatively rapid gradients in pH, temperature, salinity, etc., across the interface between these two hydrothermal systems probably account for the concentration of copper around the boundary zone between the potassium silicate and phyllic zones. With this second stage of alteration, a Lowell-Guilbert model deposit comes into being.

Supporting evidence for these two stages in the formation of Lowell–Guilbert model deposits comes from the work of Taylor & Fryer (1982, 1983), who have used REE geochemistry to demonstrate that the meteoric-hydrothermal fluids had the potential to remobilize and reconcentrate copper and molybdenum, giving rise to the second stage of hypogene leaching and enrichment at both Bakirçay and Santa Rita, New Mexico.

(d) *Fluid inclusion evidence.* This has been comprehensively reviewed by Roedder (1984) who emphasized the remarkable uniformity of the three characteristics of high maximum temperatures (as high as 725°C), high salinities (up to 60 wt % alkali chlorides) and evidence of boiling for all deposits. He considered that the extensive fluid inclusion data favours a magmatic hypothesis of origin for the mineralizing fluids. The highest temperature inclusions, which are rich in copper and other metals, characterize the central portions of porphyry systems (e.g. Chivas & Wilkins 1977, Easthoe 1978), and their distribution patterns mimic the zonal alteration-mineralization patterns. Fluid inclusion temperatures and salinities decrease both away from the central portions and with time of formation. All this evidence agrees well with the concept depicted in Fig. 15.11.

(e) *Source of the copper — is it leached from the magma or the wall rocks of the intrusion?* Economic porphyry copper deposits are more likely to develop if the parent magma is enriched in copper; can this happen during calc-alkaline differentiation? Feiss (1978) and Mason & Feiss (1979) suggested that, as copper prefers octahedral sites in crystallizing magmas, be they in melt or crystalline phases, then magmas with more octahedral sites will contain more copper in the melt at the retrograde boiling stage and thus be more likely to give rise to porphyry copper-type mineralization. Now the number of octahedral sites is proportional to the alumina:alkalies ratio and a study of mineralized versus barren intrusions in porphyry copper belts showed that the former are characterized by a higher ratio. This is one mechanism that can lead to the concentration of copper into certain magmatic differentiates.

Another is the enrichment that may occur when copper behaves as an incompatible element early in the development of magmatic systems as a result of the relatively poor development of magnetite and augite during fractional crystallization, because these minerals abstract copper from melts when they crystallize. The discovery, in a volcano, of mafic to intermediate lavas that are relatively poor in magnetite and augite, may therefore indicate the presence of a porphyry copper deposit in its roots, and such lavas, which are enriched in copper, have been recognized in Central America (Eilenberg & Carr 1981). It therefore seems very likely that some calc-alkaline magmas may be enriched in copper during differentiation, and this could help explain why only some of a series of similar intrusions become porphyry copper deposits.

If we pursue the possibility that the copper (and other wanted metals) are derived from the host intrusion, then what evidence supports this hypothesis? Only indirect evidence we must admit because although we do not have any radioactive tracers to monitor the movement of copper, molybdenum, gold and silver, we do nonetheless have fairly compelling evidence. For example, Anderson (1980) analysed the metal distribution of porphyry deposits in the Canadian Cordillera and showed that it correlates strongly with the regional time-space distribution of the host intrusions and not with the rock types hosting those intrusions. This conclusion was confirmed in a comprehensive statistical study by Griffiths and Godwin (1982), indicating that the fundamental processes governing the metallization of porphyry copper deposits are probably subduction-controlled rather than upper crustal, a conclusion supported by the work of Sillitoe & Hart (1984), which showed that the lead isotopes in Colombian porphyry copper deposits probably came from subducted pelagic sediment. The final concentration of the copper into a stockwork is of course an upper crustal process, but further evidence that it is derived from the host intrusion and not the country rocks comes from the work of Hendry *et al.* (1985), who have shown that the deep level and temporal equivalents of North American porphyry copper intrusions are depleted in copper compared with similar barren intrusions. This work suggests that copper is abstracted not only from apical portions of the porphyry host but also from deeper parts of the parent intrusions.

Porphyry molybdenum deposits

GENERAL DESCRIPTION

These have many features in common with porphyry copper deposits; some of these features have been touched on above and a useful summary account has been given by White *et al.* (1981). Average grades are 0.1–0.45 MoS_2 (molybdenum grades are *more usually* given as MoS_2) and one deposit produces by-product tin and tungsten. Host intrusions vary from quartz monzodiorite through granodiorite to granite. Stockwork mineralization is more important than disseminated mineralization and the orebodies are associated with simple, multiple or composite intrusions or with dykes or breccia pipes. There are three general orebody morphologies (Ranta *et al.* 1984, see Fig. 15.12), and tonnages range from 50–1500 Mt. Sections through the two biggest deposits are given in Figs. 15.13, 15.14. The molybdenite occurs in (i) quartz veinlets carrying minor amounts of other sulphides, oxides and gangue, (ii) fissure veins, (iii) fine fractures containing molybdenite paint, (iv) breccia matrices and, more rarely, (v) disseminated grains. Supergene enrichment, which can be very important in porphyry coppers, is generally absent or minor.

White *et al.* (1981) have divided the porphyry molybdenums of North America into Climax and Quartz Monzonite Types and Laznicka (1985) has termed the second type the granodiorite-quartz monzonite class. Confusion is building up here which can only be resolved by strict adherence *and reference* to one igneous rock classification scheme. White *et al.* appear to have used the term quartz monzonite as defined in time honoured, North American usage, the equivalent rock name in Europe being adamellite. Following the IUGS classification (Streckeisen 1976), now in all recent textbooks on igneous rocks, it must be stated that the majority of porphyry molybdenum deposits are associated with granites, including all Climax types and many 'quartz monzonite types'. Some 'quartz monzonite types' are associated with granodiorite and a few with quartz monzodiorite. Probably none is associated with quartz monzonites as defined in the IUGS classification!

This is all very confusing for the student *and* the mineral explorationist selecting targets in a new concession area. It is unfortunately a difficulty that will only slowly diminish over the years as those in the applied field begin to discipline themselves more strongly in their use of up-to-date nomenclature. The reader should remember that this is a problem that permeates the whole

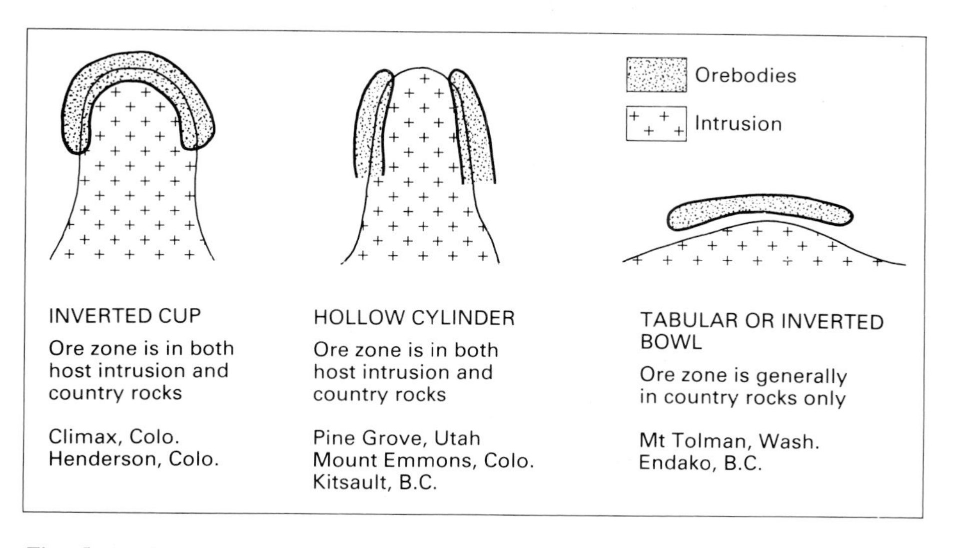

Fig. 15.12. Porphyry molybdenum orebody morphologies.

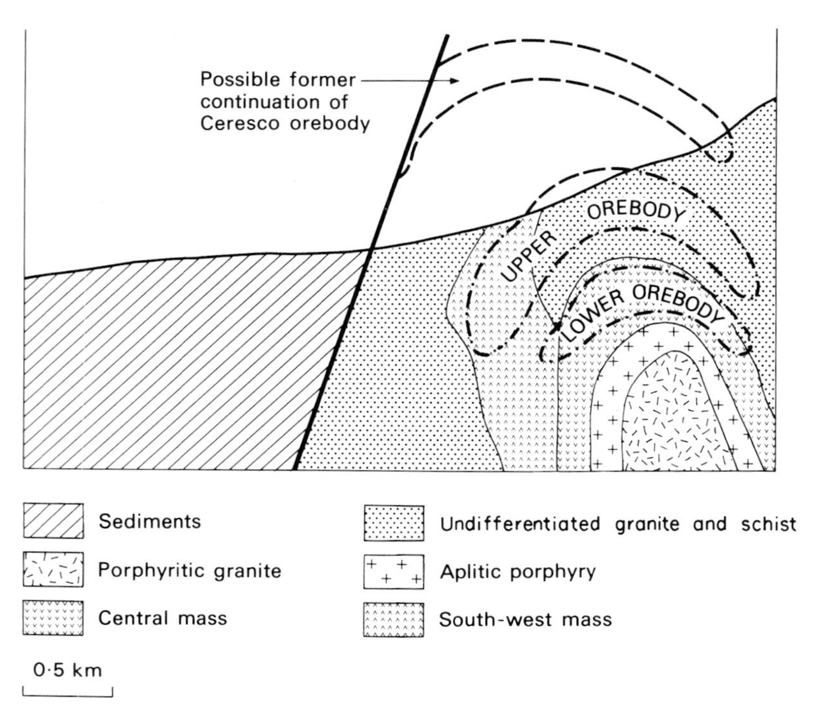

Fig. 15.13. Generalized geological section through the Climax molybdenum mine, Colorado. (Modified from Hall *et al.* 1974).

of the science and not just the subject of porphyry molybdenums! An interesting little exercise in this connexion is to superimpose the IUGS classification boundaries and approved rock names on Fig. 2 of Griffiths & Godwin (1983).

The principal differences between these two types are that the Climax type generally have high trace or accessory contents of tin and tungsten, intense silicification associated with their wall rock alteration, and are low in copper compared with the other type which lies in the transition zone to molybdenum-bearing porphyry coppers. Perhaps it would be better if the second type was named after a well known deposit of this type such as Endako — *if* there is merit in this division for, as Laznicka (1985) has remarked, there appear to be many deposits outside North America that are transitional between these two types.

HYDROTHERMAL ALTERATION

The alteration patterns are very similar to those found in porphyry copper deposits, with potassic alteration and silicification being predominant. The most detailed study is on the Urad and Henderson deposits (Wallace *et al.* 1978) where, associated with the Henderson orebody, there is a central

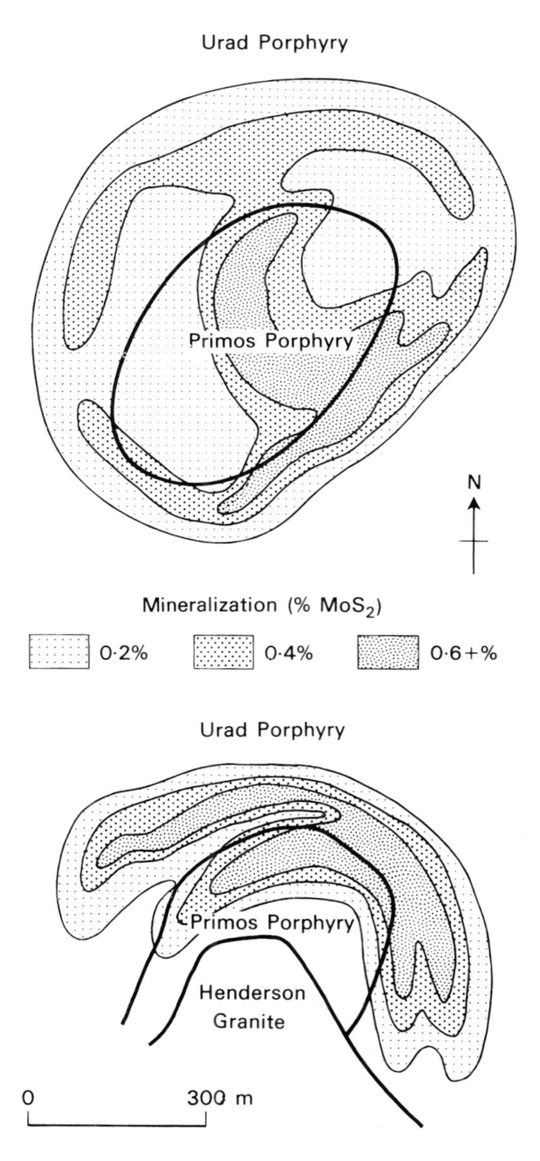

Fig. 15.14. Plan (8000 ft or 2438.4 m level) and section of the Henderson molybdenum orebody, Colorado. (Modified from Wallace *et al.* 1978).

potassic zone carrying secondary potash feldspar and biotite. Succeeding this are quartz-topaz, phyllic, argillic and propylitic zones. There is a silicified zone which lies largely within the potassic zone. The Henderson orebody is roughly coincident with the potassic and silicified zones. A prominent pyrite zone, carrying 6–10% pyrite, is developed around the Henderson orebody and a similar less distinct zone is present at Climax. Peripheral pyrite zones, which appear to coincide mainly with the phyllic zone, have been reported from a number of other porphyry molybdenums.

The close spatial association of the orebodies of most deposits with the potassic zone of alteration suggests a magmatic source and Wallace *et al.* (1978) argue that, for the Henderson and Climax deposits, this must have been unexposed master reservoirs which fed the columns of exposed host intrusions. They consider that the volumes of the host intrusions are far too small to have supplied the large tonnage of molybdenum which is present in the deposits. The source of the molybdenum and its host magma may have been continental crust, as many workers have suggested (e.g. Hollister 1975, Stein 1985) but Westra & Keith (1981) and Knittel & Burton (1985) have advanced evidence indicative of a mantle origin, which is surely a very likely explanation for the deposits in the Philippines.

Porphyry tin and porphyry tungsten deposits

Much primary tin has been won in the past from stockworks at Altenberg and Zinnwald in Germany, from many deposits in Cornwall, England, and from deposits in New South Wales, Tasmania and South Africa. Eroded stockworks in Indonesia and Malaysia have provided much of the alluvial cassiterite in those countries. Exploitation of such deposits in recent times has been mainly in Germany and New South Wales. At Ardlethan in New South Wales a quartz-tourmaline stockwork occurs in altered granodiorite carrying secondary biotite and sericite. With a grade of 0.45% tin this deposit is economic. The majority of tin stockworks in the world, however, only run about 0.1% tin and they are not at present mineable at a profit.

Tin stockworks in the countries mentioned above belong to the plutonic environment. Recently, Sillitoe *et al.* (1975) and Grant *et al.* (1980) have described porphyry tin deposits from the subvolcanic section of the Bolivian tin province south of Oruro and have shown that these deposits have much in common with porphyry copper deposits — large volumes of rock grading 0.2–0.3% tin. There is pervasive sericitic alteration that grades outwards into propylitic alteration and pyrite halos are present in two deposits. Major differences include the absence of a potassic zone of alteration, the association with stocks having the form of inverted cones rather than upright cylinders and the presence of swarms of later vein deposits.

Porphyry tungsten (usually W-Mo) deposits have now been described from various parts of the world especially North America (e.g. Noble *et al.* 1984, Davis & Williams-Jones 1985) and China (Clarke 1983). Hydrothermal alteration is not always present as pervasive zones and when present may not show a regular concentricity. Grades are low: at Mount Pleasant, New Brunswick, 45 Mt running 0.2% W and 0.1% Mo, and at Logtung, Yukon, 162 Mt running 0.13% WO_3 and 0.052% MoS_2. Geological settings are very variable: subvolcanic quartz-feldspar porphyry at Mount Pleasant, diorite to biotite granite in China, and granite, but principally its contact aureole, at Logtung.

16

Stratiform Sulphide and Oxide Deposits of Sedimentary and Volcanic Environments

In this chapter we are concerned with a class or classes of deposit whose origin is at present hotly debated. As was pointed out in Chapter 2, there are many types of stratiform deposit. This chapter is mainly devoted to sulphide deposits and the related oxide deposits can only be mentioned *en passant*. The latter do not include bedded iron and manganese deposits, placer deposits and other ores of undoubted sedimentary origin, which are dealt with in Chapter 19. The related oxide deposits are in tin and uranium (and possibly iron) with which may be grouped certain tungsten deposits.

Concordant deposits referred to in Chapter 2 which belong to this class include the Kupferschiefer of Germany and Poland; Sullivan, British Columbia; the Zambian Copperbelt, and the large group of volcanic-associated massive sulphides. There appears to be a possible gradation in type and environment from deposits such as the Kupferschiefer, which are dominantly composed of normal sedimentary material and which occur in a non-volcanic sedimentary environment, through deposits such as Sullivan, which are richer in sulphur and may have some minor volcanic formations in their host succession, to the volcanic massive sulphide deposits which are mainly composed of sulphides and occur in host rocks dominated by volcanics. Stanton (1972) considered that these and other deposits formed a 'spectrum of occurrence' and treated them as one class. Other workers, e.g. Barnes (1975), Solomon (1976), Gustafson & Williams (1981), Franklin *et al.* (1981), Sangster (1983) and Eckstrand (1984) have felt it better to divide these deposits into two classes, the first class being those developed in a sedimentary environment where sedimentary controls are important, and the second being the volcanic-associated massive sulphide deposits in which exhalative processes were important during genesis. Such a division introduces difficulties when dealing with deposits showing only a weak link with volcanism but where exhalative processes may have been important, e.g. Sullivan. However, this division will be followed in this chapter as the author feels that deposits such as the Kupferschiefer and the Zambian Copperbelt are sufficiently different from the volcanic massive sulphide deposits to warrant some differentiation.

Stratiform sulphide deposits of sedimentary affiliation

GENERAL CHARACTERISTICS

The majority of these deposits occur in non-volcanic marine or deltaic environments. They are widely distributed in space and time, i.e. from the Proterozoic to the Tertiary, and can vary in tonnage from several hundred millions down to subeconomic sizes. In shape, they are broadly lensoid to stratiform with the length at least ten times the breadth. There is often more than one ore layer present. Feeder zones (cf. Fig. 2.13) have been identified below some deposits and may be present below many more, but as mining operations rarely penetrate into footwalls on a large scale, they will probably never be seen. The degree of deformation and metamorphism varies with that of the host rocks, suggesting a pre-metamorphic formation. They are frequently organic-rich, particularly those in shales, and usually contain a less complex and variable suite of minerals and recoverable metals than volcanic massive sulphide deposits. The sulphides have a small grain size so that fine and often costly grinding is necessary to liberate them from the gangue. They may show a shore to basinward zoning of Cu+Ag→Pb→Zn (Barnes 1975), and in those deposits dominated by copper mineralization the zoning is typically as follows: barren (hematite but no sulphide)→chalcocite →bornite→chalcopyrite, both outwards and upwards from the centre of mineralization. It is possible and probably valid to erect various sub-classes (cf. Eckstrand 1984), by emphasizing differences in metal ratios, geological environments and so on. Space does not permit this treatment in detail here, but attention must be drawn to the tendency for copper and lead-zinc deposits to be separate from each other and to have markedly different metal ratios.

The geological settings of these deposits are mostly intracratonic and the majority do not appear to be directly related to orogenic events or plate margin activity. Regional settings include (1) first marine transgressions over continental deposits (Kupferschiefer, Zambia, White Pine, (2) carbonate shelf sequences (Ireland), (3) fault controlled, sedimentary basins (Selwyn Basin, Yukon; Belt-Purcell Basin, British Columbia). Some of these environments appear to be aulacogens (cf. Elmore 1984).

Economically both the copper and lead-zinc deposits are of great importance on a world-wide scale; indeed sediment-hosted, stratiform copper deposits are second only to porphyry coppers as producers of the metal.

SOME EXAMPLES OF COPPER DEPOSITS

(a) *The European Kupferschiefer.* This is probably the world's best known copper-rich shale. It is of late Permian age and has been mined at Mansfeld (East Germany) for almost 1000 years. The Kupferschiefer underlies about 600 000 km^2 in Germany, Poland, Holland and England (Fig. 16.1). Copper concentrations greater than 0.3% occur in about 1% and zinc concentrations greater than 0.3% in about 5% of this area. Thus, although all the Kupferschiefer is anomalously high in base metals, ore grades are only encountered in a few areas. The most notable recent discoveries have been in southern Poland where deposits lying at a depth of 600–1500 m have been

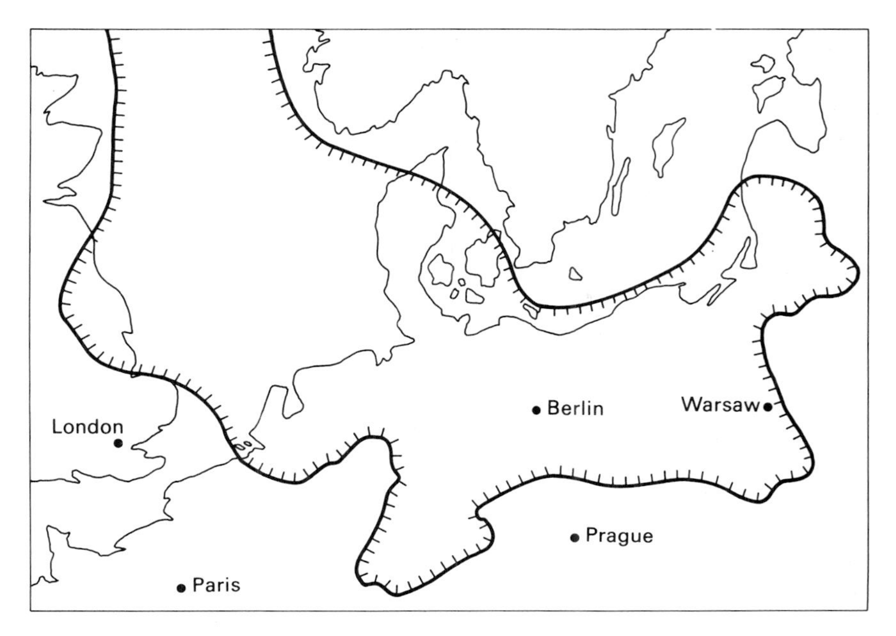

Fig. 16.1. Extent of the Zechstein Sea in Central Europe. The Kupferschiefer occurs at the base of the Upper Permian (Zechstein).

found during the last two decades. Here, the Kupferschiefer varies from 0.4–5.5 m in thickness. Average copper content is around 1.5% and reserves at 1% Cu amount to some 1500 Mt making Poland the leading copper producer in Europe. The area underlain by these deposits is approximately 30 × 60 km.

The Kupferschiefer consists of thin alternating layers of carbonate, clay and organic matter with fish remains which give it a characteristic dark grey to black colour. The Kupferschiefer is the first marine transgressive unit overlying the non-marine Lower Permian Rotliegendes, a red sandstone sequence, and it is succeeded by the Zechstein Limestone which in turn is overlain by a thick sequence of evaporites. The Kupferschiefer and the Zechstein evaporates may represent a tidal marsh (sabkha) environment which developed as the sea transgressed desert sands.

The copper and other metals are disseminated throughout the matrix of the rock as fine-grained sulphides (principally bornite, chalcocite, chalcopyrite, galena, sphalerite), commonly replacing earlier calcite cement, lithic fragments and quartz grains as well as other sulphides. Typical features of the mineralization are shown in Fig. 16.2. A zone of superposed diagenetic reddening known as the Rote Fäule facies transgresses the stratigraphic horizons. Copper mineralization lies directly above the Rote Fäule and the copper zone is overlain, in turn, by lead-zinc mineralization. This relationship to the Rote Fäule has meant that the delineation of this facies is the most important feature of the search for new orebodies. The Rote Fäule-copper zones are

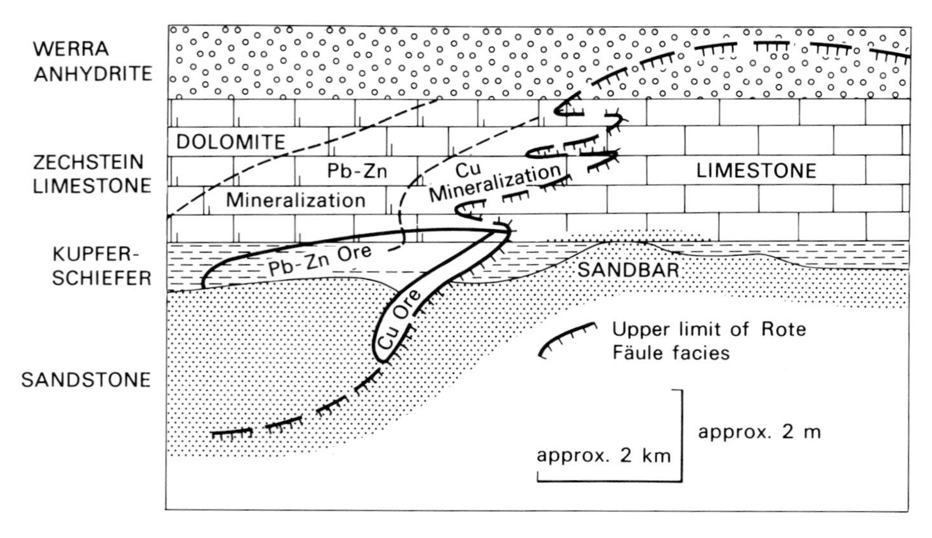

Fig. 16.2. Diagrammatic section through orebodies in the basal Zechstein with the Rote Fäule facies alteration gently transgressing the bedding above an area of sandbars formed by marine reworking of the Rotliegendes. Sulphide mineralization occurs in the unoxidized zone adjacent to the Rote Fäule with copper nearest to it and lead-zinc further away. (After Brown 1978).

coincident with underlying highs in the buried basement, and the metal zoning dips away from the highs toward the basin centres.

(b) *The Zambian Copperbelt.* This is part of the larger Central African Copperbelt of Zambia and Shaba (Zaïre) which produce about 17% of the western world's copper. In 1984 Zambia produced 531 000 t of copper and the mill grade at the principal mines varied from 1.49–2.81%, with appreciable by-product cobalt from some. The industry in Zambia is contracting rapidly. The harsh realities are that the mines are old, with declining grades and an average fifteen years life ahead of them, and the country simply does not have the cash for investment to increase their efficiency and operate them profitably. Shaba produced about 500 000 t copper in 1984 from ore averaging around 4% Cu+Co.

Almost all the copper mined in 1984 came from restricted horizons within the late Proterozoic Katangan sediments of the Lufilian Arc (Fig. 16.3). The Katangan rests unconformably on a granite-schist-quartzite basement and the lowermost Katangan sediments fill in the valleys of the pre-Katangan land surface. Most mineralization in Zambia and south-eastern Shaba occurs in the Ore Formation which lies a few metres above the level at which the pre-Katangan topography became filled in. Shale or dolomitic shale forms the host rock for about 60% of the mineralized ground, and the shale orebodies form a linear group to the south-west of the Kafue Anticline (Fig. 16.4). Arkose-arenite hosted ores occur mainly to the north-east of the anticline, e.g. Mufulira (see page 23). The footwall succession consists of quartzites, feldspathic sandstones and conglometrates of both aquatic and aeolian origin.

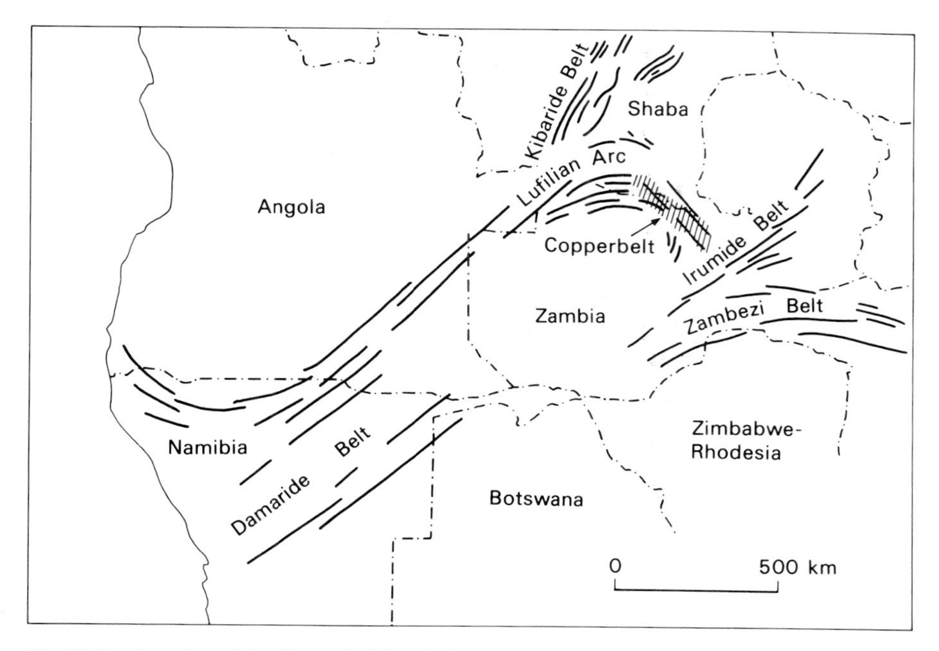

Fig. 16.3. Location of the Copperbelt in relation to the main tectonic trends of Central Africa. (After Raybould 1978).

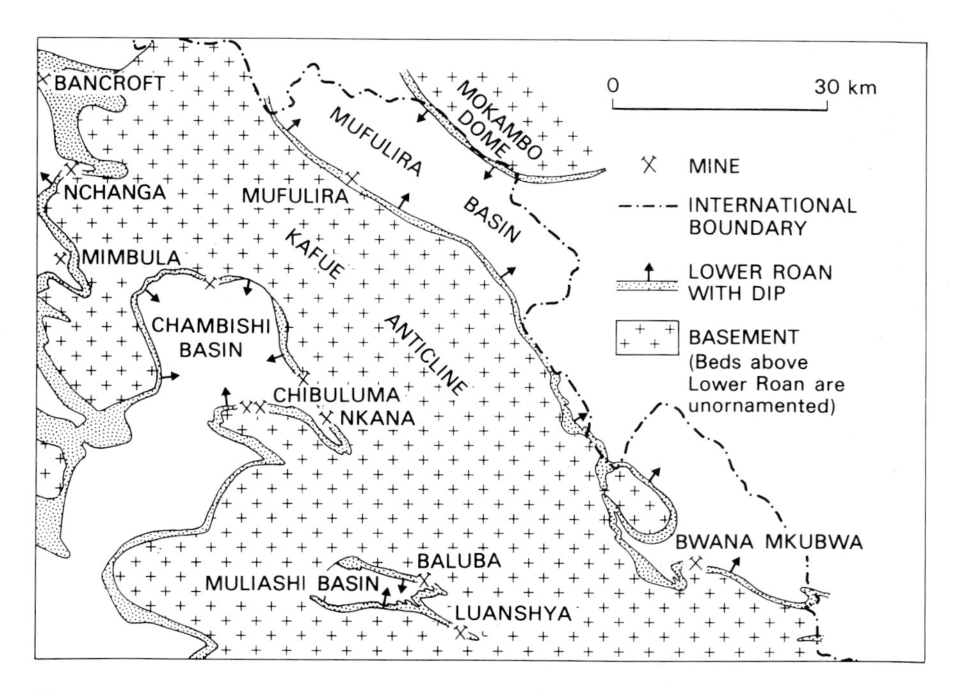

Fig. 16.4. Location map for the Zambian Copperbelt showing the regional geology. (Modified from Fleischer *et al.* 1976).

The Ore Formation, generally 15–20 m thick, is succeeded by an alternating series of arenites and argillites which, with the rocks below them, make up the Lower Roan Group. All the rocks and their contained copper minerals have suffered low to high grade greenschist facies metamorphism and many of the so-called shales are biotite-schists. In places they are tightly folded (Fig. 16.5).

Copper, together with minor amounts of iron and cobalt, occurs mainly in the lower part of the Ore Formation as disseminated bornite, chalcopyrite and chalcocite. In places, mineralization passes for short distances into the underlying beds. Both the upper and lower limits of mineralization are usually sharply defined. The sulphide minerals show a consistent zonal pattern with respect to the strandline from barren near-shore sediments to chalcocite in shallow water, to bornite with carrollite and chalcopyrite, then chalcopyrite and finally pyrite (in places with sphalerite) in the deeper parts of marine lagoons and basins.

(c) *The White Pine copper deposit, northern Michigan.* A different type of copper deposit is found in Precambrian strata at White Pine (Fig. 16.6). This district supplies about 5% of the copper mined in the USA. Ore grade is about 1.2% copper as chalcocite and native copper which occur in the Nonesuch Shale, deposited about 1000 Ma ago. This formation is about 150–200 m thick but only the lower 8–15 m contain significant copper mineralization (Burnie *et al.* 1972). The name 'Nonesuch Shale' is misleading because much of the formation is siltstone or sandstone. Copper mineralization is almost invariably confined to individual lithogical units and the content changes with sedimentary facies variations. The Nonesuch Shale conformably overlies the Copper Harbor Conglomerate, the upper beds of which are also locally cupriferous.

The Nonesuch Shale has been divided into three subzones. A basal Cupriferous Zone contains chalcocite and native copper, but only traces of bornite, chalcopyrite and pyrite. Next comes a transition zone with a gradation in the mineralogy from chalcocite through bornite and chalcopyrite to pyrite; lead-zinc minerals also occur. This is overlain by copper-poor pyritous shale and siltstone petrologically similar to the copper-bearing rocks. Within the cupriferous zone, maximum copper concentrations are found in siltstone and shale units immediately overlying sandstones which are usually copper-poor.

GENESIS

Ideas concerning the origins of these deposits show in general a similar evolution from epigenetic, to syngenetic, to what might be termed 'just epigenetic'. For example the original White Pine Mine worked ore from both the Nonesuch Shale and the Copper Harbor Conglomerate along the White Pine Fault and, because the mine was on a structural feature, it was assumed that the ore was epigenetic and deposited by hydrothermal solutions which rose up the fault. The discovery that the ore persisted along the base of the Nonesuch Shale over many tens of km^2 led to a concept of a syngenetic origin.

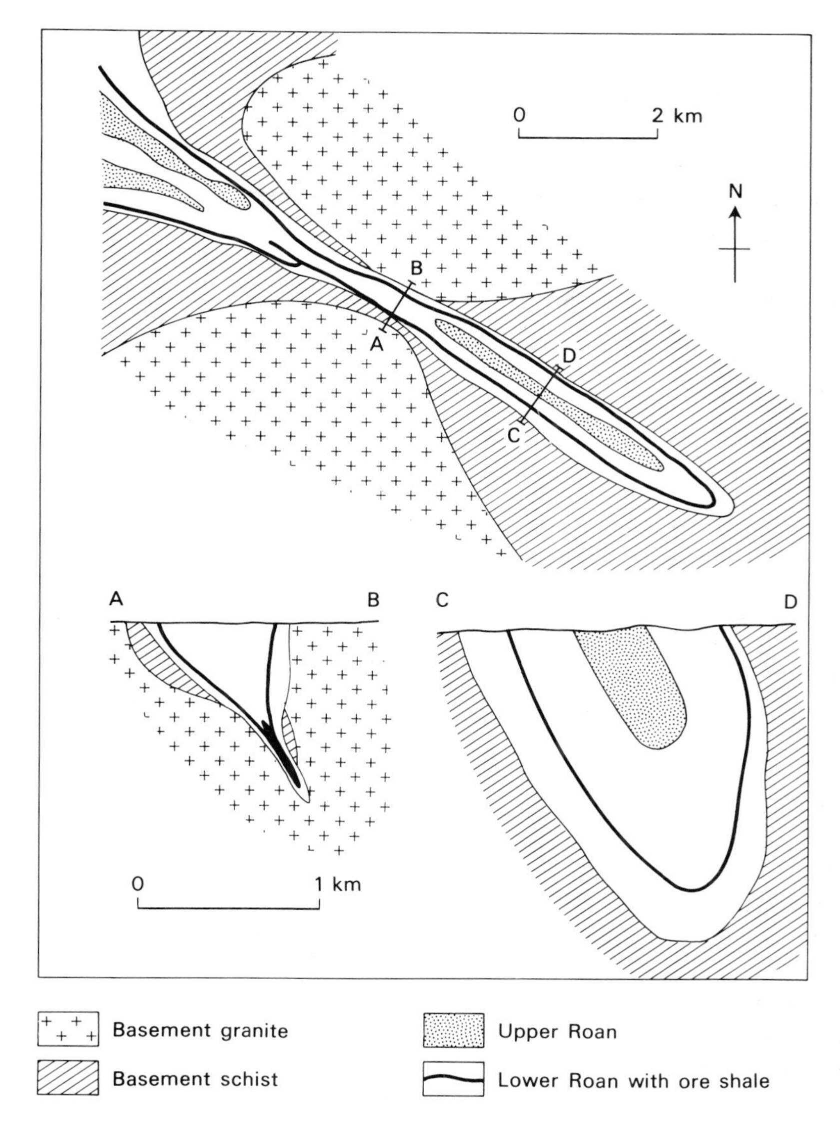

Fig. 16.5. Sketch map and sections of the Luanshya Deposit, Zambia. (Modified from Dixon 1979).

Further study showed that ore in the locality cuts across bedding, leading to a more sophisticated concept of genesis — that copper-bearing solutions circulated through beds which contained syngenetic pyrite, and that the copper replaced the iron.

For the Zambian deposits many workers have proposed and still propose a syngenetic origin, e.g. Fleischer *et al.* (1976) who cite many lines of evidence, particularly slumped ore horizons, for mineralization having occurred during

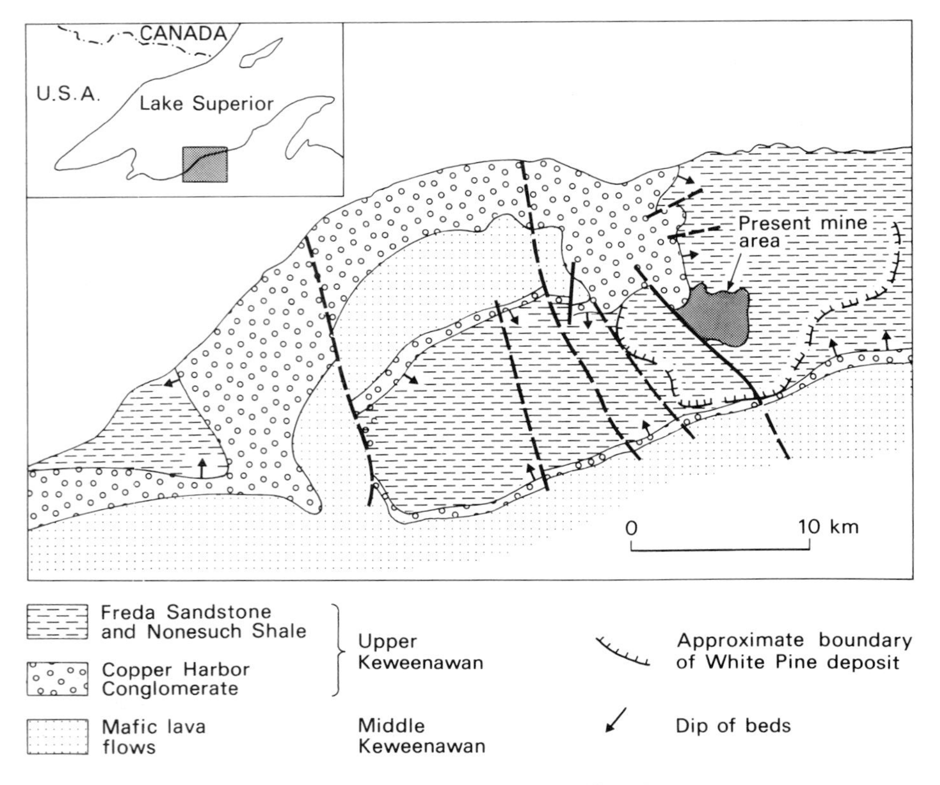

Fig. 16.6. Geological sketch map of the area containing the White Pine copper deposit. (Modified from Burnie *et al.* 1972 and White 1971).

sedimentation. These authors postulated that the supply of copper came from springs or streams draining a hinterland of aeolian sands and terrestrial red beds, and that it was selectively precipitated by hydrogen sulphide in bodies of standing water to form the zone of mineralization described above. Binda (1975) described detrital grains of bornite and bornite-bearing rock fragments from Mufulira and suggested that clastic sedimentation played a role (of yet unknown importance) as an ore-forming process in the Copperbelt. From an epigenetic viewpoint, Raybould (1978) has pointed out that the Zambian Copperbelt and other major Proterozoic stratiform copper and lead-zinc deposits such as McArthur River, Northern Territory; Mount Isa, Queensland; Sullivan, British Columbia; White Pine, Michigan, etc., appear to have developed in intracratonic and cratonic-margin rift systems of mid to late Proterozoic age. He suggested that the mineralization resulted directly from deep-seated processes coeval with the rifting and not from surface weathering. Annels (1979) has suggested that diagenetic metal chloride-enriched brines were responsible for the mineralization at Mufulira.

Since then, work on deposits of this type all over the world has led to the conclusion by most observers that copper and associated metals have been added to their host rocks after sedimentation and after at least very early syndiagenetic accumulations of sulphate and sulphide were formed, some of which, particularly pyrite, have been replaced by later copper and cobalt minerals. Features such as this and the slightly transgressive nature of the mineralization have now been reported from many deposits. Brown (1978) and Chartrand & Brown (1985) illustrate the former observations well and the latter feature has been described above. The proposition that the Central African Copperbelt sedimentation and mineralization took place in a rift zone has been elaborated by, among others, Annels (1984), who suggested that the mineralization was due to hydrothermal leakage from the bounding fractures, probably into relatively unconsolidated sediment. Gustafson & Williams (1981) suggested that the unifying feature for all sediment-hosted, stratiform, base metal deposits was that each was developed in a structural situation that permitted heated basinal brines to be moved to a shallow site of sulphide deposition. The reader who wants a quick overall view of the state of play in the investigation of these deposits cannot do better than read the abstracts on pp. 177–214 of *Can. Mineral. 24* (1986).

SOME EXAMPLES OF LEAD-ZINC DEPOSITS

These too can form giant orebodies, or groups of stacked orebodies (Table 16.1). They appear, as Russell *et al.* (1981) and Large (1983) *inter alia* have suggested, to form a distinctive group of ores formed in local basins on the sea floor as a result of protracted hydrothermal activity accompanying continental rifting. The host rocks are generally shales, siltstones and carbonates. The ores are finely layered and often interbanded with rock material (Figs. 2.14, 16.7), with both being affected by soft sediment deformation suggestive of a synsedimentary origin of the sulphides. At Silvermines and Tynagh, fossil hydrothermal chimneys have been found which are similar to those known from present day hydrothermal vents on the East Pacific Rise and other constructive plate margins (Banks 1985). These discoveries, together with the evidence of feeder zones and other observations described above, suggest that some (probably all) of these deposits have been formed by hydrothermal solutions venting into restricted basins on the sea floor. The environment was not, however, that of the deep ocean, but more like that of the Gulf of California, where sulphide and baryte deposits are forming today in the axial rifts of a region of active sedimentation (Lonsdale & Becker 1985). However, the sea depths are still much greater than those postulated for the formation of these giant lead-zinc deposits — about 50–200 m. Russell *et al.* (1981) suggested that the exhalations were formed by convection cells similar to those sketched in Fig. 4.12, which dissolved base metals from the rocks they traversed (see Chapter 4). Such cells would only have to penetrate a few km into the oceanic crust in order to be heated to and sustained at the necessary temperatures, if supplied with the latent heat of crystallization from a magma chamber (Cann & Strens 1982).

Table 16.1. Size and grade of some sediment-hosted, stratiform, lead-zinc deposits (Gustafson & Williams 1981 and other sources.)

Country	Deposit	Tonnage of ore in Mt (Reserves and past production)	Grade			By-products	Age
			Cu%	Pb%	Zn%		
Australia	Broken Hill	180	0.2	11.3	9.8	Ag 175 g t^{-1}	L. to M. Prot.
	McArthur River	237	0.2	4.1	9.2	Ag 41 g t^{-1}	M. Prot.
	Mount Isa	88.6	0.06	7.1	6.1	Ag 160 g t^{-1}	M. Prot.
Canada	Howard's Pass	100	—	1.5	6.0	—	Sil.
	Sullivan	160	—	6.6	5.9	Ag 68 g t^{-1} Sn, Cd, Cu, Au	M. Prot.
Germany	Meggen	60	0.2	1.3	10.0	Baryte	Dev.
	Rammelsberg	30	1.0	9.0	19.0	Ag 103 g t^{-1} Baryte	Dev.
Ireland	Navan	70	—	2.6	10.1	—	Carb.
	Silvermines	18.4	—	2.8	7.4	Ag 21 g t^{-1} Baryte	Carb.
	Tynagh	12.3	0.4	4.9	4.5	Ag 58 g t^{-1}	Carb.
RSA	Gamsberg	93.5	—	0.6	7.4	—	M. Prot.

For an intracratonic situation, like that of Ireland, it is of course necessary to postulate much deeper penetration of sea water. Russell *et al.* suggested a penetration of about 10 km if the deposit was mainly Pb-Zn, with a depth of about 15 km being necessary to permit the leaching of copper. This is a promising model, as it not only accounts for the formation of these deposits, but also for their distribution, and hence the prediction of the approximate locations of other possible deposits. The reader is strongly recommended to study Russell *et al.* (1981) or further developments of this hypothesis in Russell (1983). Unfortunately, Brown & Williams (1985) have thrown a spanner into the works by suggesting that Russell's cells reaching 10–15 km down into a thick Lower Palaeozoic crust in Ireland are an untenable hypothesis, as recent seismic work has shown that the depth to the Precambrian basement is only 4 km. They therefore favour mineralization by basinal brines, as described in Chapter 4.

Volcanic-associated massive sulphide deposits

Some attention has already been paid to these deposits in Chapters 2 and 4. This section will therefore be used to amplify and add to what has already been written. The reader is consequently recommended to read pp. 26–27 and 63–69 before reading this section.

GENERAL CHARACTERISTICS

Some of the more important features have been covered in the above sections

Fig. 16.7. Photographs of ore specimens from Mount Isa, Queensland. The light coloured material is sulphide and the dark is silicate. Note the slump folds, associated fracturing and synsedimentary faults affecting both sulphide-rich and silicat-rich layers.

of this book and only additional material is included here. The mineralogy of these deposits is fairly simple, the major minerals, in order of abundance, being: pyrite, pyrrhotite, sphalerite, galena, chalcopyrite (bornite and chalcocite are occasionally important), minor arsenopyrite, magnetite and tetrahedrite-tennantite. The gangue is principally quartz, but occasionally, carbonate is developed, and chlorite and sericite may be locally important.

The geochemical division into iron, iron-copper, iron-copper-zinc and iron-copper-zinc-lead deposits has already been touched on, but it must be emphasized that while we may find pyrite deposits without any appreciable copper, copper is never found on its own. Similarly, if we find lead, we will have zinc and at least accessory copper too. With zinc will come copper and perhaps lead. Some of the names commonly given to these different types have been mentioned on p. 63. Although for a number of years the dominantly iron-copper-zinc deposits of the Canadian Archaean were considered by many workers to be a variant of the Kuroko-type, it is now generally agreed that they are best considered as a separate type, which Hutchinson (1980) termed Primitive. A summary of these different types is given in Table 16.2.

Stanton (1978) objected to the division of these deposits into various types, and considered these ores to be part of one continuous spectrum, showing a progressive geochemical evolution which accompanies that of the calc-alkaline rocks in island arcs. It is certainly difficult to assign some deposits to particular types except on a purely geochemical basis. For example, the deposits of West Shasta, California, which are believed to have formed in a Devonian bimodal suite of island arc volcanics (South & Taylor 1985, Lindberg 1985), are, like Kuroko deposits, intimately associated with rhyolite lavas and domes, but they are a Cu-Zn (+Au-Ag) type and lack significant lead. Thus although they have many of the properties of Kuroko-type deposits, a case could be made out for correlating them with the Primitive-type, of which it must be remarked that some deposits, e.g. Kidd Creek, Ontario, do carry recoverable lead. Yes, reader, the situation is confusing! — we may be dealing with a continuous spectrum or spectra that we have not yet recognized, but meanwhile mining geologists will continue to use the terms listed in Table 16.2, because in the present state of our knowledge, they imply useful summary descriptions of the deposits, helpful in formulating exploration models.

It is important to note that precious metals are also produced from some of these deposits, indeed in some Canadian examples of the Primitive type they are the prime product. Both Besshi and Kuroko types may also produce silver and gold whilst the Cyprus type may have by-product gold.

The vast majority of massive sulphide deposits are zoned. Galena and sphalerite are more abundant in the upper half of the orebodies whereas chalcopyrite increases towards the footwall and grades downward into chalcopyrite stockwork ore (Fig. 2.19). This zoning pattern is only well developed in the polymetallic deposits. As the number of mineral phases decreases, so the zonation tends to become obscure and may not be in evidence at all in Cyprus type deposits.

Table 16.2. Volcanic-associated massive sulphide deposit types. (Modified from Hutchinson 1980).

Type	Volcanic rocks	Clastic sedimentary rocks	Depositional environment	General conditions	Plate tectonic setting	Known age range
Besshi (=Kieslager) $Cu-Zn\pm Au\pm Ag$	Within plate (intraplate) basalts	Greywackes and other turbidites	Deep marine sedimentation with basaltic volcanism	Rifting	Epicontinental or back-arc	Early Proterozoic Palaeozoic
Cyprus $Cu(\pm Zn)\pm Au$	Ophiolitic suites, tholeiitic basalts	Minor or absent	Deep marine with tholeiitic volcanism	Tensional, minor subsidence	Oceanic rifting at accreting margin	Phanerozoic
Kuroko $Cu-Zn-Pb$ $\pm Au\pm Ag$	Bimodal suites, tholeiitic basalts, calc-alkaline lavas and pyroclastics	Shallow to medium depth clastics, few carbonates	Explosive volcanism, shallow marine to continental sedimentation	Rifting and regional subsidence, caldera formation	Back-arc rifting	Early Proterozoic Phanerozoic
Primitive $Cu-Zn\pm Au\pm Ag$	Fully differentiated suites, basaltic to rhyolitic lavas and pyroclastics	Immature greywackes, shales, mudstones	Marine, < 1 km depth. Mainly developed in greenstone belts	Major subsidence	Much debated: fault-bounded troughs, back-arc basin?	Archaean—early Proterozoic

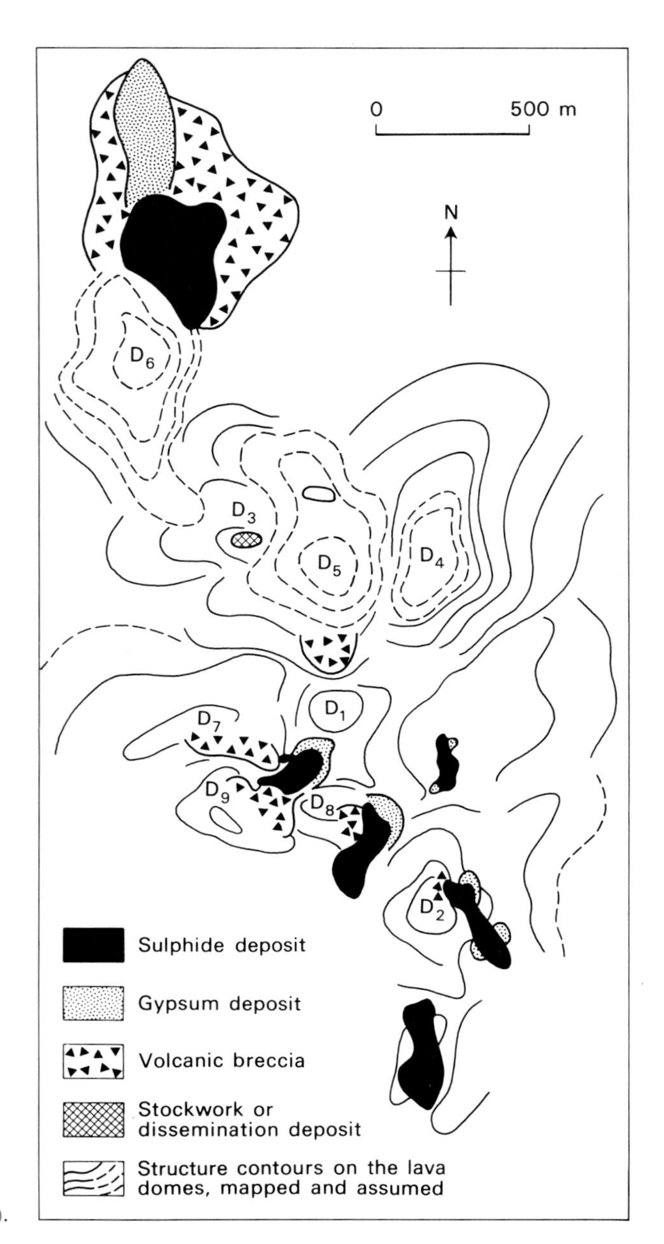

Fig. 16.8. Distribution of dacite lava domes and Kuroko deposits, Kosaka District, Japan. (Modified from Horikoshi & Sato 1970).

Textures vary with the degree of recrystallization. The dominant original textures appear to be colloform banding of the sulphides with much development of framboidal pyrite, perhaps reflecting colloidal deposition. Commonly, however, recrystallization, often due to some degree of metamorphism, has destroyed the colloform banding and produced a granular ore. This may show banding in the zinc-rich section, whereas the chalcopyrite-pyrite ores are rarely banded. Angular inclusions of volcanic host rocks are occasionally present and soft sediment structures (slumps, load casts) are sometimes seen. Graded bedding has also been reported from some deposits (Sangster & Scott 1976, Ohmoto & Skinner 1983).

Wall rock alteration is usually confined to the footwall rocks. Chloritization and sericitization are the two commonest forms. The alteration zone is pipe-shaped and contains within it and towards the centre the chalcopyrite-bearing stockwork. The diameter of the alteration pipe increases upward until it is often coincident with that of the massive ore. Metamorphosed deposits commonly show alteration effects in the hanging wall. This is probably due to the introduction of sulphur released by the breakdown of pyrite in the orebody. The sulphur, by reacting with pore solutions, could give rise to extensive hydrogen ion production.

The close association with volcanic domes has already been noted. The Kuroko deposits of the Kosaka district, Japan, are a good example (Fig. 16.8). All the Japanese Kuroko deposits are associated with Miocene volcanics and fossiliferous sediments developed along the eastern margin of a major geosyncline. Mineralization occurred during a limited period of the Middle Miocene over a strike length of 800 km in the Green Tuff volcanic region.

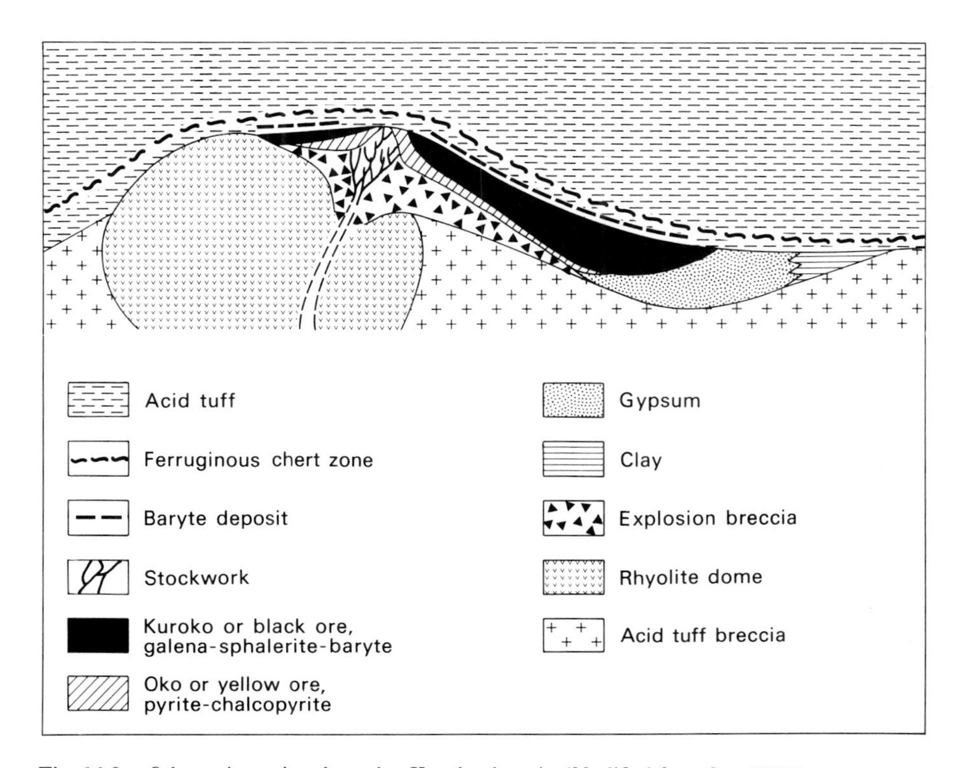

Fig. 16.9. Schematic section through a Kuroko deposit. (Modified from Sato 1977).

Within this region more than a hundred Kuroko-type occurrences are known, but most are clustered into eight or nine districts. Their host rocks are acid pyroclastic flows which are also centred on domes, particularly those showing evidence of explosive activity (Fig. 16.8). Kuroko ores have a consistent stratigraphical succession of ore and rock types, and an idealized deposit (Fig. 16.9) contains the following units:

(1) hanging wall: upper volcanic and/or sedimentary formation;
(2) ferruginous quartz zone: chiefly hematite and quartz;
(3) baryte ore zone;
(4) Kuroko or black ore zone: sphalerite-galena-baryte;
(5) Oko or yellow ore zone: cupriferous pyrite ores; about this level, but often towards the periphery of the deposit, there may be the Sekkoko zone of anhydrite-gypsum-pyrite;
(6) Keiko or siliceous ore zone: copper-bearing, siliceous, disseminated and/or stockwork ore;
(7) footwall: silicified rhyolite and pyroclastic rocks.

GENESIS

The genesis of these deposits is discussed on pp. 63–69.

Volcanic-associated oxide deposits

IRON DEPOSITS

Solomon (1976) suggested that stratiform oxide ores such as the magnetite-hematite-apatite ores of Kiruna and Gällivare, northern Sweden, are oxide end members of the massive sulphide group — a view echoed by Parák (1985). These ores sometimes contain as much as 15% P in the form of apatite, and magnetite (±hematite) -apatite ores are often referred to as Kiruna-type ores, but some in the type area are virtually free from apatite. There does indeed appear to be a range of ore types from massive sulphide deposits containing minor magnetite, through magnetite-pyrite ores with minor chalcopyrite and trace sphalerite such as Savage River, Tasmania (Coleman 1975), to stratiform iron oxide deposits which may, or may not, carry appreciable phosphorus. Some possibly related deposits, e.g. in Iran and the USSR, appear to have the form of mineralized pipes. These oxide deposits can be immense. The Kiruna orebody crops out over a strike length of 4 km, extends down dip for at least 1 km and is 80–90 m thick. Phosphorus-poor ore runs 67% Fe and phosphorus-rich 2–5% P and about 60% Fe. Production is now entirely from underground workings and at 18 Mt p.a. makes Kiruna the world's largest underground mine. At Savage River, ore reserves in 1975 were 93 Mt in one of many deposits in a zone up to 23 km long. Among the pipe deposits there are fourteen in the Bagq district of Iran together totalling > 1000 Mt (Förster & Jafarzadeh 1984), and the single deposit of Korshunovsk in Irkutsk, USSR carries 428 Mt of ore (Sokolov & Grigor'ev 1977). Here we only have space for a few remarks on the geology and genesis of these deposits. They all occur in volcanic or volcanic-sedimentary terrane, but there is no suggestion of some correlation of deposit type with volcanic rock type, as may be the case with the massive sulphides. This is because Savage River is associated with metamorphosed tholeiitic basalts, Kiruna with keratophyres, Iron Mountain, Missouri with andesites and the Bafq district and Cerro de Mercado, Mexico with rhyolites.

A sketch map of the Kiruna area is given in Fig. 16.10. The stratiform and concordant nature of the orebodies and their volcanic setting are apparent. The ores are both massive in part and well banded elsewhere — a banding that in places looks very much like bedding, sometimes with cross stratification. At Luossavaara, what appears to be a stockwork underlies the orebody (Fig. 16.11) and explosive volcanic activity has formed a hanging wall breccia containing fragments of the ore. All this evidence suggests that these are exhalative deposits or oxide lava flows that have undergone some reworking at the surface. The latter hypothesis was put forward by Williams (1969) in view of the interesting occurrences of undoubted magnetite-hematite-apatite lava flows in Chile (Park 1961), but the bulk of the evidence favours an exhalative origin for Kiruna and similar stratiform deposits. It remains to be seen if the pipe deposits formed from similar mineralizing fluids that were trapped within the crust like those at the Great Bear Lake deposits in Canada (Hildebrand 1986).

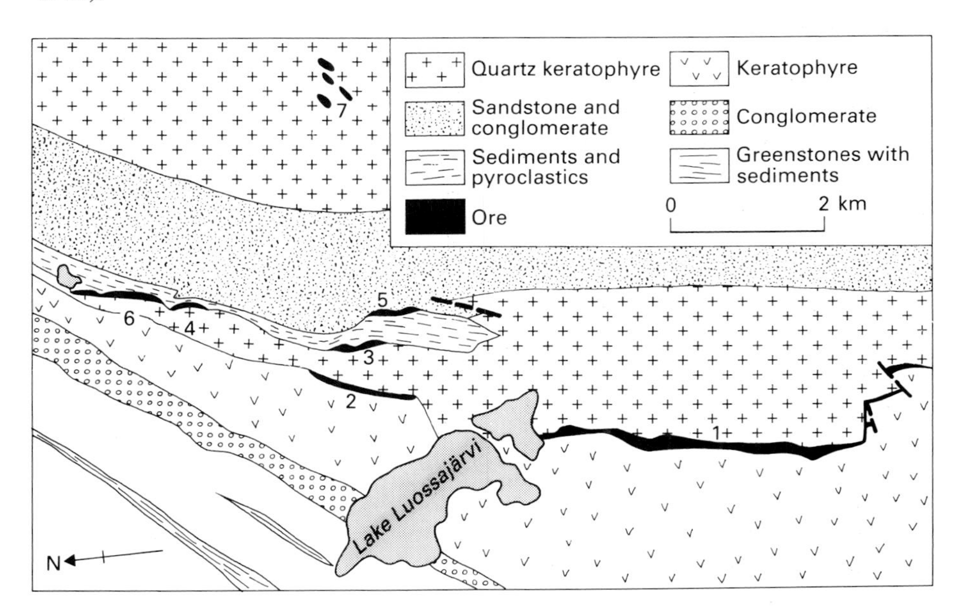

Fig. 16.10. Sketch map of the geology of the Kiruna area showing the location of the iron deposits. 1 — Kirunavaara; 2 — Luossavaara; 3 — Rektorn; 4 — Henry; 5 — Haukivaara; 6 — Nukutusvaara; 7 — Tuollovaara. The formations dip and young eastwards. (After Parák 1975, 1985).

Fig. 16.11. Stockwork of magnetite veins cutting keratophyre in the footwall of the Luossavaara iron orebody.

Fig. 16.12. Banded scheelite ore, Felbertal, Austria, showing folding of both scheelite-bearing and silicate layers. Photograph taken using ultraviolet light and supplied through the courtesy of Prof. Dr R. Höll (scale in cm).

Another and older genetic hypothesis for these deposits, that they are magmatic segregations, appears to be losing favour — but it still has some persuasive advocates (cf. Frietsch 1978).

Other possible exhalative oxide deposits include the Rexspar uranium deposit, British Columbia (Preto 1978), certain uranium deposits in Labrador (Gandhi 1978), and some tin ores. Exhalative tin ores were first described from the Erzgebirge of East Germany (Baumann 1970), but the most important deposits so far found are probably those of Changpo in Dachang, China (Tanelli & Lattanzi 1985). Exhalative tungsten ores have also come into prominence during recent years particularly as a result of the pioneering work of Höll & Maucher (1976) who have traced Sb-W-Hg mineralization along the Eastern Alps through the whole length of Austria. The mineralization is stratiform, occurs in Lower Palaeozoic inliers and is associated with metatholeiites. The scheelite mineralization is mostly present in fine-grained metachert bands that probably represent an original exhalite and which is very finely banded (Fig. 16.12). The scheelite bands appear to be synsedimentary and have suffered the same deformation and metamorphism as the enclosing rocks; one of the world's largest tungsten mines (Felbertal, to the south of Salzburg), exploits this mineralization. Mineralization of this type has now been reported from a number of areas of the world e.g. The Italian Alps (Brigo & Omenetto 1983), Broken Hill, New South Wales (Barnes 1983), Sante Fe, New Mexico (Fulp & Renshaw 1985), Sangdong, South Korea (Fletcher 1984) and La Codosera, Spain (Arribas & Gumiel 1984). The general volcanic association appears to be with basalts.

The mercury end member brings us back to sulphide mineralization in the form of cinnabar — the principal mineral of the famous Almaden mercury deposits of Spain (Saupé 1973).

17

The Vein Association and Some Other Hydrothermal Deposits

Introduction

Vein, manto and pipe deposits have already received considerable attention in some earlier chapters. Their morphology and nature were described in Chapter 2 when such subjects as pinch-and-swell structure, ribbon ore shoots, mineralization of dilatant zones along faults, vein systems and orebody boundaries were outlined. The reader may find it convenient to review what is written on these subjects on pp. 11–16. Similarly in Chapter 3, in a discussion of precipitation from aqueous solutions, crustiform banding, (that characteristic texture of veins) was described, as were fluid inclusions and wall rock alteration, whose relationship to veins is most marked. It is generally agreed that vein filling minerals were deposited from hydrothermal solutions, and in Chapter 4 some consideration was given to the origin and nature of hydrothermal solutions including metamorphic processes such as lateral secretion. The relevant sections are on pp. 45–56 and 56–63. In Chapter 5, paragenetic sequence and zoning were discussed with special reference to vein deposits; see pp. 76–81.

If this chapter had been written forty or so years ago it would probably have been the largest chapter in the book. However, since the 1940s the importance of veins has been steadily diminishing. The reasons for this are twofold. Firstly, a number of deposits formerly thought to belong to this association have been recognized as belonging to other ore classes. For example the Horne orebody of Noranda, Quebec was, for decades, believed to be a hydrothermal replacement pipe but is now considered to be a deformed volcanic massive sulphide deposit. The second and more important reason is economic. Taking copper as an example, only those ores containing more than 3% Cu were economically workable until the nineteenth century and such high metal levels were generally only reached in hydrothermal veins, so exploration and mining concentrated on this class of orebody. Today, rocks containing 0.5% Cu can be economically exploited by large scale mining methods and consequently veins do not hold their pre-eminent position any more. They are, however, still important for their production of gold, tin, uranium and a number of other metals and industrial minerals such as fluorspar and baryte. Of course thick, high grade, base metal veins are still well worth finding and a good example is the El Indio Mine high in the Chilean Andes, where massive sulphide veins can be over 10 m thick with average grades of 6–12% Cu, 8–10 g t^{-1} Au and 160 g t^{-1} Ag (Walthier *et al.* 1985). As a result of their past importance, studies of vein ores have had a very profound influences on theories of ore genesis almost up to the present day, an influence out of all proportion to their true value.

Vein deposits and the allied, but less frequent, tubular orebodies show a great variation in all their properties. For example, thickness can vary from a few millimetres to more than a hundred metres and as far as environments are concerned, veins can be found in practically all rock types and situations, though they are often grouped around plutonic intrusions as in Cornwall, England. Mineralogically they can vary from monomineralic to a mineral collector's paradise such as the native silver-cobalt-nickel-arsenic-uranium association of the Erzgebirge and Great Bear Lake, Northwest Territories.

Kinds of veins

The majority of veins have probably formed from uprising hydrothermal solutions which precipitated metals under environments extending from high temperature-high pressure near magmatic, to near surface low temperature-low pressure conditions. A few veins are pegmatitic and represent end stage magmatic activity; a few are clearly volcanic sublimates.

Mineralogically, gangue minerals can be the dominant constituents as in auriferous quartz veins. Quartz and calcite are the commonest, with quartz being dominant when the host rocks are silicates and calcite with carbonate host rocks. This suggests derivation of gangue material from the wall rocks. Sulphides are most commonly the important metallic minerals but in the case of tin and uranium oxides are predominant. Here and there, native metals are abundant, especially in gold- and silver-bearing veins.

Zoning

This important property has already been considered. It is so commonly developed within single veins or groups of veins (district zoning) that some further mention of it must be made here. The base metal veins of Cornwall show some of the best and earliest studied examples of orebody and district zoning (Fig. 5.6 and Table 5.1), which has had important economic implications. For example, after the recognition of zoning in this tinfield, probably some time before de la Beche wrote about it in 1839, a number of abandoned copper mines were deepened and the delighted owners found tin mineralization beneath the copper zone. This tin had not been discovered earlier because there can be a barren zone of as much as 100 m or so between the tin and copper zones. On the other hand, as at the old Dolcoath Mine, there may be an overlap and a mine which commenced life as a copper producer could in time become a copper-tin producer and lastly a tin producer. Such was the history of this, the richest individual tin vein that has been worked anywhere in the world. In all, Dolcoath produced 80 000 t Sn and 350 000 t Cu and it is an interesting exercise to calculate the value of this metal production at present day prices! As can be seen from Table 5.1, the zoning generally takes the form of a progressive change in composition with depth. Gradual changes in metallic minerals are accompanied by somewhat less pronounced changes in the gangue minerals with quartz being present at all levels. This zoning has been attributed to a slow fall in temperature with increasing distance from the granite intrusions, though the zonal boundaries, if they are isothermal surfaces, are not parallel to the granite contacts. This discrepancy is, however, simply due to their adopting a compromise attitude between that of the source of heat and the cooling surface (ground level).

Some important vein deposit types

As may be imagined from the variable nature of veins, the definition of types would be a fruitful field for the classification-mad scientist who delights in pigeon-holing nature, even if he can find only one example for some of his holes! Eckstrand (1984) lists eight vein deposit types in his valuable volume and five of these are briefly discussed here. The first three will be considered together, for economy of space and because they are all gold vein types.

As gold is very much in mineral explorationists' sights at the time of writing, the student is strongly recommended to study and compare Eckstrand's mineral deposit types 8.2, 11 and 15. The first and least important is the turbidite-hosted vein gold found in the Archaean (e.g. Yellowknife, Northwest Territories) and in the Phanerozoic (e.g. Nova Scotia; Ballarat, Bendigo, Castlemaine in Australia). The veins may be concordant with the development of saddle reefs as well as being thoroughly discordant and developed in reverse faults, ladder veins etc. On a world-wide basis this has not yet been established as an important deposit type. Volcanic-associated vein (11) and intrusion-associated gold (15) are much more important. These are among the big gold producers of the Archaean greenstone belts and are found and sought for in this terrane all over the world. Many famous names belong here, such as the Golden Mile at Kalgoorlie, Western Australia, the Kolar Goldfield of India and the Kirkland Lake and Timmins areas of Ontario. Tonnages may exceed 60 Mt, e.g. Hollinger, Ontario, and grades are mainly from 7–17 g t^{-1}. They are usually developed in greenschist facies terranes and are associated with tholeiitic pillowed basalts and komatiites and their pyroclastic equivalents, or with felsic to mafic intrusives. As the structural control of mineralization changes from more widely spaced faults or fractures, to closely spaced minor fractures, these deposits shade into one of the types of disseminated gold deposits described below. Good descriptions of vein gold deposits are to be found in Hodder & Petruk (1982), Colvine (1983) and Foster (1984).

The controls on the location of Archaean gold deposits are becoming better known, and many vein sets have been recognized as being commonly strata-bound and often confined to competent less basic basaltic units in metamorphosed tholeiitic-komatiitic sequences. The physical properties of these rocks have favoured hydraulic fracturing and fluid access, and their mineralogy and geochemistry have controlled gold deposition within the veins. These considerations led Phillips *et al.* (1983) to single out rocks of this type in greenschist to transitional amphibolite facies areas as favourable exploration targets.

Vein uranium has a world-wide significance. The classic vein deposits of Joachimsthal in the Erzgebirge (*Jáchymov* in Czech), from which the Curies separated radium, produced large tonnages of uranium concentrates for USSR consumption during the two decades following the explosion of the first atomic bombs. These are the famous Ag-Co-Ni-As-U veins described by Georgius Bauer, who wrote under the Latin name of Agricola, in his book *De Re Metallica* (1556). Agricola's works laid the foundation of modern mining science and of the geology of ore deposits. Vein uranium is important throughout the Hercynian Massifs of central and western Europe (see Fig. 5.5), and indigneous deposits make a significant contribution to the French nuclear industry. The veins are frequently emplaced in two-mica granites and the vein material is regarded as having been leached out of these 'fertile' granites by circulating hydrothermal solutions (Cuney 1978, Leroy 1978, 1984).

Vein tin deposits are described below and in various other parts of this book e.g. pp. 51–52, 77–79, 86–89. Apart from placer tin deposits, they are the world's most important source of tin.

Some examples

THE LLALLAGUA TIN DEPOSITS, BOLIVIA

Bolivia possesses the greatest known reserves of tin outside the countries of south-east Asia, most of her reserves being in vein and disseminated deposits. There is a long history of mining dating back to the Spanish colonial days of the sixteenth century when most of the important tin deposits were discovered and were mined initially for the fabulously rich silver ores that had formed, partly by supergene enrichment processes, in the upper parts of the vein systems. Some of these deposits are still being mined, mainly for tin ores beneath the silver-rich zones.

The Bolivian tinfield extends along the Andean ranges east of the high plateau of Bolivia from the Argentine border northwards past Lake Titicaca and into Peru (Fig. 17.1), where many new tin deposits may be found (Kontak & Clark 1985). North-west of Oruro the deposits are mainly tin and tungsten veins associated with granodioritic batholiths. The batholiths range in age from Triassic to Miocene, with the Miocene ones being best mineralized.

South of Oruro there is a tin-silver association spatially related to high level subvolcanic intrusions. At some of these volcanic centres both the intrusives

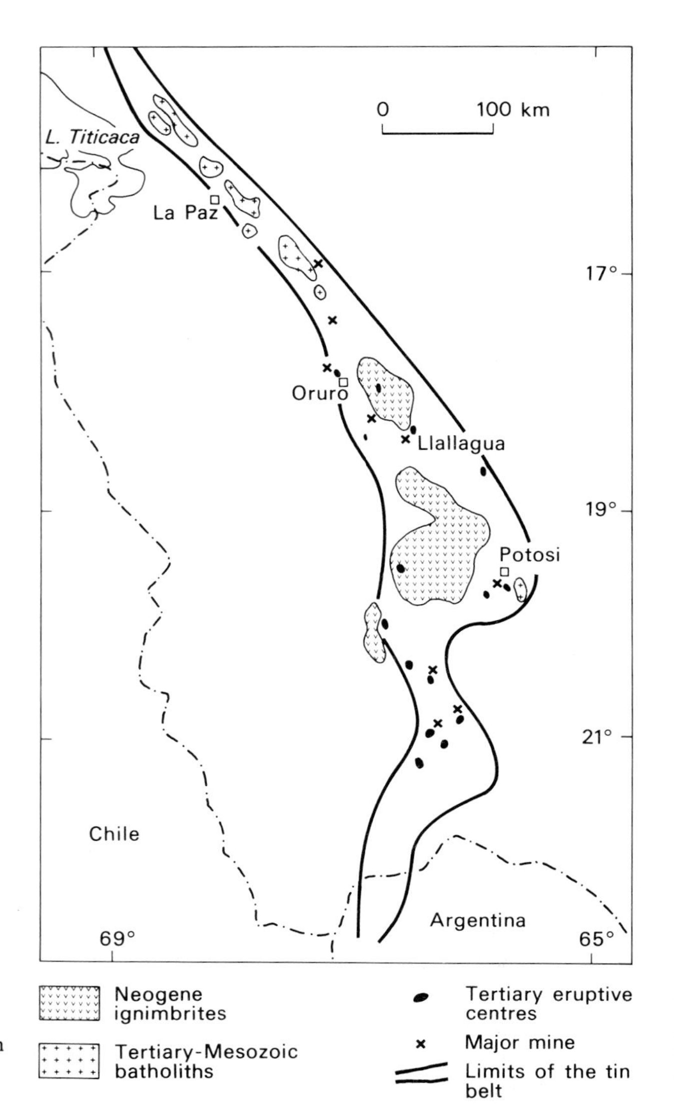

Fig. 17.1. The Bolivian Tin Belt. (After Grant *et al.* 1977).

and the coeval volcanics are preserved, at others erosion has removed the volcanics completely leaving only the intrusives. This is the case at Llallagua, the world's largest tin mine working primary tin deposits which is estimated to have produced over 500 000 t of tin since the beginning of this century. The mine occurs in the Salvadora Stock which occupies a volcanic neck cutting the core of an anticline in Palaeozoic rocks. The stock, which is made up of xenolithic and highly brecciated porphyry, narrows with depth. The original texture and rock composition (probably quartz-latite) are obscured by pervasive alteration. The ubiquitous brecciation is important because it has produced an increased permeability of the stock, which has given rise to alteration and mineralization independent of the geometry of the later vein systems. This

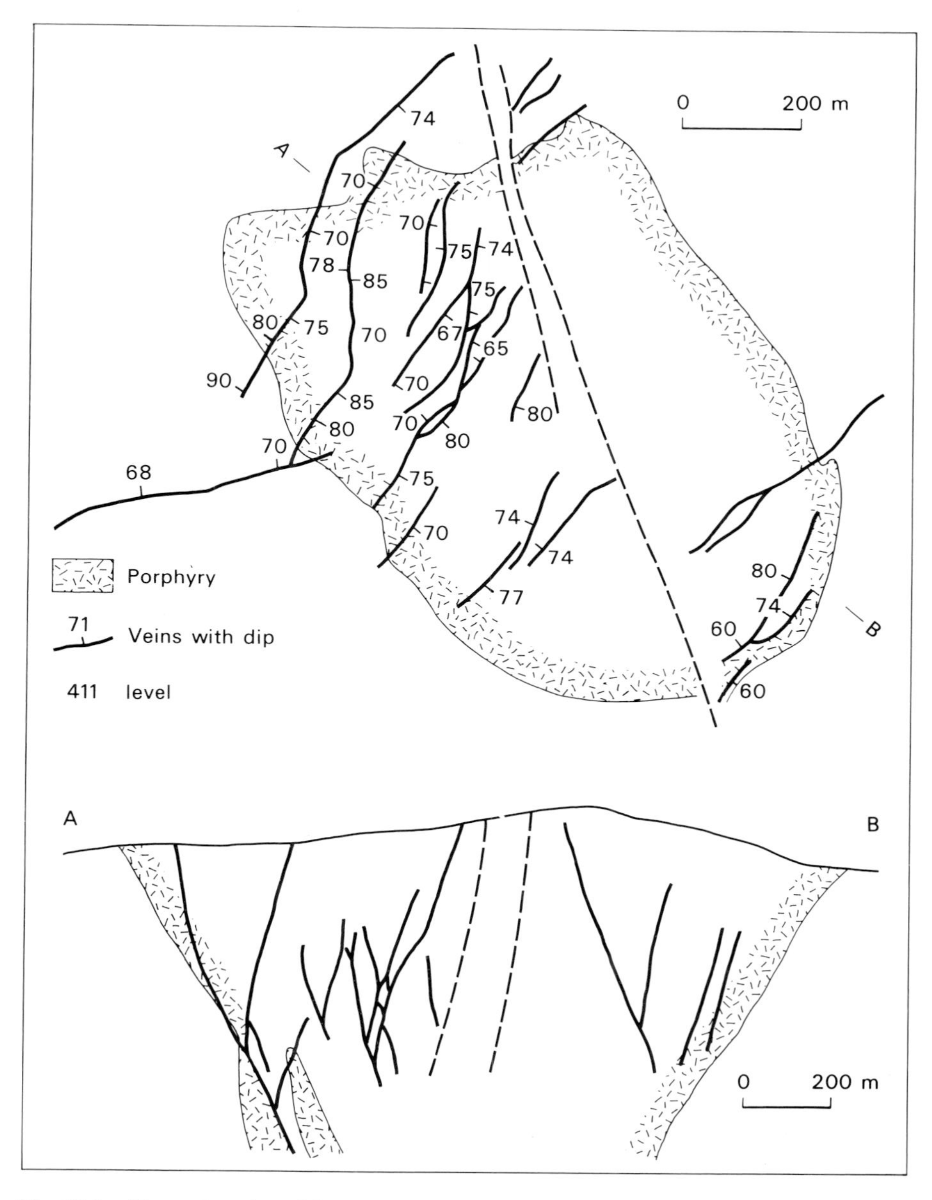

Fig. 17.2. Plan and section of the major veins at Llallagua, Bolivia. (Modified from Turneaure 1960).

dispersed mineralization may be of high enough grade to permit bulk mining methods to be used, i.e. it may be a porphyry tin deposit (Grant *et al.* 1977). The alteration has produced a host rock consisting of primary quartz, tourmaline, sericite and secondary quartz.

There is a network of veins in and around the stock, some of which are shown on Fig. 17.2. The major veins trend about 030° and appear to form part of a conjugate system of normal faults. They are typified by the San Jose type which generally has a dip of 45–80°, a good width and strike persistence, and an average width of about 0.6 m, although widths of up to 1.8 m are known. These are the richest veins and can contain up to 1 m of solid cassiterite. Clay gouge is very common. The Serrano vein type is much thinner (average 0.3 m), nearly vertical and impersistent. These veins can be as richly mineralized as the San Jose type, but with such narrow veins dilution with country rock material occurs during mining. There is little or no clay gouge. These veins may have formed in vertical tension gashes associated with the normal faulting.

The first stage of mineralization consisted of the formation of crusts of quartz followed by bismuthinite and cassiterite, and many high grade veins are composed almost entirely of these three minerals. Wolframite and tourmaline also being to this early stage (Turneaure 1960). This was followed by a stage of sulphide mineralization, principally pyrrhotite and franckeite, the pyrrhotite being later largely replaced by pyrite, marcasite and siderite. Arsenopyrite, sphalerite, stannite and rare chalcopyrite were also formed during this later stage of sulphide deposition. The discerning reader will have noted that in this vein system there is present, over a very restricted vertical range, a host of minerals which in Cornwall are developed in a zonal sequence spread out vertically over hundreds of metres (Table 5.1). The concentration of low and high temperature minerals into the same 'zone' is called telescoping and is thought to be due to the existence of high temperatures in near surface host rocks induced by the volcanic activity and a blanket of hot volcanic deposits. Fluid inclusion work by Grant *et al* (1977) has shown that the pervasive alteration and early vein growth, including cassiterite deposition, took place at temperatures (uncorrected for pressure) of about 400–350°C. The temperature dropped to 300°C and lower during the sulphide deposition.

BUTTE, MONTANA

This is one of the world's most famous vein mining districts. From 1880–1964 Butte produced 300 Mt of ore yielding 7.3 Mt Cu, 2.2 Mt Zn, 1.7 Mt Mn, 0.3 Mt Pb, 20 000 000 kg Ag, 78 000 kg Au, together with significant amounts of bismuth, cadmium, selenium, tellurium and sulphuric acid. Because of this wealth of mineral production from a very small area (little more than 6 × 3 km), with more than a score of mines, Butte has been aptly called the richest hill on earth, for the monetary value of its production has only been exceeded by the much larger Witwatersrand Goldfield of South Africa. Cut-and-fill stoping of the veins was the main mining method up to 1950, then it was joined by the block caving of veined ground, and in 1955 large scale, open pit working of low grade porphyry-type mineralization commenced. Reserves are still extensive, of the order of 10 Mt of high grade vein copper and silver ore and 500 Mt of low grade copper mineralization.

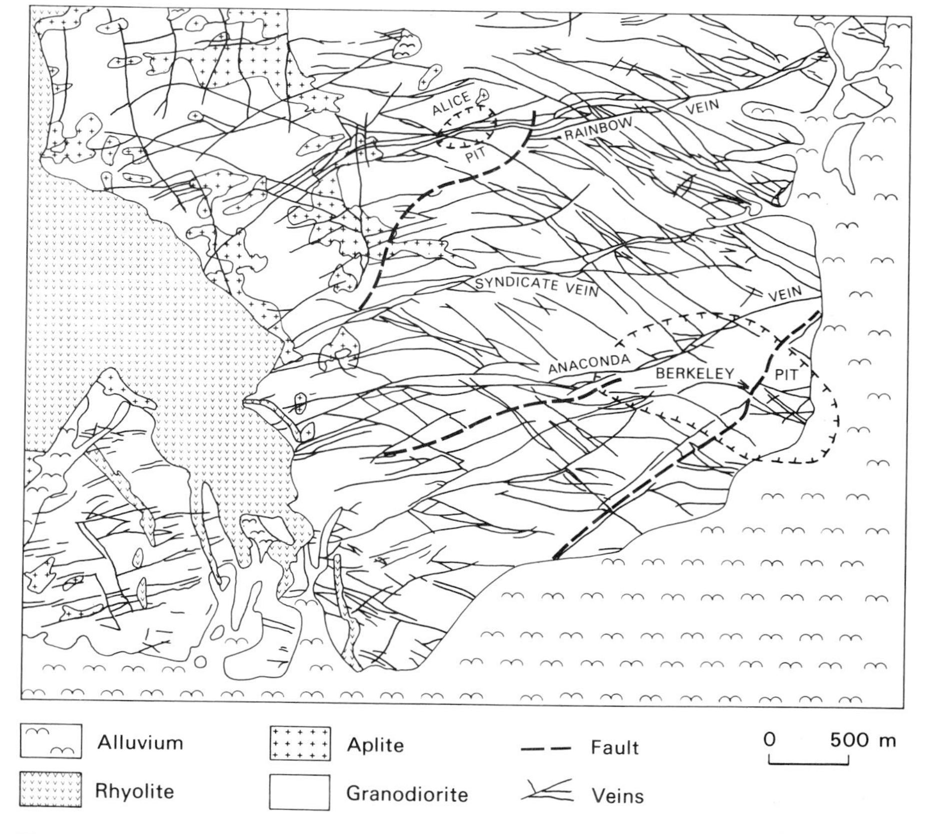

Fig. 17.3. Surface geology and veins of the Butte district, Montana. (Modified from Meyer *et al.* 1968).

The Butte field is in the south-western corner of the Cretaceous Boulder Batholith. Soon after the emplacement of the intrusion large parts of the area were covered by rhyolitic and dacitic eruptions (Fig. 17.3). The veins occur in a granodiorite which has given a radiometric age of 78 Ma. There are two mineralization stages: pre-main and main. The pre-main stage consists of small quartz veins carrying molybdenite and chalcopyrite found in the deeper central parts of the mineralized zone. They are bordered by alteration envelopes carrying potash feldspar, biotite and sericite. The biotite has been dated at 63 Ma.

The main mineralization stage occurs in several vein systems of which the most important are the easterly trending Anaconda and the later north-westerly trending Blue veins. The Anaconda veins are the major producers in the western third of the mineralized zone and also in the eastern third, where they divide into myriads of closely spaced south-easterly trending minor veins. This is called horse-tailing and gives rise to porphyry-type mineralization suitable for mass mining methods. The Anaconda veins are the largest and most

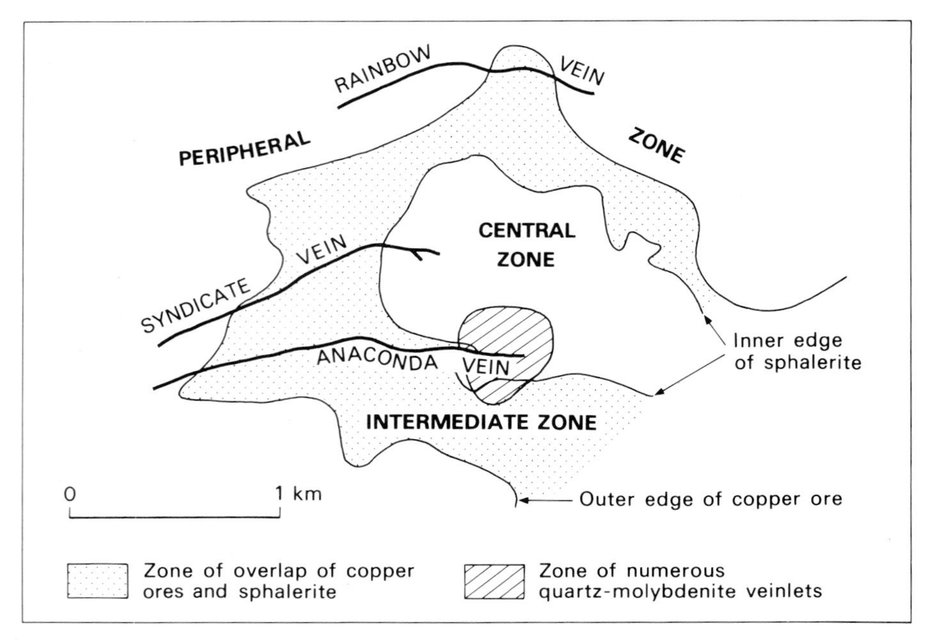

Fig. 17.4. Zoning on the 2800 ft (853.44 m) level at Butte, Montana. (Data from Meyer *et al.* 1968).

productive, averaging 6–10 m in thickness with local ore pods up to 30 m thick. The Blue veins usually offset the Anaconda veins with a sinistral tear movement and sometimes ore has been dragged from the Anaconda into the Blue veins by this fault movement. Individual oreshoots persist along strike and in depth for hundreds of metres.

All the veins contain similar mineralization and this is strongly zoned (Fig. 17.4). There is a Central Zone of copper mineralization which at depth contains the pre-main stage quartz-molybdenite veins and which is particularly rich in chalcocite-enargite ore. This gradually gives way outwards to ore dominated by chalcopyrite and containing minor amounts of sphalerite — the Intermediate Zone. The Peripheral Zone is principally sphalerite-rhodochrosite mineralization with small quantities of silver. All the veins are bordered by zones of alteration, which usually consists of sericitization next to the vein followed outwards by intermediate argillic alteration and then by propylitization. In deeper parts of Butte, advanced argillic alteration is present next to the veins. Sericite produced by this alteration dates at 58 Ma, distinctly later than the biotite of the quartz-molybdenite mineralization. The zoning suggests that the main mineralization was effected by hydrothermal solutions which passed upwards and outwards during a long period of time through a steadily evolving fracture system.

Brimhall (1979) has demonstrated that the pre-main stage mineralization consisted of widespread, zoned, low grade, disseminated copper-molybdenum mineralization of porphyry type formed at 600–700°C. During the later main stage mineralization a geothermal system, involving hydrothermal fluids of meteoric origin and possibly set in motion by a younger intrusive system, achieved temperatures of 200–350°C and leached much of the disseminated copper, redepositing it to form the rich vein orebodies.

Unconformity-associated uranium deposits

Deposits of this type have only come into prominence during the last fifteen years or so. The first to be discovered was the Rabbit Lake deposit in Saskatchewan in 1968, and the initial discoveries in the East Alligator River Field of Australia were made in 1970. As a result there is no agreed name for this deposit type, and in addition to the above title from Eckstrand (1984) they have been called *inter alia* 'unconformity vein-type deposits' and 'Proterozoic vein-like deposits.' Some are in no way vein-like, e.g. the McClean deposits (Figs. 2.7 & 2.8), and a name omitting the word vein is to be preferred. Since 1970 more deposits have been found and are now known to include the largest (Jabiluka, Northern Territory, 200 000 t contained U_3O_8), highest grade (Cigar Lake, Sask, 8.49% U) and the most valuable uranium deposits (Key Lake 86 000 Mt running 2.6% U_3O_8) in the world. The Athabaska Basin in Saskatchewan and the Northern Territory of Australia host the main deposits so far found and these contain about half the western world's low cost uranium reserves (Dahlkamp 1984). The Key Lake Mine started up in October 1983 at a capital cost of £282 million, and, at its planned production level of 5443 t U_3O_8 p.a., its output is 12% of the total from the western world.

Orebodies of this deposit type range from very small up to more than 50 Mt (Jabiluka II) and grade from 0.3 to over 5%. They may also have many metal by-products, e.g. Jabiluka II has 15 ppm Au and Key Lake contains large quantities of nickel. Other elements of note are Co, As, Se, Ag and Mo. The geological environment is that of middle Proterozoic, sandstone dominated sequences unconformably overlying older metamorphosed Proterozoic basement rocks. Most of the mineralization occurs at or just below the unconformity, where it is intersected by faults passing through carbonaceous schists in the basement (Fig. 17.5). The orebodies tend to be tubular (Figs. 2.7, 2.8) to flattened cigar-shaped. The high grade orebodies grade outwards into stratiform disseminations and fracture fillings and in Saskatchewan the mineralization may continue above the unconformity (Fig. 17.5). Important zones of wall rock alteration are developed around the orebodies, with the development of clay minerals, sericite and chlorite, and these zones greatly broaden the exploration target. A palaeoregolith is present at the unconformity below the Athabaska Group sandstone.

Although most of the Saskatchewan deposits are found along the basin rim (Fig. 17.6), the presence of the Cluff Lake Mine and other deposits in the Carswell Circular Structure, which brings the unconformity to the surface, indicates that other orebodies probably await discovery in deeper parts of the basin. The basin is now 1750 m deep at its centre but, on the basis of fluid

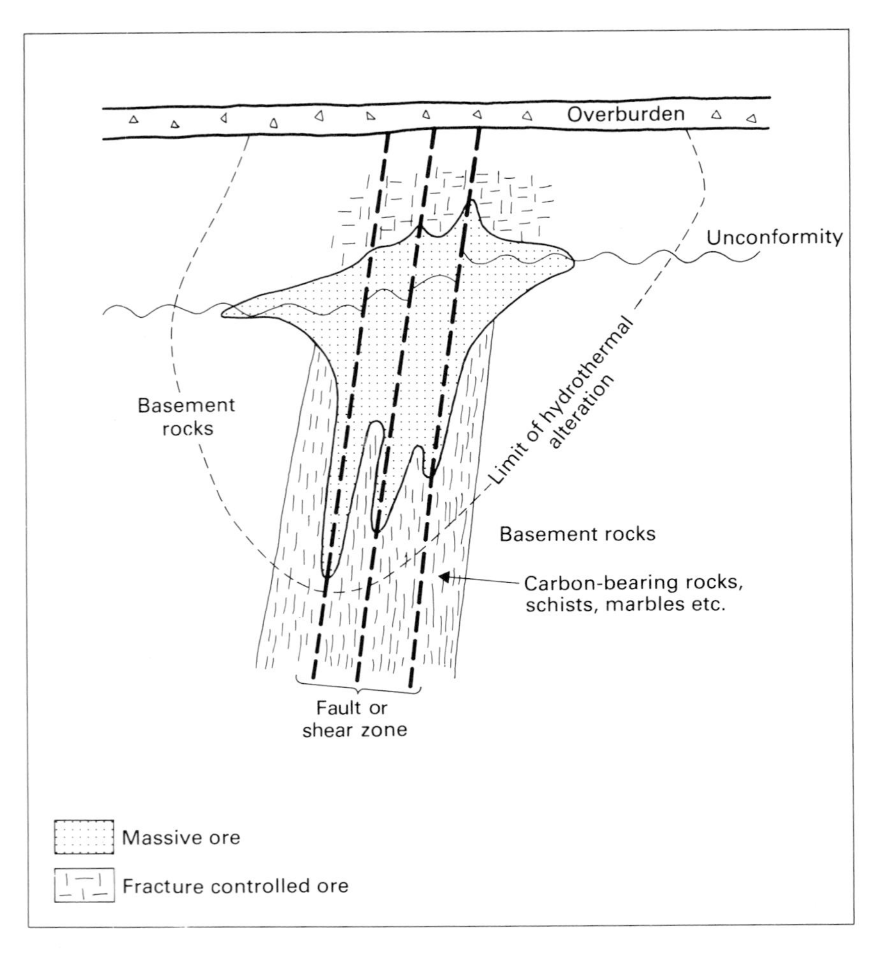

Fig. 17.5. Generalized diagram of an unconformity-associated uranium deposit. (After Clark *et al.* 1982).

inclusion studies, Pagel (1975) and Pagel & Jafferezic (1977) postulated that the basin sediments were once at a depth of 3500 m. These and other fluid inclusion studies indicate that both the uranium deposits and the Athabaska sandstones have been permeated by brines and CO_2-rich fluids at about 160–200°C. Stable isotopic H and O data (Pagel *et al.* 1980) suggest that these were one and the same fluid. (Similar fluids have been found in inclusions in the Australian deposits.) Isotopic studies by Bray *et al.* (1982) of the McClean Lake deposit permit a positive correlation of the $\delta^{13}C$ from graphite in the host metasediments and from siderite intergrown with uranium mineralization, suggesting that the carbon of the siderite has been derived from the graphite. Similarly, their work on sulphur isotopes showed that the sulphur in the ore zone sulphides could have been derived from the basement rocks. All the deposits have a largely epigenetic aspect, and radiometric dating at Key Lake

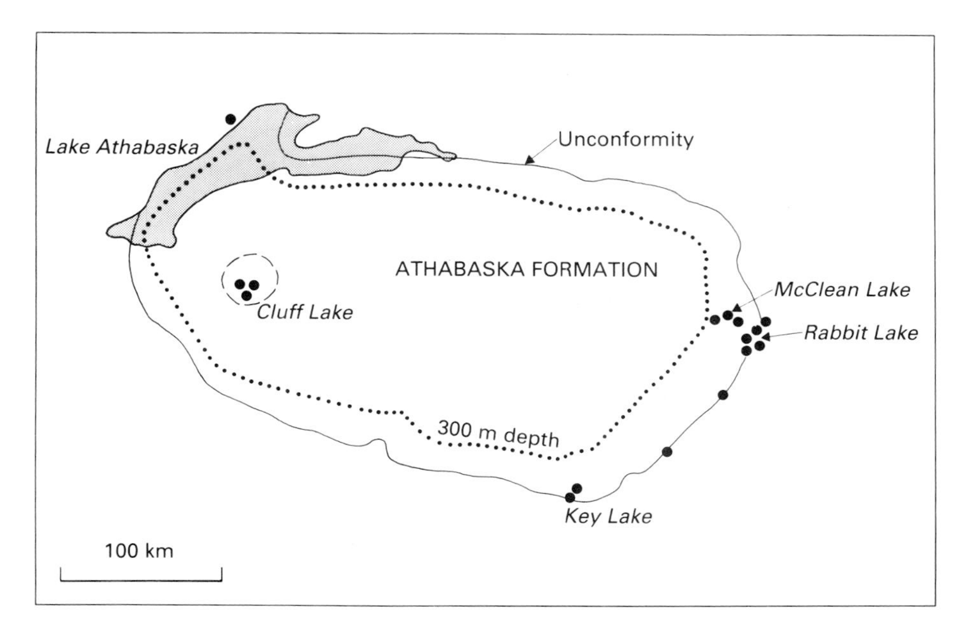

Fig. 17.6. Outline of the Athabaska Basin and distribution of the associated uranium deposits. The basement rocks outside the basin are older Proterozoic and Archaean in age. (After Clark *et al.* 1982).

(Trocki *et al.* 1984) shows that the mineralization is about 300 Ma younger than the Athabaskan sediments and their major period of diagenesis.

All this evidence has led most geologists to postulate uranium deposition from low temperature (100–200°C) geothermal systems driven by regional heating events (cf. Gustafson & Curtis 1983). The uranium is thought to have travelled as uranyl complexes [such as $UO_2(CO_3)_3{}^{4-}$] with reduction of U^{6+} due to carbonaceous material in the schists playing an important part in the mineralization process and involving reactions such as:

$$2UO_2(CO_3)_3^{4+} + 8H^+ + C \rightarrow 2UO_2 + 7CO_2 + 4H_2O$$

The source of the uranium is still conjectural. For the Narbarlek deposit, Ewers *et al.* (1983) showed that it could have been derived from basement granite or metamorphics, or the overlying Kombolgie Formation. In both the Canadian and Australian fields there are uranium-rich rocks and uranium deposits present in the basement. Source rocks for the uranium are not therefore a problem in elucidating the genesis of these deposits.

With such characteristic features marking their geological environment, an exploration model for these deposits may be readily drawn up and favourable areas for exploration outlined. The student may care to try out such an exercise using the data given above and apply his model to the Canadian Shield using a 1:5 000 000 geological map. He can then compare his ideas with those of Clark *et al.* (1982). Of course, similar favourable areas can be outlined elsewhere and uranium deposits of this type have been found in France (George *et al.* 1983). An interesting summary of exploration for these deposits has been related by Ullmer (1985).

Disseminated gold deposits

The range of deposit subtypes that can be described under this title is considerable. The best comprehensive reference is Boyle (1979, pp. 295–331) who recognized four subtypes but wrote that: 'These deposits cover a wide spectrum.' I will follow his classification.

DISSEMINATED AND STOCKWORK GOLD-SILVER DEPOSITS IN IGNEOUS INTRUSIVE BODIES

Some of these are referred to as porphyry gold deposits. Orebodies of this general type are commonly in the range 5–15 Mt with grades around 8–16 ppm, but there are many larger deposits of lower grade. The mineralization occurs in highly fractured zones of irregular outline which have been healed by veinlets, veins and stringers of auriferous quartz. These zones are marked by considerable hydrothermal alteration of the host rock which varies with the rock type. In granitic rocks the common alteration is sericitization, silicification, feldspathization and pyritization; in intermediate and basic rocks, carbonatization, sericitization, serpentinization and pyritization.

Deposits of this subtype occur in orogenic belts in all the continents and range from Archaean to Phanerozoic in age. Good potted descriptions of eleven deposits are given by Boyle, who points out that there is a marked relationship in many gold belts between albites, albite-porphyries, quartz-feldspar porphyries and gold. The exact nature of this relationship is uncertain and may vary from deposit to deposit. A good summary of these relationships as seen in Ontario can be found in Marmont (1983).

DISSEMINATED GOLD-SILVER OCCURRENCES IN VOLCANIC FLOWS AND ASSOCIATED VOLCANICLASTIC ROCKS

Orebodies of this subtype have only recently been outlined and an increase in the price of gold in real terms will bring many more of these deposits into the class of potential orebodies. The mineralization occurs in large, diffuse volumes of alteration in rhyolites, andesites or basalts. Some are closely associated with known gold deposits. Reported gold values are mainly in the range 0.02–0.03 ppm but some are much richer, e.g. Round Mountain, Nevada at 66 ppm. They are probably related to Carlin-type deposits which are described below.

DISSEMINATED DEPOSITS IN TUFFACEOUS ROCKS AND IRON FORMATIONS

Gold deposits in tuffs and other pyroclastic rocks are common in the greenstone belts of the Precambrian (Woodall 1979). A well described example is the Madsen Mine, Ontario (Durocher 1983). The orebodies took the form of echelon ore zones hosted by heterogenous, sheared and highly altered tuffs occurring in the lower parts of a tholeiitic-komatiitic sequence (Fig. 17.7). The orebodies, which were delineated by assay boundaries and had an average grade of about 8 ppm, were localized by rolls (open folds) in the hanging wall and footwall contacts. Orebodies at the Triton Gold Mine, Western Australia (Campbell 1953) occurred in a similar geological environment, with the same structural controls governing their location, and had a similar grade.

The spatial association of many Archaean gold deposits with ultramafic lavas suggests that komatiites may be the ultimate source rock of much gold, a point of importance in designing exploration programmes and one which has been well aired by Keays (1984).

A common factor of the greenstone belts of Archaean cratons is the presence of several large gold deposits (>50 t Au) in areas containing numerous

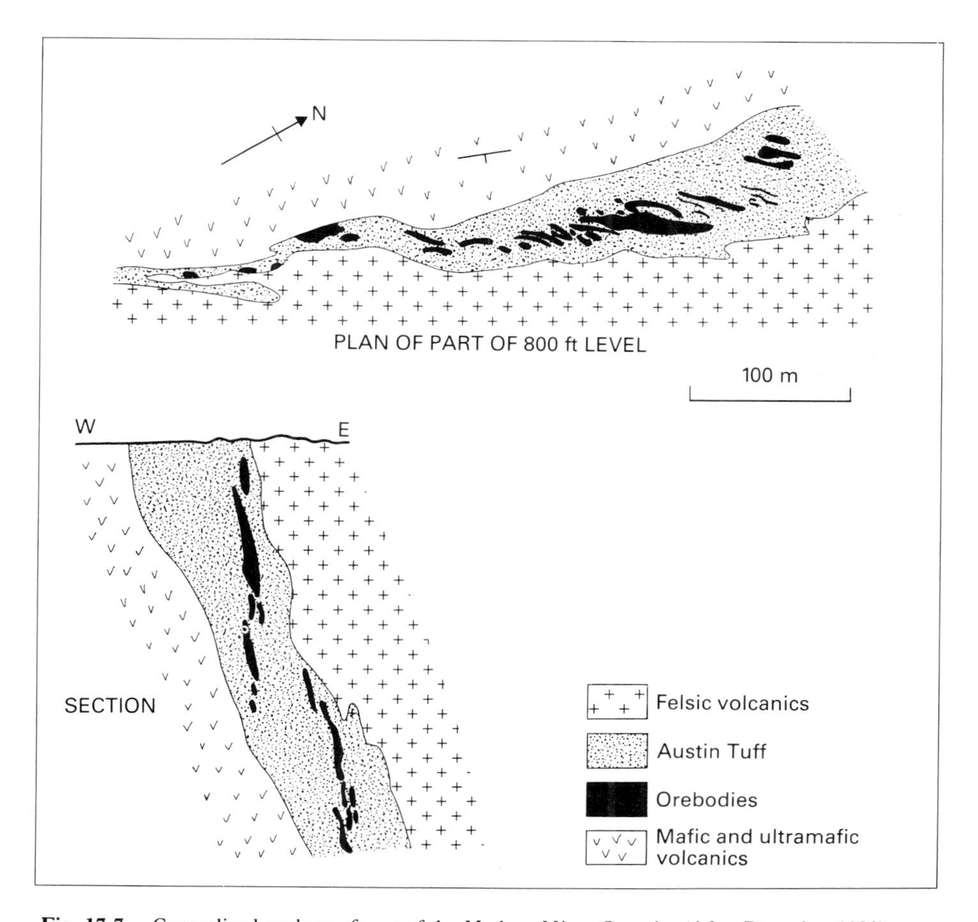

Fig. 17.7. Generalized geology of part of the Madsen Mine, Ontario. (After Durocher 1983).

smaller deposits. These large and many smaller deposits occur in various facies of iron formation of which lean sulphide-bearing carbonate and silicate facies are perhaps the most important. Sawkins & Rye (1974) termed these 'Homestake-type' deposits after the famous large and long lived mine in South Dakota, and included in this group Morro Velho, Brazil and the Kolar Goldfield of India. Many of the Zimbabwean deposits are of this type, as well as many in the Murchison Province, Western Australia.

The Homestake Mine has produced about 1100 t Au since it began production in 1877, and is the biggest gold mine in the USA. The orebodies occur only in the Homestake Formation, a thin (<100 m) auriferous, quartz-sideroplessite schist unit within a thick sequence of metasediments that only contains minor metabasaltic rocks. The deposit has suffered complex polyphase folding and low to medium grade metamorphism that gave rise to considerable redistribution and concentration of the gold (and sulphides), such that the orebodies are now spindle-shaped zones which appear to be localized in part by dilatant zones formed by superposition of F_2 on F_1 folds (Fig. 17.8). The orebodies consist of quartz, chlorite and ankerite with pyrrhotite, arsenopyrite and gold. Most of the chlorite appears to be altered cummingtonite or sideroplessite and much of the quartz takes the form of metamorphic segregations, veins and stringers, such that for many decades the deposit was considered to be epigenetic.

Stable isotope studies have shown that the sulphide sulphur and the oxygen of the quartz are indigenous to the Homestake Formation and the inference is that the gold too is syngenetic. A model for the formation of these deposits was proposed by Fripp (1976), involving circulating sea water, similar to that shown in Fig. 4.12 but with the saline solutions debouching into basins in which banded iron formation is being precipitated. Kerrich & Fryer (1979)

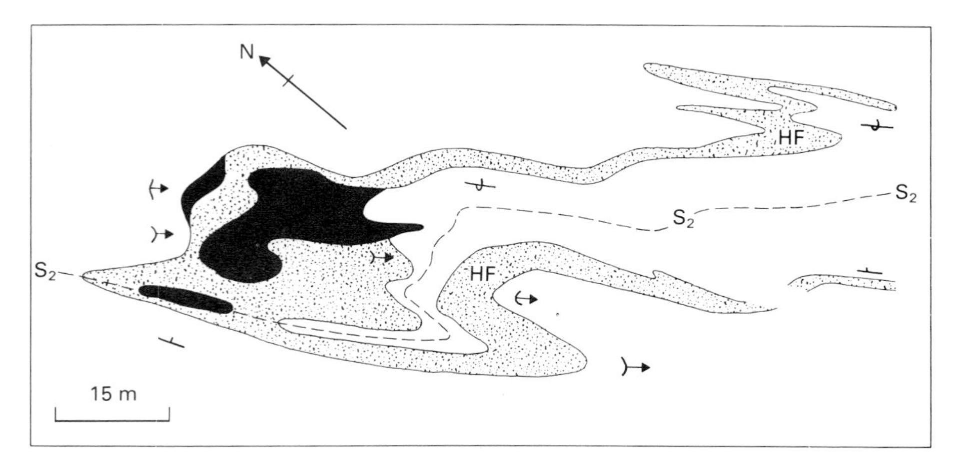

Fig. 17.8. Superposed folds, 2300 ft level, Homestake Mine, S. Dakota. An early fold with axial surface S_2 has smaller folds superposed on its limbs. Note development of orebodies in thickened hinge zones of later folds. Ore-bearing zones in black. HF = Homestake Formation. (Compiled from several sources).

have suggested that the fluids were of metamorphic origin but again suggest venting into a basin in which BIF was forming, cp. Fig. 4.8. On the other hand, Phillips, Groves & Martyn (1984) have suggested that some, perhaps all, of these deposits are epigenetic; but they side with Kerrich and Fryer in advocating a metamorphic source for the mineralizing fluids. These contrasting views lead to rather different exploration philosophies. A possible sequence of events, which ignores the origin of the mineralizing solutions, is shown in Fig. 17.9.

CARLIN-TYPE DEPOSITS

There are at least nine different names for this deposit type in the English

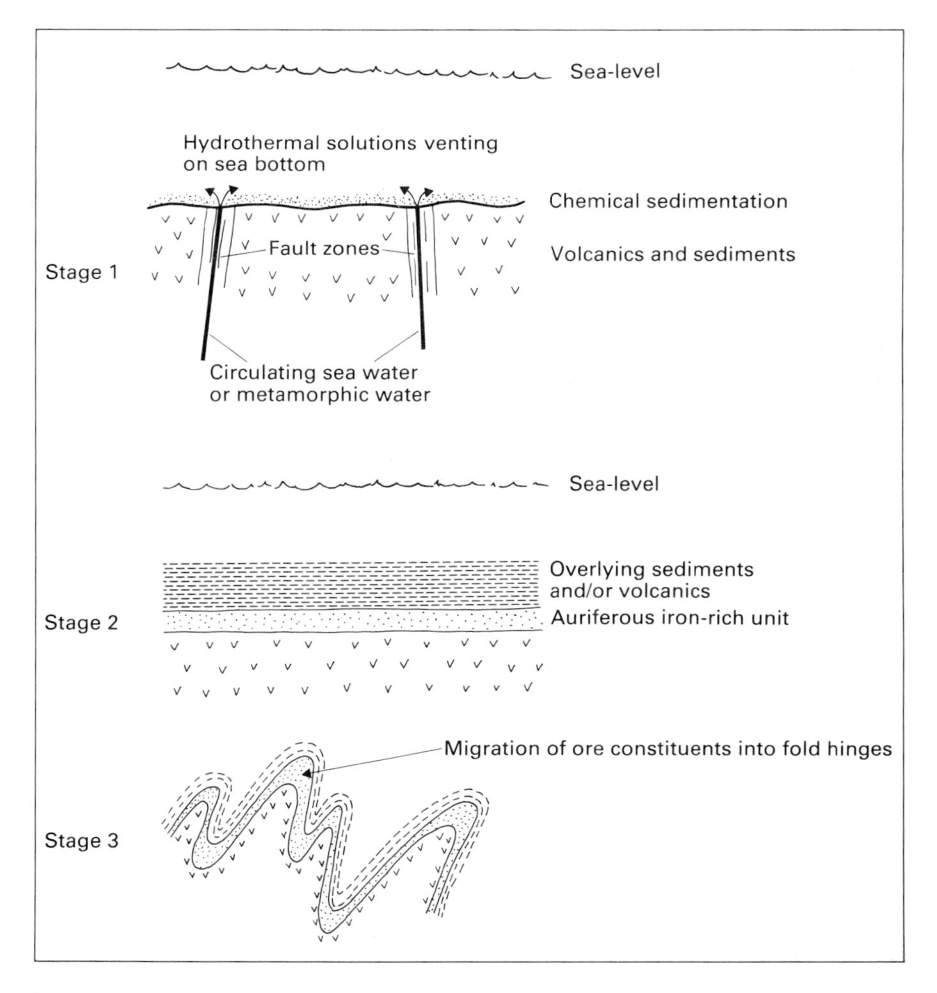

Fig. 17.9. Generalized model for the formation of Homestake-type deposits. (Modified from Sawkins & Rye 1974).

literature. For this writer Carlin-type is the least wanting as far as the likelihood of misleading the reader is concerned, and it has the virtue of using the name of the first deposit of this type to be found. (The Carlin Mine, Nevada, came into production in 1965). Nevertheless, we must not put these deposits into a strait-jacket and expect all deposits to reproduce the exact features of the type deposit. This would constrain our exploration and scientific study of these deposits and they are, at the time of writing, an important exploration target.

The deposits at present being exploited or proved up are mainly located in Nevada, Utah, Idaho and California. Their grade varies from 2.7–13.4 ppm Au and mercury and silver are often won as by-products, although in some Au/Ag can be very high. Orebody tonnages are in the range 5–80 Mt and resources can be very large, e.g. Round Mountain, Nevada, with 186 Mt grading 1.7 ppm Au (Tooker 1985). Orebody shapes vary from broadly tabular to quite irregular and they have assay boundaries. They are bulk-mineable deposits. New finds are being made all the time (*vide* Burger 1985) and there are many regions which could host such deposits; the young island arcs of the Western Pacific being favoured by Sawkins (1984) for exploration.

The known deposits have been classified as carbonate-hosted, volcanic hot spring-hosted and volcanic-hosted (Tooker 1985). These descriptions can, however, mislead the unwary reader, for Carlin and some of the other deposits in the carbonate-hosted category occur in carbonaceous siltstones which did, or do, contain appreciable quantities of carbonate minerals. In the primary unoxidized ores calcite has been removed, and pyrite and fine-grained silica, together with gold, arsenic, thallium, antimony and mercury, have been introduced (Radtke 1985). Other host rocks in this group may be designated limestones. Host rocks in the hot spring group vary from volcaniclastic to a mixture of these with interlayered sediments. The volcanic associated deposits have been developed at or near the surface. Perhaps the most important rock property, regardless of petrography, is permeability, allowing the establishment of circulating geothermal systems such as those described in Chapter 4 (cp. Fig. 4.3). Faults too have played an important part as circulation channels. Fluid inclusion studies indicate temperatures of the mineralizing fluids of 160–250°C in most cases and stable isotope studies suggest that all the introduced materials were probably leached from the underlying rocks (cf. Rye 1985). Only free gold is present (i.e. no combined gold as tellurides) and the grain size is less than 1 μm. Wall rock alteration in the form of silicification, jasperoid formation, argillization, propylitization and alunitization is common. A schema for the formation of Carlin-type deposits is given in Fig. 17.10.

From the regional tectonic point of view these deposits occur in complex terrane. Central and western Nevada is a case in point, consisting as it does of several geological terranes of distinct stratigraphical and structural domains that have been accreted throughout much of the Phanerozoic to the North American Craton. Sawkins (1984) suggested that at their time of formation the environment was either that of rear-arc rifting or a principal arc setting, and he favours the latter on the basis of early to mid Tertiary age determinations.

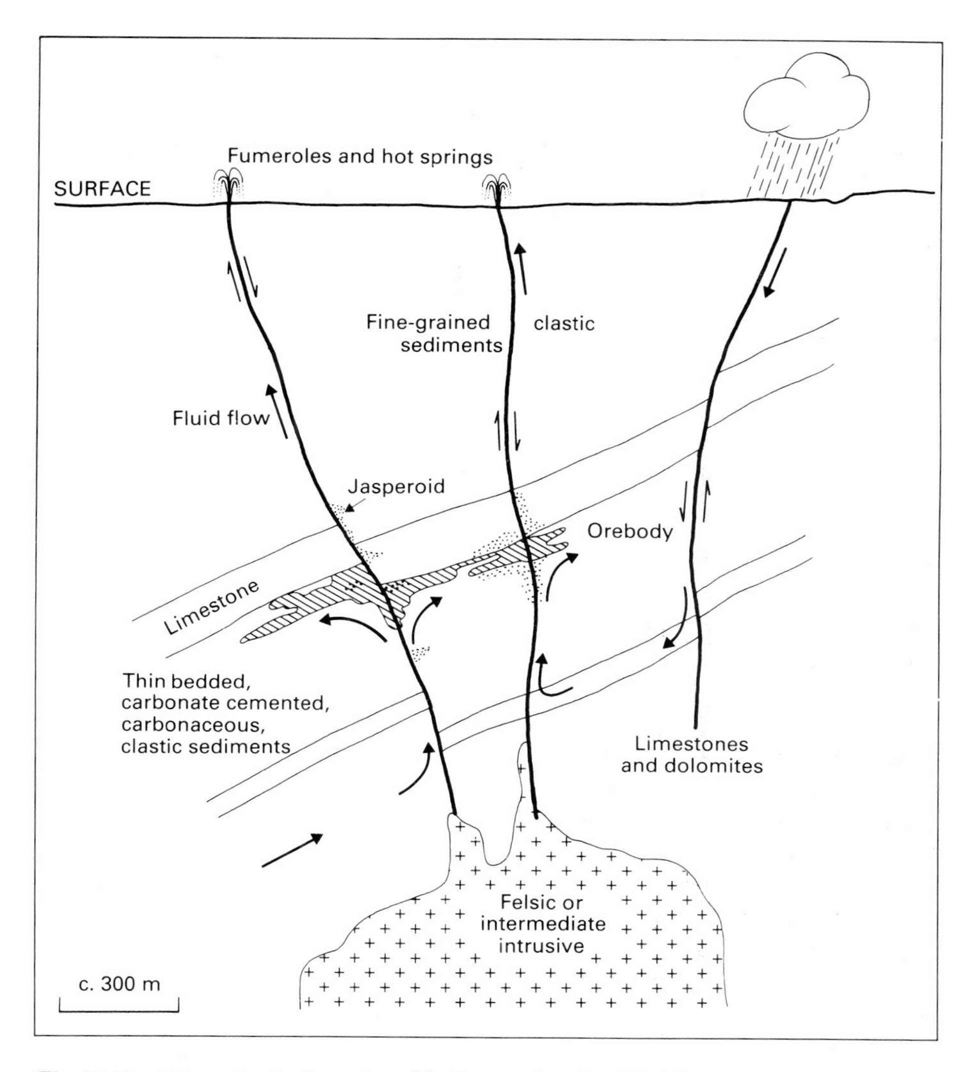

Fig. 17.10. Schema for the formation of Carlin-type deposits. (After Sawkins 1984).

Silberman (1985) has shown that many geothermal systems in the general area of Nevada are only 0.5–3.0 Ma old, which favours the rear-arc rifting regime such as the Taupo Volcanic Zone of North Island, New Zealand or an older analogue like the Caledonian marginal basin of Wales. Ivosevic (1984) also favours this environment. Even more difficult to place in its tectonic environment is the Cinola deposit of the Queen Charlotte Islands, Birtish Columbia (Champigny & Sinclair 1982) which occurs in the suspect terrane of Wrangellia (Windley 1984).

Finally, it seems likely that carbon may have played a role in precipitating the gold, judging by its presence, or former presence, in many deposits;

Springer (1983) has speculated on the potential of Archaean carbonaceous schists as hosts of large low grade deposits of gold.

Note added in press

In a comprehensive summary of sediment-hosted deposits, Bagby & Berger (1986) have put forward a twofold division into jasperoidal and Carlin types. The reader will find this paper to be important and rewarding reading.

18

Strata-bound Deposits

Introduction

The term strata-bound is applied, irrespective of their morphology, to those deposits which are restricted to a fairly limited stratigraphical range within the strata of a particular region. For example, the vein, flat and pipe lead-zinc-fluorite-baryte deposits of the Pennine orefields of Britain are restricted to the Lower Carboniferous and therefore spoken of as being strata-bound. To take a very different example, the stratiform deposits of the Zambian Copperbelt are all developed at about the same stratigraphic horizon in the Roan Series (Chapter 16) and these too may be described as strata-bound. Clearly, stratiform deposits can be strata-bound but strata-bound ores are not necessarily stratiform. Two important associations will be dealt with here: carbonate-hosted base metal deposits and sandstone-uranium-vanadium-base metal deposits.

Carbonate-hosted base metal deposits

These are important producers of lead and zinc and also, sometimes principally, of fluorite and baryte. Copper is important in some fields, notably that of Central Ireland.

DISTRIBUTION IN SPACE AND TIME

Most of the lead and zinc produced in Europe and the United States comes from this type of deposit. In Europe there are important fields in Central Ireland, the Alps, southern Poland and the Pennines of England. In the United States there are the famous Appalachian, Tri-State (south-west Missouri, north-east Oklahoma and south-east Kansas), south-east Missouri and Upper Mississippi districts. There are also important fields in north Africa (Tunisia and Algeria) and Canada.

There are very few deposits of this type in the Precambrian. Substantial deposits first appear in the Ordovician of south-east Missouri and important deposits occur in all systems, except the Silurian, up to the Cretaceous, as can be seen from the following examples:

Ordovician: South-east Missouri (Old and New Lead Belts);
Devonian: Pine Point, Northwest Territories, Canada;

Carboniferous: Central Ireland; British Pennines; Tri-State;
Permian: Trento Valley, Italy;
Triassic: Eastern Alps (Austria and northern Yugoslavia);
Jurassic: Southern Poland (also in Triassic and Devonian);
Cretaceous: Northern Algeria and Tunisia.

ENVIRONMENT

We are not concerned in this group of deposits with skarns and the skarn environment, but with deposits which occur mainly in dolomites and, to a lesser extent, in limestones. The most important feature is the presence of a thick carbonate sequence because thin carbonate layers in shales seldom contain important deposits of this type (Sangster 1976). The fauna and lithologies of the limestone and dolomite hosts show that they were mostly formed in shallow water, near shore environments of warm seas, and a plot of major carbonate-hosted deposits on palaeolatitude maps shows a grouping of these deposits in low latitudes (Dunsmore & Shearman 1977). The warmer climates of low latitudes encourage the development of reefs and so the frequent, but by no means universal, association of these deposits with reefs (e.g. south-east Missouri) and carbonate mudbanks (e.g. Ireland) is not surprising. The occurrence of reefs and carbonate mudbanks is related to ancient shorelines and sea bed topographies. Nowadays, along the shorelines where carbonate deposition is occurring we often find arid zones, as in the Persian Gulf, with desert prograding supratidal flats or coastal sabkhas. There, gypsum and anhydrite are precipitated from marine-derived groundwaters to form evaporites. These may be of considerable significance. The isotopic composition of the sulphur or sulphides from a number of carbonate-hosted deposits suggests origination from sea water sulphate, particularly sea water of the same age as the limestone country rocks. Sulphate evaporites are known to be interbedded with the limestones in relatively close regional proximity to many carbonate-hosted deposits (Dunsmore & Shearman 1977). Thus, as Stanton (1972) has emphasized, the primary regional control of such deposits is palaeogeographical.

Environments such as those described above developed in the past along the margins of marine basins which formed in stable cratonic areas such as the Devonian Elk Point Basin of western Canada in which the Pine Point deposits occur. Carbonate-hosted lead-zinc deposits also occur in a very different environment — in the failed arms (aulacogens) of the triple junctions of rifted continental areas as in the Benue Trough and the Amazon Rift Zone, and on the flanks of embryonic oceans as down the flanks of the Red Sea and the Tethys Ocean (Alps). In this environment too there is an important development of evaporites. A negative but important point is that these lead-zinc deposits are remote from post-host rock igneous intrusives which might be the source for mineralizing solutions.

Returning to the cratonic basin-shelf sea environment we may note other regional controls which are well exemplified by the Mississippi Valley region and the British deposits. In these regions, the orefields are present in positive areas of shallow water sedimentation and separated from each other by shale-rich basins. Such positive areas in the British Isles are often underlain by older granitic masses, and Evans & Maroof (1976) suggested that these very competent rocks fractured easily to produce channelways for uprising solutions which, on reaching the overlying limestones, gave rise to the mineralization. In addition, a large number of deposits are clearly related spatially to faults, sometimes of a regional character (Figs. 2.12, 5.7) up which the ore solutions may have passed.

Sangster (1976) has divided carbonate-hosted base metal deposits into two major types: (i) Mississippi Valley, and (ii) Alpine. Other workers do not make this distinction and refer to all low temperature carbonate-hosted deposits as being of Mississippi Valley-type. Sangster contends that the first type, which is strata-bound, was emplaced after lithification of the host rocks and was largely controlled by pre-ore structures, whilst he considers the Alpine-type to be stratiform and synsedimentary. The author of this book feels that this is a difficult distinction to sustain. For example, Sangster classifies the Irish deposits as being Alpine in type although they show a complete spectrum from thoroughly epigenetic types through to stratiform, perhaps synsedimentary orebodies (Evans 1976a), and thus what seems to be a single orefield consisting of related deposits would have to be split into two different types. However, because this distinction is made by some workers on a global basis, the author has avoided using the term Mississippi Valley-type for all carbonate-hosted base metal deposits.

A number of workers, e.g. Rickard *et al.* (1975) and Sawkins (1984) have suggested that a variant of this deposit class is the sandstone-hosted, lead deposit type (Bjrlykke & Sangster 1981), exemplified by Laisvall, Sweden and Largentière, France. There is thus a great deal of confusion as to what are the common properties of this class of deposit and Sangster (1983b) has suggested that this is because we are trying to classify the unclassifiable, that the differences between individual ore districts outweigh the similarities, both numerically and in significance.

OREBODY TYPES AND SITUATIONS

As has been made clear by the above discussion, the orebodies are very variable in type. In the British Pennines, vein orebodies with ribbon ore shoots occupying normal faults are the main deposit type in the northern field (Fig. 2.4). In the southern Pennines, veins are again the most important orebodies but there they occupy tear (wrench) faults. The orebodies in the Tri-State field are in solution and collapse structures, caves and underground channelways connected with karst topography. In Ireland, the orebodies vary from stockwork brecciation zones to stratiform deposits (Fig. 2.12). At Pine Point, the ores are in interconnected small-scale solution cavities which Dunsmore (1973) has suggested are the result of the dissolution of carbonate rocks by corrosive

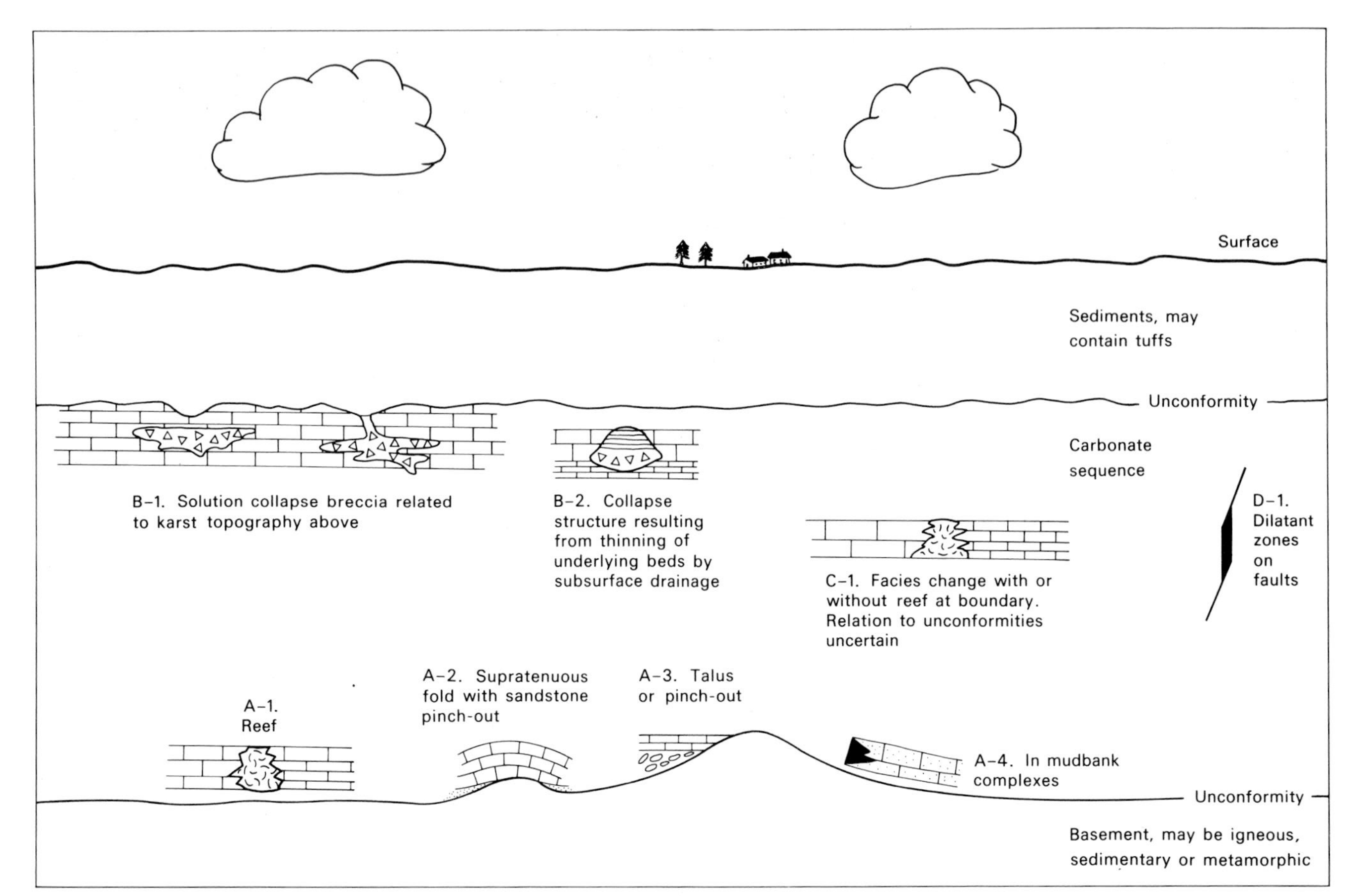

Fig. 18.1. Idealized vertical section illustrating the range of geological situations in which carbonate-hosted base metal deposits are known to occur. (Modified from Callahan 1967).

fluids generated by a reaction between petroleum and sulphate ion. Some of the geological situations in which these deposits occur are shown in Fig. 18.1. They may be listed as follows:
(1) Above unconformities in environments such as reef and facies changes (A-1), supratenuous folds (A-2); above the pinch-outs of permeable channel-way horizons (A-3); above or in mudbank complexes (A-4);
(2) Below unconformities in solution-formed open spaces (caves, etc.) related to a karst topography predating the unconformity (B-1), or in collapse structures formed by the dissolution of underlying beds by subsurface drainage (B-2);
(3) At a facies change in a formation, or between basins of deposition (C-1);
(4) In regional fracture systems (D-1).

GRADE, MINERALOGY AND ISOTOPIC CHARACTERISTICS

Average ore grades range mainly from 3–10% combined Pb + Zn with individual orebodies running up to 50%. Tonnages generally range from a few tens of thousands up to 20 Mt. As an example, we can look at the Navan Mine in Ireland, the largest zinc mine in Europe. This has a number of closely spaced orebodies whose proven, probable and possible reserves total 70 Mt grading 12% Zn + Pb. Zn/Pb = 5/1, the cut-off grade is 4% and the annual production of ore is 2.25 Mt.

The characteristic minerals of this ore association are galena, sphalerite, fluorite and baryte in different ratios to one another varying from field to field. Pyrite and especially marcasite may be common and chalcopyrite is important in a few deposits. Calcite, dolomite, other carbonates and various forms of silica usually constitute the main gangue material. Colloform textures are common in some ores. High trace amounts of nickel seem to be characteristic of these ores (Ixer & Townley 1979).

Sulphur isotopic abundances have been studied in both the sulphide and sulphate minerals. These tend to vary from field to field (Fig. 18.2) but generally show a range of positive $\delta^{34}S$ values. This range may be explained in terms of fractionation as a function of mineral species, temperature or chemical environment or by the mixing or sulphur from different sources (Heyl *et al.* 1974). These authors have suggested that a comparison with values for crustal rocks (Fig. 18.2) indicates a crustal source for the sulphur of these deposits. Thus, the Pine Point values suggest that the sulphur was derived from marine evaporites (Rye & Ohmoto 1974). Silvermines, however, appears to be a notable exception, for sulphur isotopic studies have suggested a meteoritic-type (mantle?) source for all the sulphur (Greig *et al.* 1971), though Coomer & Robinson (1976) have suggested that some of the sulphur was derived from sea water. Work on the Tynagh Pb-Zn-Cu-Ag orebody (also in Ireland) has shown, however, that most of the sulphur was derived from sea water and only a small part from a deep-seated source (Boast *et al.* 1981).

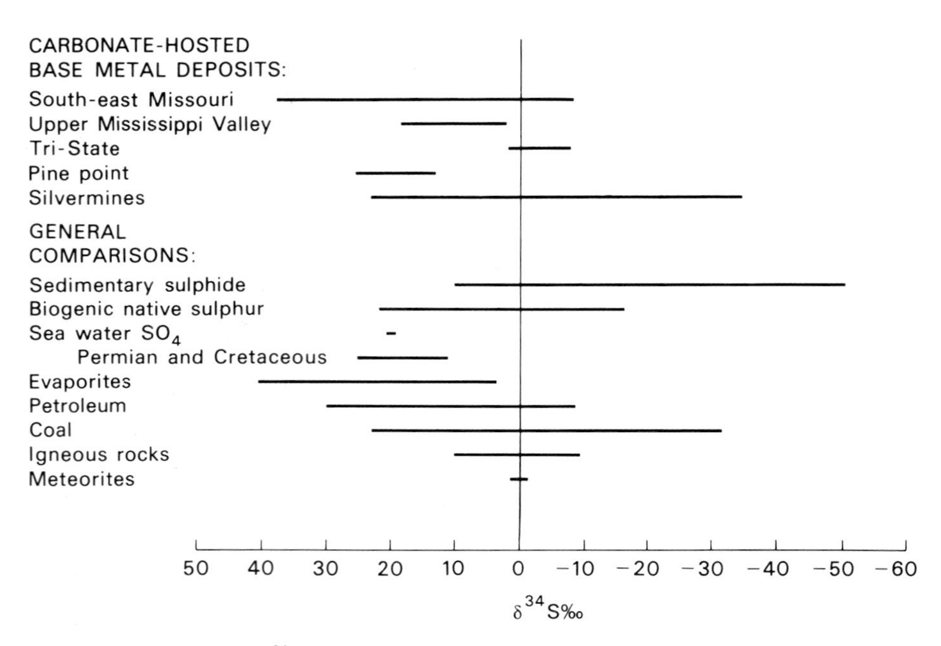

Fig. 18.2. The range of $\delta^{34}S$ for some carbonate-hosted base metal deposits and the range for major sources of sulphur that could have contributed to the ore deposits. (Modified from Heyl *et al.* 1974).

Like sulphur, lead isotope abundances are variable in nature. This variation results from the addition to existing lead of varying amounts of ^{206}Pb, ^{207}Pb and ^{208}Pb produced by the radioactive decay of uranium and thorium. Two distinct categories of lead, ordinary lead and anomalous lead, have been recognized. Ordinary lead has isotopic ratios which increase steadily with time so long as it remains in uniform source rocks (probably the mantle) in contact with constant amounts of uranium and thorium. Once it has been removed from its source and separated from the elements producing radiogenic lead ($^{206-208}Pb$), its isotopic composition is fixed and if a mathematical model is assumed for the rate of addition of radiogenic leads in the source region then a model lead age can be calculated. Generally, such ages are in reasonable agreement with other radiometric age determinations, but sometimes they are grossly incorrect and such leads are defined as anomalous. Some give negative ages, i.e. the model lead age says that they have not yet been formed! These leads must have had a more complicated history, presumably within the crust, during which they acquired extra amounts of radiogenic lead. They are sometimes called J-type leads after Joplin, Missouri from where the first examples to be found were collected. Other leads, on a simple mathematical model, are older than the rocks in which they occur. These could be leads which were first removed from the mantle, then 'stored' in some older rocks before being remobilized and redeposited in younger rocks.

Leads from the various fields of the Mississippi Valley have been found to

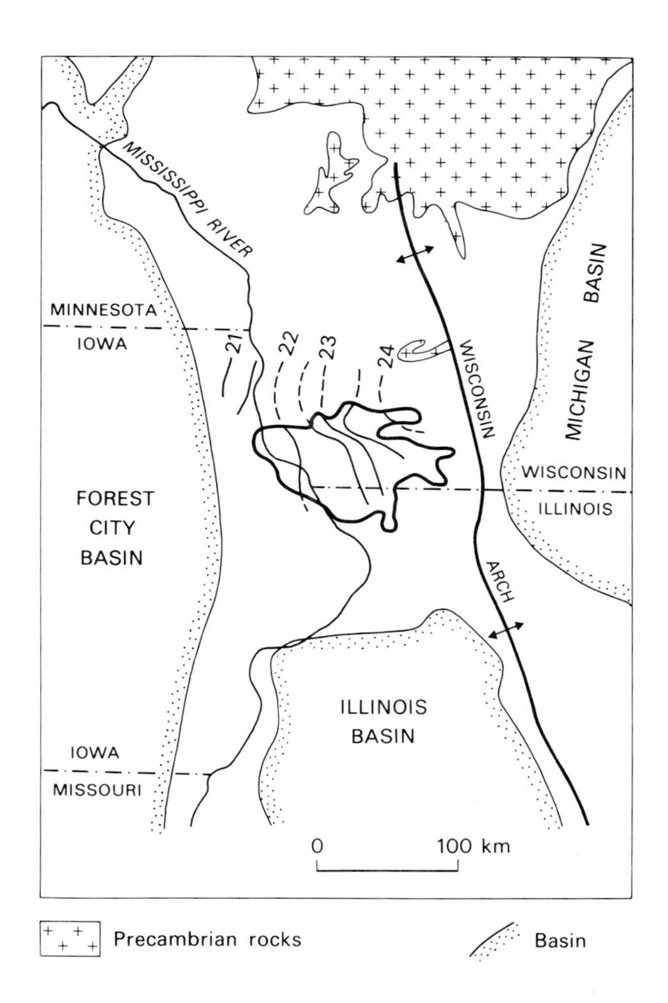

Fig. 18.3. Variation of $^{206}Pb/^{204}Pb$ ratios in the Upper Mississippi Valley Mining District and the location of neighbouring sedimentary basins. (Based on Heyl *et al.* 1966).

be notably enriched in radiogenic lead (i.e. having $^{206}Pb/^{204}Pb$ ratios of twenty or greater) compared to ordinary lead. They are all J-type leads. The strange thing is that although J-type lead is ubiquitous in the Mississippi Valley fields, most similar carbonate-hosted deposits such as those of Pine Point, the British Pennines, central Ireland and southern Poland contain ordinary lead. These facts suggest that the Mississippi Valley lead was derived from a crustal source relatively high in uranium and thorium, which could have provided it with anomalous amounts of radiogenic lead. A highly probable source of this nature would be the Precambrian basement (Heyl *et al.* 1974) because of the conspicuous similarity between the source age of the lead, as determined from the slope of the lead-lead isochron, and the accepted age of the basement rocks (1450 Ma) in the Upper Mississippi Valley district. On the other hand, work in the south-east Missouri district suggests that all the lead may have been derived by hydrothermal solutions passing through the permeable Cambrian Lamotte Sandstone.

The Upper Mississippi Valley district not only contains the most radiogenic lead yet known from a carbonate-hosted deposit anywhere in the world, it also shows an isotopic zoning (Fig. 18.3). The lead becomes progressively more radiogenic from west to east. Heyl *et al* (1966) have pointed out that this could be produced by addition of highly radiogenic lead leached from the basement to more normal lead carried by basinal brines moving up dip into the district from the neighbouring Forest City and Illinois Basins (Fig. 18.3).

The fact that the lead in other carbonate-hosted ore districts (Pine Point, etc.) is ordinary lead has caused most workers to postulate a deep-seated uniform source (e.g. Greig *et al.* 1971 for Ireland), and it has been pointed out that there are a number of profound faults in the nearby Canadian Shield which, when projected south-westwards, pass under the ore zone at Pine Point.

ORIGIN

There is little doubt that the majority of deposits of this class have been formed from epigenetic hydrothermal solutions. A few show syngenetic features and the occasional deposit shows evidence of both epigenetic and syngenetic deposition, e.g. the Mogul Mine at Silvermines, Ireland, (Fig. 2.12) where the lower ore along the fault is epigenetic and yields meteoritic values for the sulphur isotopic ratios but the upper orebody appears to be syngenetic and to mark the time of ore formation. Some of the mechanisms of transportation and deposition of lead and zinc in hydrothermal solutions have been discussed in Chapter 4. The source(s) of these solutions and their metallic constituents is very problematical. For most fields, lead isotopic studies suggest a deep source for the metals but in some cases basinal brines may have played an important role. Jackson & Beales (1967) have argued strongly for such a mechanism and this may have been the case for the Upper Mississippi Valley district. Again, the source of sulphur may have been partially deep-seated (Ireland?) or from marine evaporites (Pine Point?) or from sea water.

Clearly Sangster (1983b) is right in recommending that we look at the differences between these deposits and do not try to force them into a strait-jacket or look for one unique model for their formation. As the reader can discern from Chapter 4, we have today four general models for the genesis of carbonate-hosted lead-zinc deposits:

(1) transport of the metals as bisulphide complexes, one fluid carrying metals and sulphur with precipitation by boiling, cooling by contact with ground water etc.;

(2) transport of metals as chloride complexes (more favoured hypothesis) and precipitation when this solution meets one carrying H_2S — mixing model;

(3) transport of metals as chloride complexes and sulphur as sulphate in the same solution, precipitation when the sulphate is reduced by encountering organic material;

(4) organometallic complexes as carriers of the metals, H_2S in the same solution with precipitation by cooling. Much more research will have to be carried out before we can decide whether one general model is applicable or whether we have a number of processes at work with different combinations being operative during the genesis of different deposits.

Sandstone-uranium-vanadium-base metal deposits

These deposits are found in terrestrial sediments, frequently fluviatile, which were generally laid down under arid conditions. As a result, the host rocks are often red in colour and for this reason copper deposits of this type are commonly referred to as 'red bed coppers'. Uranium-rich examples are called Colorado Plateau-type, carnotite-type, Wyoming roll-front-type, Wyoming geochemical cell uranium-type, western states-type, or sandstone uranium-type. The last term is now in common use.

In deposits of this type one or two metals are present in economic amounts, whilst the others may be present in minor or trace quantities. Thus copper

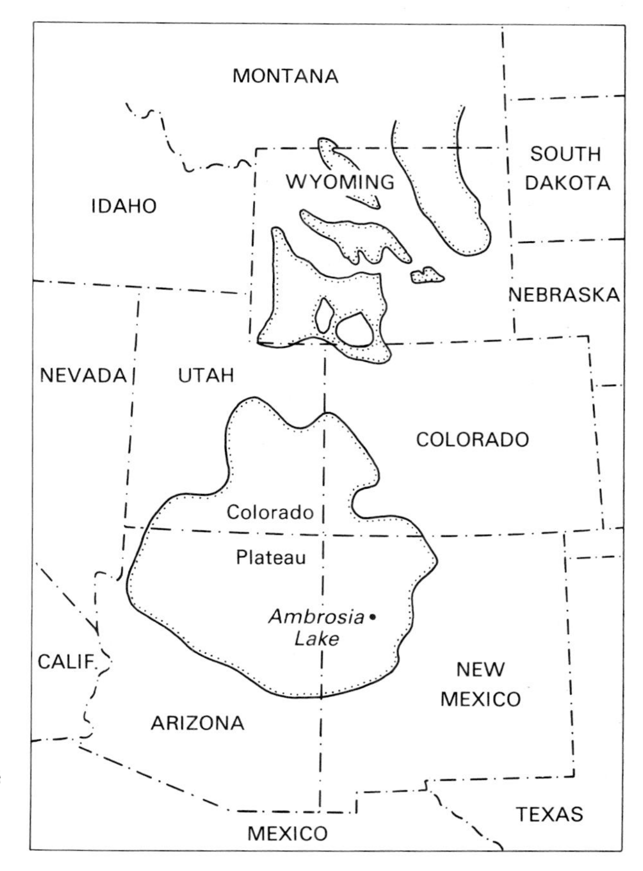

Fig. 18.4. Map showing the Colorado Plateau and Wyoming basins. (After Rackley 1976).

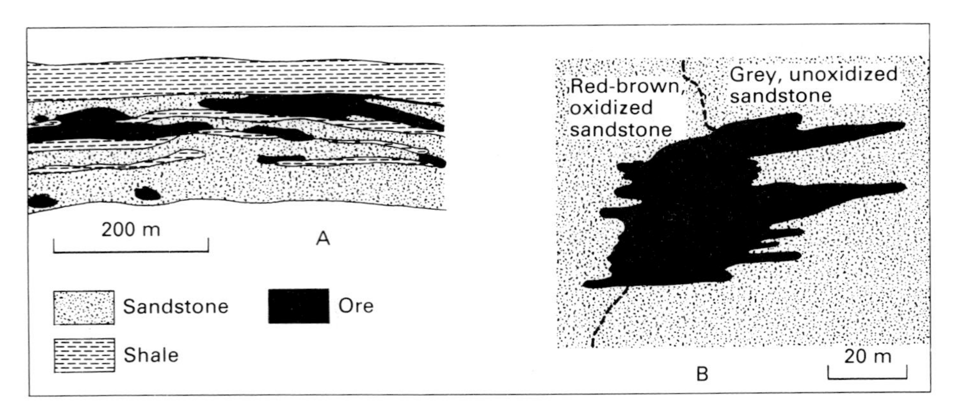

Fig. 18.5. Cross sections through two forms of sandstone uranium-type deposits in the Ambrosia Lake Field, New Mexico. A = blanket or peneconcordant (after Dahlkamp 1978) and B = stack or tectolithologic (after Dixon 1979).

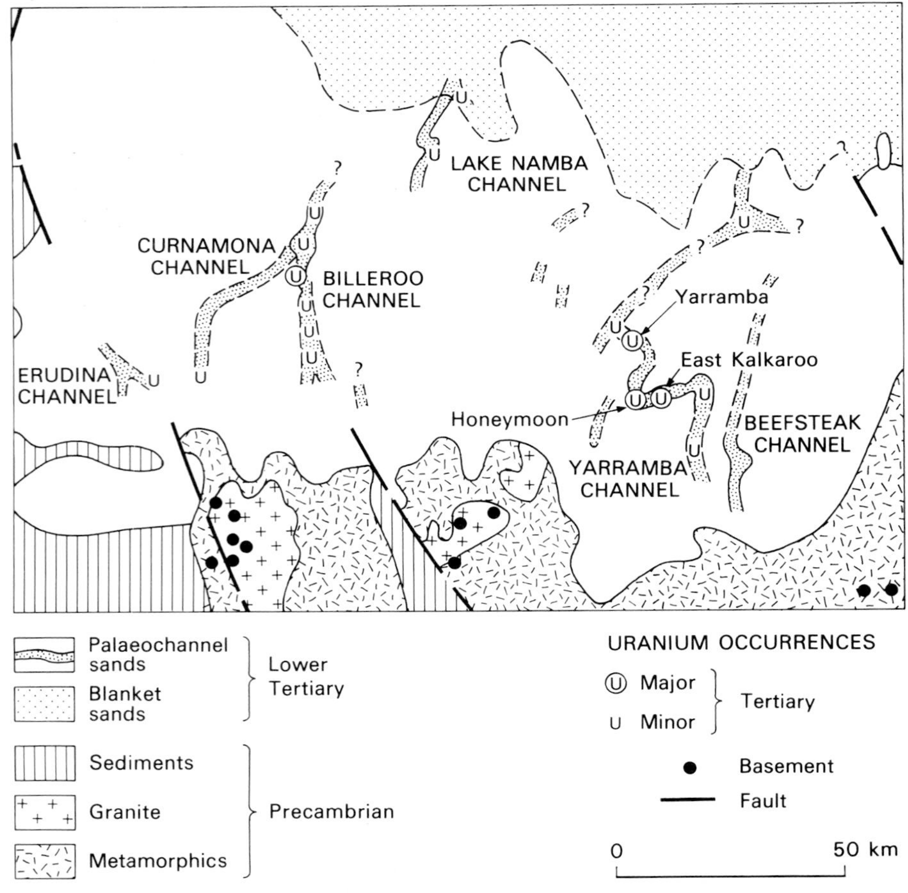

Fig. 18.6. Fossil stream channels with uranium mineralization in the Tertiary of South Australia. Note the many uranium occurrences in the basement which may have been the source of the uranium. (Redrawn from Brunt 1978).

mineralization (with chalcocite, bornite and covellite) is widespread in red bed successions though it is not often up to ore grade (Chapter 2, page 23) and the same applies to silver and lead-zinc mineralization. Uranium mineralization (±vanadium) may be accompanied by trace amounts of the above metals but usually occurs as separate deposits.

SANDSTONE URANIUM-TYPE DEPOSITS

Uranium deposits of this general type are widespread in the USA and they have provided over 95% of its domestic production of uranium and vanadium. From the global point of view they probably constitute a quarter of the non-communist world's reserves, and they are now known in many parts of the world (Dahlkamp 1978). In the USA, these deposits are well developed in the Colorado Plateau region and in Wyoming (Fig. 18.4).

Metals occurring in these deposits in significant quantities are: uranium, vanadium, copper, silver, selenium and molybdenum. A deposit may contain any one or more of these metals in almost any combination except vanadium and copper which are usually mutually exclusive. Amounts of uranium, vanadium and copper vary enormously within and between deposits and many orebodies fluctuate so much in grade that a single overall average figure is not informative. Generally, grades vary from 0.1–1% U_3O_8, but can be locally much higher with such phenomena as whole tree trunks entirely replaced by uranium minerals. The usual range in mineral deposits is 1000 to 10 000 t of

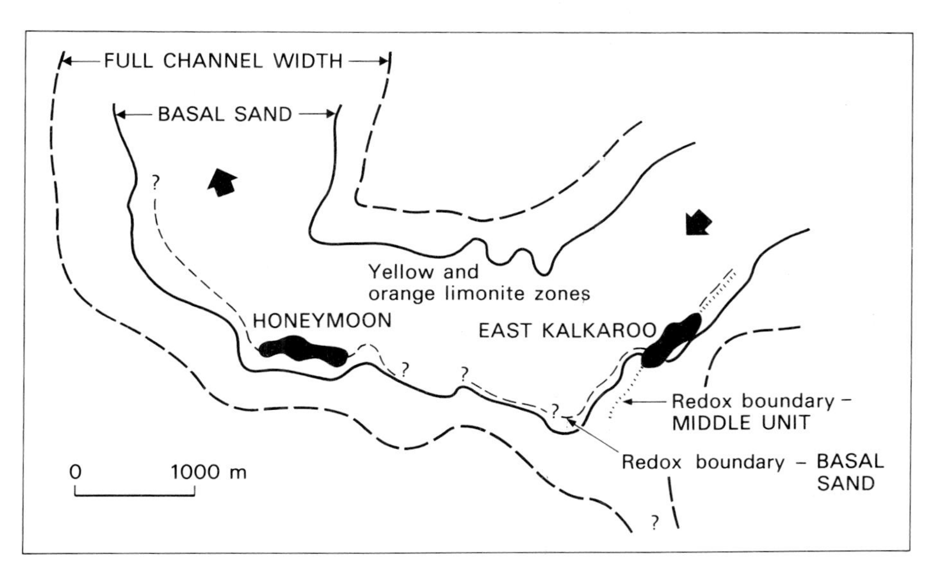

Fig. 18.7. Part of the Yarramba channel showing the position of two major uranium occurrences. Both deposits occur in embayments along the same channel bank. Patchy ore grade mineralization is also present along the redox boundaries away from the deposits. The opposite bank is barren as the yellow limonitic oxidation zone extends up to the pinchout of the sand units. (Redrawn from Brunt 1978).

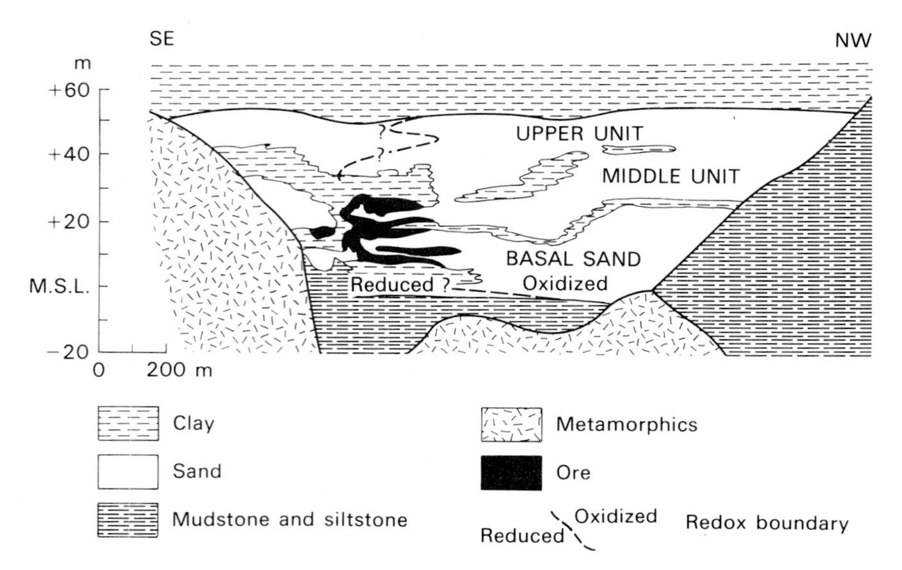

Fig. 18.8. Cross-section across the Yarramba channel showing the East Kalkaroo deposit. Note the roll-front configuration of the ore. (Redrawn from Brunt 1978).

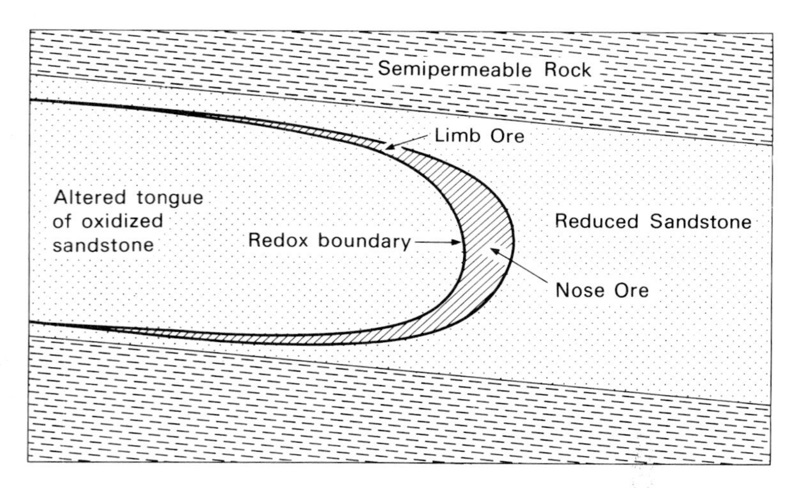

Fig. 18.9. Idealized cross section of a roll-front uranium deposit. (Redrawn from Reynolds & Goldhaber 1978).

contained U in ores grading 0.1–0.2% U. Many deposits contain less than 1000 t U, but some contain more than 30 000 t U. Some American deposits carry up to 1.5% V_2O_5 and some up to 0.2% Mo, but Mo in many cases is deleterious (Eckstrand 1984).

Most of the orebodies are similar. Small irregular pods are common and are sporadically distributed within a favourable rock unit. The larger deposits form mantos hundreds of metres long, about a hundred metres wide and a few metres thick. They can be mined by open pit or underground methods whilst the smaller bodies can be exploited by *in situ* leaching. The elongate orebodies follow buried stream courses or lenses of conglomeratic material. The most common forms of deposit are termed (i) roll-front (Fig. 18.9); (ii) blanket or peneconcordant (Fig. 18.5A) and (iii) stack or tectolithologic (Fig. 18.5B), which are often related to permeable fault zones. These different morphologies can be related to the flow of mineralizing fluids through the host rock. The deposits are epigenetic, in the sense that they were formed in their present position after the host sediment was deposited — how much later is very debatable. The typical orebody represents an addition of less than 1% of ore minerals which are accommodated in pore spaces where they form thin coatings on the detrital grains, whilst, in the case of high grade deposits, they may entirely fill the pore spaces. The disseminated form and microscopic size of the ore minerals increase the susceptibility to subsequent oxidation and remobilization by both alteration and weathering. The principal primary uranium minerals are pitchblende and coffinite $[(USiO_4)_{1-x}(CH)_{4x}]$. Vanadium, if present, is generally in the form of roscoelite (vanadium mica) and montroseite [VO(OH)].

Sedimentological studies have shown that the usual immediate host rocks of these ores are fossil stream channel deposits. These consist of linear formations of permeable sandstone and conglomeratic sandstone enclosed by relatively impermeable rocks — shales, mudstones, etc. (Figs. 18.6–18.8). During deposition, climatic conditions were warm to hot and seasonably humid. Abundant vegetation grew in the depositional area and animals burrowed and mixed dead vegetation with the sediment. Frequent reworking by the streams incorporated sufficient organic material into the sediment to produce reducing conditions when it decayed. The sands deposited under these conditions were organic-rich, pyritic, light to dark grey, and the associated clays were light to medium grey or green, pyritic and commonly carbonaceous. Petrographic studies show that the sediment was often derived from a granitic source area and during weathering of the granite its trace content of uranium would be oxidized to the hexavalent state and taken into solution. This uranium would migrate through the basin of deposition to be lost in the sea unless it came into contact with reducing conditions in the organic-rich sediment a short distance beneath the sediment-water interface, in which case it would enrich the sediment (Rackley 1976). In some areas, acid tuffs rather than granites appear to have been the source rocks for the uranium. As the oxygen-rich waters encroached upon the reducing environment an irregular tongue-shaped zone of oxidized rock was formed. The interface, or redox boundary, between the oxidized and reduced rocks has, in cross-section, the shape of a crudely crescentic envelope or roll, the leading edge of which cuts across the host strata (Fig. 18.9) and points down dip towards reduced ground that still contains authigenic iron disulphides. The reduced ground is generally grey in colour, the oxidized ground is drab yellow to orange or red due to the development of limonite and hematite by the alteration of the sulphides. Upon encountering reducing conditions, the uranium became reduced to the insoluble tetravalent state and was precipitated. Continuous or episodic introduction of oxygenated groundwater resulted in continuous or episodic solution and redeposition of uranium and migration of the redox interface down the palaeoslope. This process can lead to ore grade accumulation at or near the concave edge of the roll and, to a lesser extent, in reduced rock near the upper and lower limbs of the roll. Later reduction or oxidation of the ore beds may materially alter the form and mineralogy of the orebodies and obscure the primary redox relationships.

Useful summary discussions of this deposit type can be found in Tilsley (1981) and Nash *et al.* (1981).

19

Sedimentary Deposits

Broadly speaking, sediments can be divided into two large groups, allochthonous deposits and autochthonous deposits. The allochthonous deposits are those which were transported into the environment in which they are deposited and they include the terrigenous (clastic) and pyroclastic classes. The autochthonous sediments are those which form within the environment in which they are deposited. They include the chemical, organic and residual classes. Table 19.1 shows the relationships between these various terms.

Table 19.1. A classification of sedimentary rocks.

Group	Class
I. Allochthonous sediments	(a) *Terrigenous deposits*—clays, siliclastic sands and conglomerates (b) *Pyroclastic deposits*—tuffs, lapillituffs, agglomerates volcanic breccias
II. Autochthonous sediments	(c) *Chemical precipitates*—carbonates, evaporites, cherts, ironstones, phosphates (d) *Organic deposits*—coal, lignite, oil shales (e) *Residual deposits*—laterites, bauxites

Some sediments are sufficiently rich in elements of economic interest to form orebodies and examples from both sedimentary groups will be described in this chapter. The student must realize, however, that the total range of sedimentary material of economic importance is much greater than can be included in this small volume. The residual deposits will be dealt with in the next chapter together with other ores in whose formation weathering has played an important role.

Allochthonous deposits

Allochthonous sediments of economic interest are usually referred to by ore geologists as mechanical accumulations or placer deposits. They belong to the terrigenous class formed by the ordinary sedimentary processes which concentrate heavy minerals. Usually this natural gravity separation is accomplished by moving water, though concentration in solid and gaseous mediums may also occur. The dense or heavy minerals so concentrated must first be freed from their source rock and must possess a high density, chemical resistance to weathering and mechanical durability. Placer minerals having these properties in varying degrees include: cassiterite, chromite, columbite, copper, garnet, gold, ilmenite, magnetite, monazite, platinum, ruby, rutile, sapphire, xenotime and zircon. Since sulphides readily break up and decompose they are rarely concentrated into placers. There are, however, some notable Precambrian exceptions (perhaps due to a non-oxidizing atmosphere) and a few small recent examples.

Placer deposits have formed throughout geological time, but most are of Tertiary and Recent age. The majority of placer deposits are small and often ephemeral as they form on the earth's surface usually at or above the local base level, so that many are removed by erosion before they can be buried. Most placer deposits are low grade, but can be exploited because they are loose, easily worked materials which require no crushing and for which relatively cheap semi-mobile separating or hydraulic mining plants can be used. Mining usually takes the form of dredging, about the cheapest of all mining methods. Older placers are likely to be lithified, tilted and partially or wholly buried beaneath other lithified rocks. This means that exploitation costs are much higher and then the deposits, to be economic, must contain unusually valuable minerals (e.g. gold) or be of high grade.

The world-wide distribution of placer deposits is very much a product of variation (both at present and in the recent geological past) of geomorphological processes acting at the earth's surface, provided primary sources are present. For example, in semi-arid to arid morphogenetic zones, fluvial processes may be more effective in liberating and transporting heavy minerals than in reworking and concentrating them to economic grades. These processes must therefore be carefully analysed before choosing areas for exploration (Sutherland 1985).

Placers can be classified in various ways but in this book the simple, traditional, genetic classification shown in Table 19.2 will be used. The traditional usage is to be found in Lindgren (1922), McKinstry (1948), Bateman (1950), Routhier (1963), Lamey (1966), Jensen & Bateman (1979) and many other textbooks. The different usage introduced by Macdonald (1983), which is very likely to be an influential book, has been adopted by Edwards & Atkinson (1986); so we are probably entering a period of confusion and it remains to be seen which set of terms becomes the standard usage of the future. Colluvial is not altogether a happy choice of term as for many writers it implies accumulation at the *base* of a cliff or slope, and it is often used as a synonym for talus. Anyone who has consulted the Oxford English Dictionary would eschew its usage forthwith!

RESIDUAL PLACERS

These accumulate immediately above a bedrock source (e.g. gold or cassiterite vein) by the chemical decay and removal of the lighter rock materials and they may grade downwards into weathered veins as in some tin areas of Shaba. In residual placers chemically resistant light minerals (e.g. beryl) may also occur.

Table 19.2. A classification of placer deposits.

Mode of origin	Class (traditional usage)	Usage in Macdonald (1983)
Accumulation *in situ* during weathering	(a) Residual placers	(a) Eluvial
Concentration in a moving solid medium	(b) Eluvial placers	(b) Colluvial
Concentration in a moving liquid medium (water)	(c) Stream or alluvial placers	(c) Fluvial
	(d) Beach placers	(d) Strandline
	(e) Offshore placers	(e) Marine placers
Concentration in a moving gaseous medium (air)	(f) Aeolian placers	(f) Desert or coastal aeolian

Residual placers only form where the ground surface is fairly flat; when a slope is present, creep will occur and eluvial placers will be generated (Fig. 19.1). Residual placers formed over carbonatites are important as producers of apatite, e.g. at Jacupiranga, Brazil; Sokli, Finland (Notholt 1979) and Sukulu, Uganda (Reedman 1984). They are sources and potential sources of niobium, zircon, baddeleyite, magnetite and other minerals. These residual placers have often formed on carbonatites which are themselves subeconomic.

ELUVIAL PLACERS

These are formed upon hill slopes from minerals released from a nearby source rock. The heavies collect above and just downslope of the source and the lighter non-resistant minerals are dissolved or swept downhill by rain wash or are blown away by the wind. This produces a partial concentration by reduction in volume, a process which continues with further downslope creep. Obviously, to yield a workable deposit this incomplete process of concentration requires a rich source. In some areas with eluvial placers, the economic material has accumulated in pockets in the bedrock surface, e.g. cassiterite in potholes and sinkholes in marble in Malaysia.

STREAM OR ALLUVIAL PLACERS

These were once the most important type of placer deposit and primitive mining made great use of such deposits. The ease of extraction made them eagerly sought after in early as well as in recent times and they have been the cause of some of the world's greatest gold and diamond rushes.

Our understanding of the exact mechanisms by which concentrations of heavy minerals are formed in stream channels is still incomplete. Rubey (1933) considered fall velocity to be the most important segregating mechanism and Rittenhouse (1943) used the concept of hydraulic equivalence to explain heavy mineral concentrations. Brady & Jobson (1973); however, showed that fall

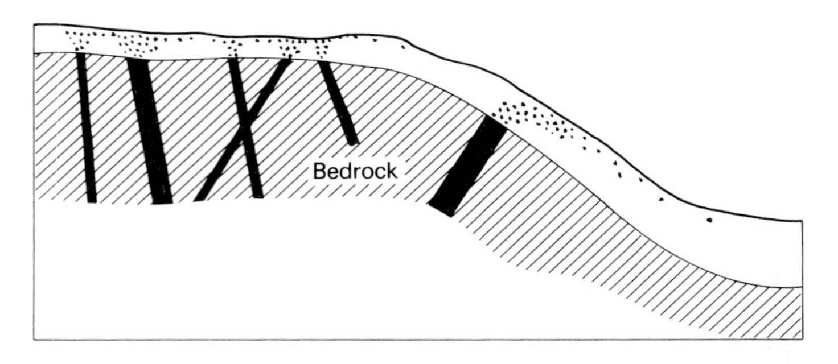

Fig. 19.1. The formation of residual (left) and eluvial (right) placer deposits by the weathering of cassiterite veins.

velocities are of little importance and they found that bed configuration and grain density are the most important factors.

It is well known that the heavy mineral fraction of a sediment is much finer grained than the light fraction (Selley 1976). There are several reasons for this. Firstly, many heavy minerals occur in much smaller grains than do quartz and feldspar in the igneous and metamorphic rocks from which they are derived. Secondly, the sorting and composition of a sediment is controlled by both the density and size of the particles, known as their hydraulic ratio. Thus a large quartz grain requires the same current velocity to move it as a small heavy mineral. Clearly, if we have a very rapid flow all grains of sand grade will be in motion, but with a slackening of velocity, the first materials to be deposited will be large heavies, then smaller heavies, plus large grains of lighter minerals. If the velocity of the transporting current does not drop any further then a heavy mineral concentration will be built up. For this reason such concentrations are developed when we have irregular flow and this may occur in a number of situations — always provided a source rock is present in the catchment area.

The first example is that of emergence from a canyon. In the canyon itself net deposition is zero. As the stream widens and the gradient decreases at the canyon exit, any heavies will tend to be deposited and lighter minerals will be winnowed away. Again, where we have fast-moving water passing over projections in the stream bed, the progress of heavy minerals may be arrested (Fig. 19.2). Waterfalls and potholes form other sites of accumulation (Fig. 19.3) and the confluence of a swift tributary with a slower master stream is often another site of concentration (Fig. 19.4) (Best & Brayshaw 1985). Most important of all, however, is deposition in rapidly flowing meandering streams. The faster water is on the outside curve of meanders and slack water is opposite. The junction of the two, where point bars form, is a favourable site for deposition of heavies. With lateral migration of the meander (Fig. 19.5), a pay streak is built up which becomes covered with alluvium and eventually lies at some distance from the present stream channel.

Obviously, placer deposits do not form in the meanders of old age rivers

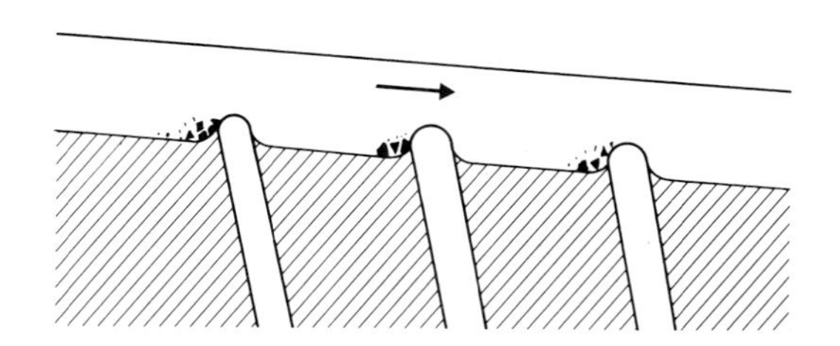

Fig. 19.2. Quartzite ribs interbedded with slate serving as natural riffles for the collection of placer gold.

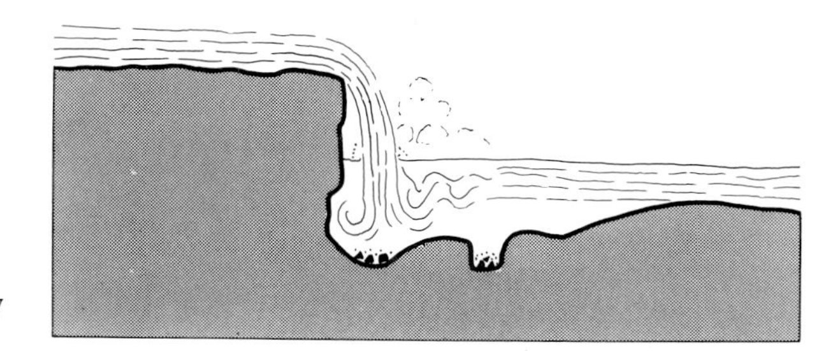

Fig. 19.3. Plunge pools at the foot of waterfalls and potholes can be sites of heavy mineral accumulations.

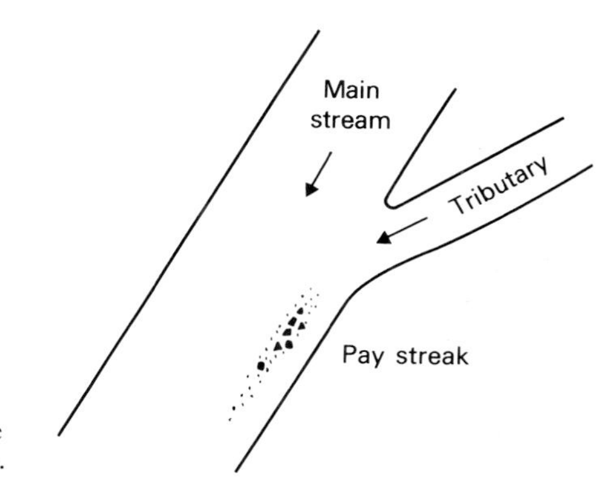

Fig. 19.4. A pay steak may be formed where a fast-flowing tributary enters a master stream.

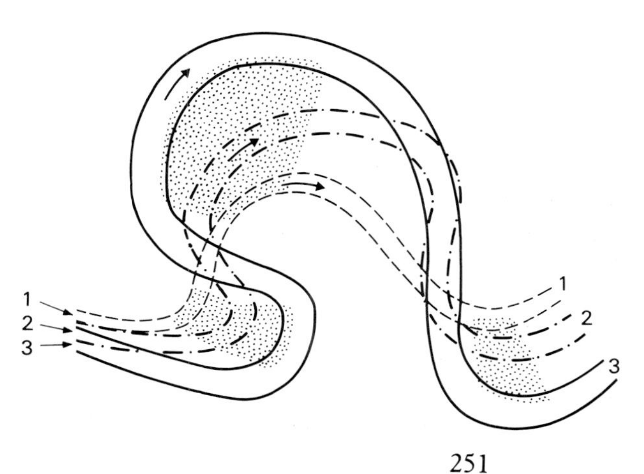

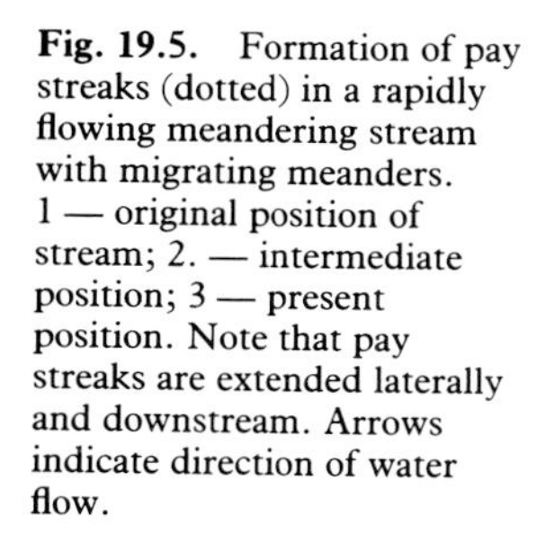

Fig. 19.5. Formation of pay streaks (dotted) in a rapidly flowing meandering stream with migrating meanders. 1 — original position of stream; 2. — intermediate position; 3 — present position. Note that pay streaks are extended laterally and downstream. Arrows indicate direction of water flow.

because current flow is too sluggish to transport heavy minerals. In the upper reaches, current flow may be too rapid and there may also be a lack of source material. The middle reaches are most likely to contain placer deposits where we have well graded streams in which a balance has been achieved between erosion, transportation and deposition. Gradients measured on a number of placer gold and tin deposits average out at a little under 1 in 175.

Much of the above discussion hinges on the concept of hydraulic equivalence in situations where the mineral grains are being transported either in a series of saltation leaps or in suspension. Reid & Frostick (1985) have pointed out, however, that in most fluvial *and littoral marine* situations the transport of sand and larger particles is largely as part of a traction carpet, in which case settling equivalent is unimportant. They consider that the important processes are (i) entrainment equivalence — larger light mineral grains stand proud on the carpet and are subject to a greater lift and drag and are entrained, and (ii) interstice entrapment — the movement of smaller heavy mineral grains into the interstices of coarser sediment, as a result of which gravels will be better traps than sand.

Much of the world's tin is won from alluvial placers in Brazil and Malaysia — but see Chapter 13. A good summary description of the principal Malaysia deposits is given in Dixon (1979).

BEACH PLACERS

The most important minerals of beach placers are: cassiterite, diamond, gold, ilmenite, magnetite, monazite, rutile, xenotime and zircon. Examples include the gold placer of Nome, Alaska, diamonds of Namibia, ilmenite-monazite-rutile sands of Travencore and Quilon, India, rutile-zircon-ilmenite sands of eastern and western Australia, and magnetite sands of North Island, New Zealand. Of course a source or sources of the heavies must be present. These may be coastal rocks, or veins cropping out along the coast or in the sea bed, or rivers or older placer deposits being reworked by the sea. Recent marine placers occur at different topographical levels due to Pleistocene sea level changes (Fig. 19.6). The optimum zone for heavy mineral separation to take place is the tidal zone of an unsheltered beach but concentration may also occur on wave-cut terraces. Some raised beaches formed during high Pleistocene sea levels contain placer ores such as at Nome, Alaska.

Important beach placers stretch for about 900 km along the eastern coast of Australia and these are particularly important for their rutile and zircon production. They occur in Quaternary sediments which form a coastal strip up to 13 km wide and usually 30–40 m thick. Placer deposits occur along the present day beaches and in the Pleistocene sands behind them (Fig. 2.16). These stabilized sands are characterized by low arcuate ridges which probably outline the shape of former bays. As the thickest heavy mineral accumulations are usually adjacent to the southern side of headlands, a reconstruction of the palaeogeography is important during exploration. Thus at Crowdy Head (Fig.

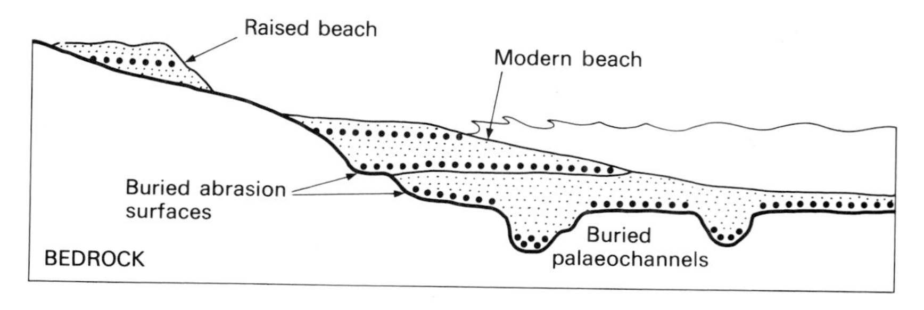

Fig. 19.6. Sketch section to illustrate some sites of beach placer deposits. Placers shown by heavy stipple. (After Selley 1976).

2.16) the deposit between A and B appears to be related to an old headland, now a bedrock outcrop at B. The point A also appears to mark the site of a former headland and further mineralization may be present to the south-west of this point.

Beach placers are formed along shorelines by the concentrating agency of waves and shore currents. Waves throw up materials on to the beach, the backwash carrying out the larger, lighter mineral grains, which are moved away by longshore drift, the larger and heavier particles thus being concentrated on the beaches. Heavy mineral accumulations can be seen on present day beaches to have sharp bases and to form discrete laminae. They are especially well developed during storm wave action. Inverse grading is present in these laminae. At the base there is a fine-grained and/or heavy mineral-rich layer which grades upwards into coarser and/or heavy mineral-poor sands. These laminations develop during the backwash phase of wave action (Winward 1975). The breaking wave moves sand into suspension and carries it beachward and as its velocity drops, its load is deposited. Then the water flow reverses and a surface layer of sand is disturbed, becoming a high particle density bed flow. During such a bed flow the smaller and denser particles sink to the bottom of the flow, producing the reverse grading and also helping to concentrate the heavies. The heavy mineral-poor sand is thus closest to the surface, waiting to be removed by the next wave. A considerable tidal variation is also important in that it exposes a wider strip of beach to wave action, which may lead to the abandonment of heavy mineral accumulations at the high water mark, where they may be covered and preserved from erosion by seaward advancing aeolian deposits.

Thus beaches on which heavy mineral accumulations are forming today include many upon which trade winds impinge obliquely, and ocean currents parallel the coast, these two factors favouring longshore drift. In addition, these beaches face large areas of ocean and so are subjected to fierce storms and large waves. Such situations are found along the eastern and western coasts of Africa and Australia where various important heavy mineral concentrations occur. But just how do such ephemeral deposits become preserved? The answer is still the subject of considerable debate. Along the present day coast of New South Wales, heavy mineral accumulations formed by storm action are rarely preserved. They are reworked and redeposited in diluted form, possibly because these beaches are now in a stable or slightly erosive stage. In the mining operations inland from the foredune of this area the Holocene and Pleistocene deposits are seen to form overlapping layers separated by heavy mineral-poor quartz sand. These layers dip south-eastwards towards the prevailing winds. This suggests that for the preservation of heavy mineral deposits either the shoreline must prograde because of a previously more abundant sediment supply than at present, or the sea level must fall to remove the heavies from the sphere of wave activity.

OFFSHORE PLACERS

These occur on the continental shelf usually within a few kilometres of the coast. They have been formed principally by the submergence of alluvial and/or beach placers (drowned placers). Well-studied examples are the tin placers of Indonesia (Batchelor 1979).

AEOLIAN PLACERS

The most important of these have been formed by the reworking of beach placers by winds, e.g. the large titanomagnetite iron sand deposits of North Island, New Zealand that are estimated to contain more than 1000 Mt of titanomagnetite, 300 Mt of which are present in the Taharoa deposit (Anon. 1983c).

FOSSIL PLACER DEPOSITS

The most outstanding examples are the gold-uranium-bearing conglomerates of the early Proterozoic. The principal deposits occur in the Witwatersrand Goldfield of South Africa, the Blind River area along the north shore of Lake Huron in Canada (only trace gold) and at Serra de Jacobina, Bahia, Brazil. Other occurrences are known in many of the other shield areas. The host rocks are oligomictic conglomerates (vein quartz pebbles) having a matrix rich in pyrite, sericite and quartz. The gold and uranium minerals (principally uraninite) occur in the matrix together with a host of other detrital minerals.

In the Witwatersrand Goldfield (Fig. 19.7) the orebodies appear to have been formed around the periphery of an intermontane, intracratonic lake or shallow water inland sea at and near entry points where sediment was introduced into the basin. Deposition took place along the interface between river systems which brought the sediments and heavies from source areas to the north and west and a lacustrine littoral system that reworked the material (Pretorius 1975, 1981). The individual mineralized areas formed as fluvial fans (Fig. 2.17) which were built up at the entry points. Each fan was the result of sediment deposition at a river mouth that discharged through a canyon and

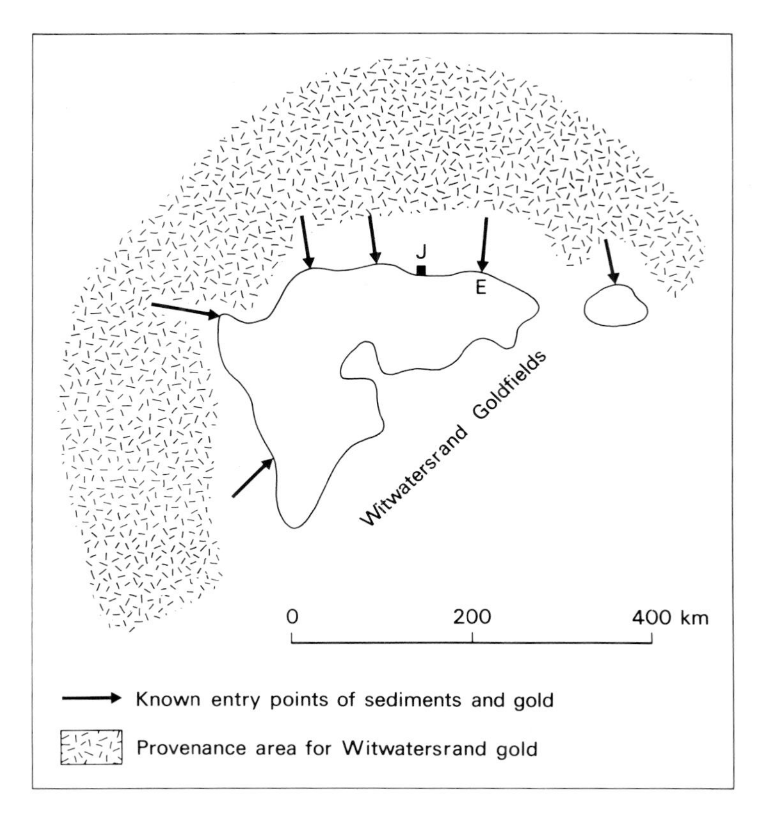

Fig. 19.7. Sketch map showing the extent of the Witwatersrand Goldfields, South Africa, the known entry points of sediment into the basin and the probable extent of the source sea. J indicates the location of Johannesburg and E the East Rand Goldfield. (Fig. 2.17). (After Anhaeusser 1976).

flowed across a relatively narrow piedmont plain before entering the basin. These processes led to the formation of the world's greatest goldfield, which between its discovery in 1886 and 1983, produced over 35 000 t of gold with an average grade of 10 g t^{-1}. Average mined grades are now below this figure.

The Blind River area has also been closely studied and here the uranium deposits (Fig. 2.18) appear to have been laid down in a fluviatile or deltaic environment, perhaps during a wet period preceding an ice age. Unlike the Witwatersrand, the host conglomerates are now at the base of the enclosing arenaceous succession and they appear to occupy valleys eroded in the softer greenstones of the metamorphic basement (Roscoe 1968).

These deposits mark an important metallogenic event in the early Proterozoic, as this type of metal concentration is rare in the Archaean and essentially absent from younger rocks. The presence of large amounts of apparently detrital pyrite and uraninite in these rocks is a problem. At the present day detrital pyrite is uncommon but known, whilst detrital uraninite is very rare but has been reported. Uraninite is not stable in water in equilibrium with atmospheres containing today's oxygen levels, and this has been taken as supporting evidence for an oxygen deficient, CO_2-rich atmosphere during much of the Precambrian. Thus an increase in oxygen content with time would be the control limiting this type of mineralization to the Archaean and early Proterozoic. This model of an anoxic Precambrian atmosphere has been challenged by a number of workers (cf. Windley 1984) and crucial evidence was presented by Simpson & Bowles (1977). They described heavy mineral concentrations in recent sediments of the Indus that contain pyrite and uraninite, the latter being strikingly similar to that of the Blind River conglomerates. Apparently, under conditions of rapid uplift, erosion, transport and burial, uraninite can persist as a detrital mineral. A useful recent discussion of the problem was given by Robinson & Spooner (1984).

Autochthonous deposits

In this section we will be concerned with bedded iron and manganese deposits. The iron deposits can be conveniently divided into the Precambrian Banded Iron Formations (BIF) and the Phanerozoic ironstones.

BANDED IRON FORMATIONS (BIF)

These form one of the earth's great mineral treasures. Besides the term BIF, these rocks are known in different continents under the terms itabirite, jaspillite, hematite-quartzite and specularite. They occur in stratigraphical units hundreds of metres thick and hundreds or even thousands of kilometres in lateral extent. Substantial parts of these iron formations are directly usable as a low grade iron ore (e.g. taconite) and other parts have been the protores for higher grade deposits (Chapter 20). Compared with the present enormous demand for iron ore, now approaching 10^9 t p.a., the reserves of mineable ore in the banded iron formations are very large indeed (James & Sims 1973). An extraordinary fact emerging from recent studies is that the great bulk of iron formations of the world was laid down in the very short time interval of 2600–1800 Ma ago (Goldich 1973). The amount of iron laid down during this period, and still preserved, is enormous — at least 10^{14} t and possibly 10^{15} t. BIF are not restricted to this period, older and younger examples being known, but the total amount of iron in them is far outweighed by that deposited during this short time interval and now represented by the BIF of Labrador, the Lake Superior region of North America, Krivoi Rog and Kursk, USSR and the Hamersley Group of Western Australia.

Banded iron formation is characterized by its fine layering. The layers are generally 0.5–3 cm thick and in turn they are commonly laminated on a scale of millimetres or fractions of a millimetre. The layering consists of silica layers (in the form of chert or better crystallized silica) alternating with layers of iron minerals. The simplest and commonest BIF consists of alternating hematite and chert layers. Note that the content of alumina is less than 1% contrasting with Phanerozoic ironstones which normally carry several per cent of this oxide. James (1954) identified four important facies of BIF.

(a) *Oxide facies*. This is the most important facies and it can be divided into the hematite and magnetite subfacies according to which iron oxide is dominant. There is a complete gradation between the two subfacies. Hematite in least altered BIF takes the form of fine-grained grey or bluish specularite. An oolitic texture is common in some examples, suggesting a shallow water origin, but in others the hematite may have the form of structureless granules. Carbonates (calcite, dolomite and ankerite rather than siderite) may be present. The 'chert' varies from fine-grained cryptocrystalline material to mosaics of intergrown quartz grains. In the much less common magnetite subfacies layers of magnetite alternate with iron silicate or carbonate and cherty layers. Oxide facies BIF typically averages 30–35% Fe and these rocks are mineable provided they are amenable to beneficiation by magnetic or gravity separation of the iron minerals.

(b) *Carbonate facies*. This commonly consists of interbanded chert and siderite in about equal proportions. It may grade through magnetite-siderite-quartz rock into the oxide facies, or, by the addition of pyrite, into the sulphide facies. The siderite lacks oolitic or granular texture and appears to have accumulated as a fine mud below the level of wave action.

(c) *Silicate facies*. Iron silicate minerals are generally associated with magnetite, siderite and chert which form layers alternating with each other. This mineralogy suggests that the silicate facies formed in an environment common to parts of the oxide and carbonate facies. However, of all the facies of BIF, the depositional environment for iron silicates is least understood. This is principally because of the number and complexity of these minerals and the fact that primary iron silicates are difficult to distinguish from low rank metamorphic silicates. Probable primary iron silicates include greenalite, chamosite and glauconite, some minnesotaite and probably stilpnomelane. Most of the iron in these minerals is in the ferrous rather than the ferric state which, like the presence of siderite, suggests a reducing environment. P_{CO_2} may be important, a high value leading to siderite deposition, a lower one to iron silicate formation (Gross 1970). Carbonate and silicate facies BIF typically run 25–30% Fe, which is too low to be of economic interest. They also present beneficiation problems.

(d) *Sulphide facies*. This consists of pyritic carbonaceous argillites — thinly banded rocks with organic matter plus carbon making up 7–8%. The main sulphide is pyrite which can be so fine-grained that its presence may be overlooked in hand specimens unless the rock is polished. The normal pyrite content is around 37%, and the banding is due to the concentration of pyrite into certain layers. This facies clearly formed under anaerobic conditions. Its high sulphur content precludes its exploitation as an iron ore; however, it has been mined for its sulphur content at Chvaletice in Czechoslovakia.

Precambrian BIF can be divided into two principal types (Gross 1970, 1980).

(a) *Algoma type*. This type is characteristic of the Archaean greenstone belts where it finds its most widespread development but it also occurs in younger rocks including the Phanerozoic. It shows a greywacke-volcanic association suggesting a geosynclinal environment and the oxide, carbonate and sulphide facies are present, with iron silicates often appearing in the carbonate facies. Algoma type BIF generally ranges from a few centimetres to a hundred or so metres in thickness and is rarely more than a few kilometres in strike length. Exceptions to this observation occur in Western Australia where Late Archaean deposits of economic importance are found. Oolitic and granular textures are absent or inconspicuous and the typical texture is a streaky lamination. A close relationship in time and space to volcanic rocks hints at a volcanic source of the iron and many regard deposits of this type as being exhalative in origin. Goodwin (1973) in a study of this deposit type in the Canadian Shield showed that facies analysis was a powerful tool in elucidating the palaeogeography and could be used to outline a large number of Archaean basins. His section across the Michipicotin Basin is shown in Fig. 19.8. Large deposits of middle Archaean age occur in the Guyanan and Liberian Shields and prior to the break-up of Gondwanaland these iron formations occupied an area of 250 000 km^2 (James 1983).
(b) *Superior type*. These are thinly banded rocks mostly belonging to the oxide, carbonate and silicate facies. They are usually free of clastic material. The rhythmical banding of iron-rich and iron-poor cherty layers, which normally range in thickness from a centimetre or so up to a metre, is a prominent feature and this distinctive feature allows correlation of BIF over considerable distances. Individual parts of the main Dales Gorge Member of the Hamersley Brockman BIF of Western Australia can be correlated at the 2.5 cm scale over about 50 000 km^2 (Trendall & Blockley 1970), and correlations of varves within chert bands can be made on a microscopic scale over 300 km (Trendall 1968).

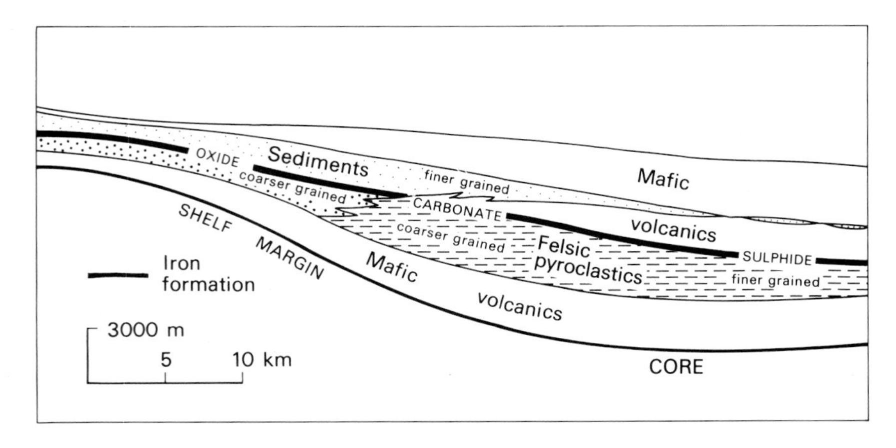

Fig. 19.8. Reconstructed stratigraphic section of the Michipicotin basin showing the relationship of the oxide, carbonate and sulphide facies of banded iron formation to the configuration of the basin and the associated rock-types. (After Goodwin 1973).

Superior BIF is stratigraphically closely associated with quartzite and black carbonaceous shale and usually also with conglomerate, dolomite, massive chert, chert breccia and argillite. Volcanic rocks are not always directly associated with this BIF, but they are nearly always present somewhere in the stratigraphic column. Superior type BIF may extend for hundreds of kilometres along strike and thicken from a few tens of metres to several hundred metres. The successions in which these BIF occur usually lie unconformably on highly metamorphosed basement rocks with the BIF, as a rule, in the lower part of the succession. In some places they are separated from the basement rocks by only a metre or so of quartzite, grit and shale and in some parts of the Gunflint Range, Minnesota, they rest directly on the basement rocks.

The development of Superior BIF reached its acme during the early Proterozoic, and Ronov (1964) has calculated that BIF accounts for 15% of the total thickness of sedimentary rocks of this age. Stratigraphic studies show that BIF frequently extended right around early Proterozoic sedimentary basins and Gross (1965) suggested that BIF was once present around the entire shoreline of the Ungava craton for a distance of more than 3200 km.

The associated rock sequences and sedimentary structures indicate that Superior BIF formed in fairly shallow water on continental shelves (Gross 1980, Goodwin 1982), in evaporitic barred basins (Button 1976), on flat prograding coastlines (Dimroth 1977) or in intracratonic basins (Eriksson & Truswell 1978). Trendall (1973) has suggested that the rhythmic microbands of the Hamersley Group so closely resemble evaporitic varves that a common origin is probable. He suggested that the banding originated by the annual accumulation of iron-rich precipitates whose deposition was triggered by evaporation from a partially enclosed basin with an average water depth of about 200 m.

There is general agreement that BIF are chemically or biochemically precipitated. Blue-green algae and fungi have been identified in the Gunflint Iron Formation of Ontario (Awramik & Barghoorn 1977) and some of these resemble present day iron-precipitating bacteria which can grow and precipitate ferric hydroxide under reducing conditions. However, there is no agreement on the source of the iron. One school considers that this was derived by erosion from nearby landmasses, another that it is of volcanic exhalative origin. One major drawback to the terrestrial derivation is that if large amounts of iron and silica were transported from the continents, large quantities of aluminous material must either have been left behind or transported and dispersed in the sea with, or not far from, the iron deposits. No such residual bauxites or aluminous sediments have been discovered. On the other hand, Miller & O'Nions (1985) give an estimate for submarine hydrothermal supply of iron to the present oceans of $\leqslant 10^{10}$ kg a^{-1}, yet it is claimed that the Hamersley BIF of Western Australia alone required $\geqslant 10^{10}$ kg a^{-1}. They concluded that a major contribution of iron from the continents did occur unless the hydrothermal iron input during the Proterozoic was overwhelmingly greater than the present day one.

The mechanism of transportation of the iron is also hotly debated. Today with an oxygen-rich atmosphere very little iron is transported to the oceans in ionic solution and most travels in colloidal solution or as particulate matter and is deposited mainly in muds. We have no true Phanerozoic analogues of the BIF of the early Proterozoic. For those who advocate a CO_2-rich, oxygen-poor early atmosphere the explanation is simple — iron could then travel as the bicarbonate in ionic solution and, since aluminium does not form a bicarbonate, the two would be separated and another genetic problem solved. With the significant development of oxygen in the atmosphere, large scale formation of BIF would cease. Unfortunately for this neat solution to several problems, there is increasing evidence of the existence of an oxygen atmosphere in the Archaean and early Proterozoic (Windley 1984).

The Lake Superior region

For an example of BIF, we can look briefly at the deposits in the United States to the west and south of Lake Superior (Bayley & James 1973). This is one of the greatest iron ore districts of the world. The western part, which is shown in Fig. 19.9, can be divided into three major units: a basement complex (>2600 Ma old); a thick sequence of weakly to strongly metamorphosed sedimentary and volcanic rocks (the Marquette Range Supergroup) and later Precambrian (Keweenawan) volcanics and sediments.

BIF is mainly developed in the Marquette Range Supergroup, but in the Vermilion district it is present in the basement. In this district there is a great thickness of mafic to intermediate volcanic rocks and sediments. BIF, mainly of oxide facies, occurs at many horizons as generally thin units rarely more than 10 m thick but one iron formation (the Soudan) is much thicker and has been extensively mined.

The remaining iron ore of this region comes from the Menominee Group of the Marquette Range Supergroup. All the iron formations of this group in the different districts are of approximately the same age. The Marquette Range Supergroup shows a complete transition from a stable craton to deep water conditions. Clastic rocks were first laid down on the bevelled basement but most of these were removed by later erosion and in many places the Menominee Group rests directly on the basement. Iron formation is the principal rock-type of this group. Despite the approximate stratigraphic equivalence of the major iron formations, they differ greatly from one district to another in thickness, stratigraphic detail and facies type. They appear to have been deposited either in separate basins or in isolated parts of the same basin or shelf area. The only evidence of contemporaneous volcanism is the occurrence of small lava flows in the Gunflint and Gogebic districts. It has been suggested *apropos* this region that volcanism appears to have been detrimental rather than conducive to iron concentration.

The same iron formation appears in the Mesabi and Gunflint districts. It is

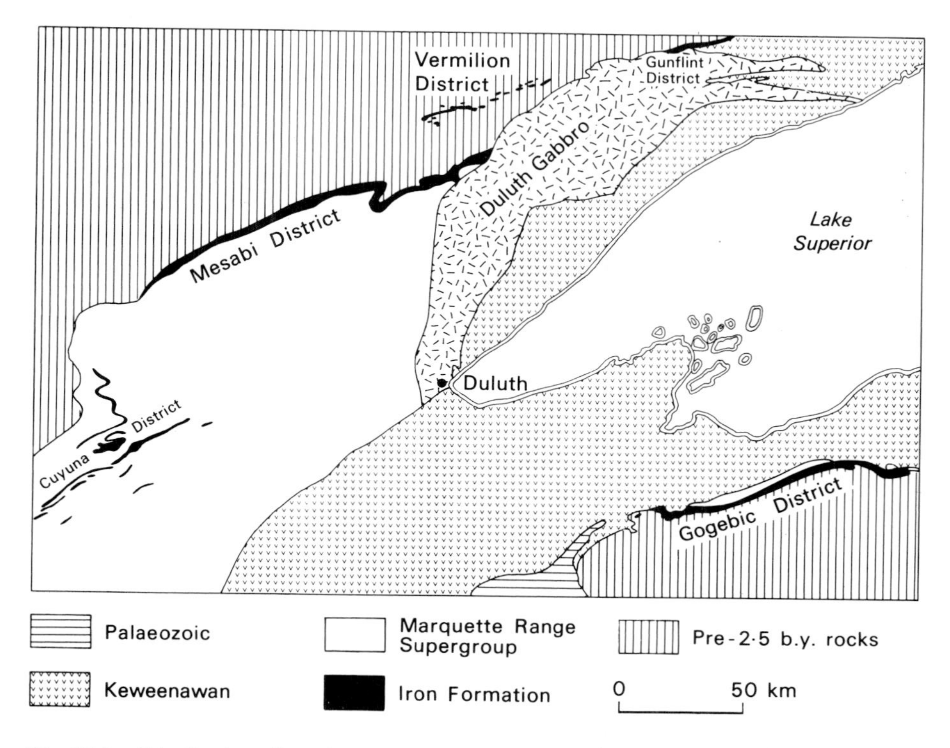

Fig. 19.9. Distribution of iron formation in Minnesota and northern Wisconsin. (After Bayley & James 1973).

100–270 m thick and consists of alternating units of dark, non-granular, laminated rock and cherty, granule-bearing irregularly to thickly bedded rock. The granules are mineralogically complex containing widely different proportions of iron silicates, chert and magnetite; some are rimmed with hematite. The iron formation of the Cuyuna district consists principally of two facies, thin bedded and thick bedded, which differ in mineralogy and texture. The first is evenly layered and laminated, the layers carrying varying proportions of chert, siderite, magnetite, stilpnomelane, minnesotaite and chlorite, while the second contains evenly bedded and wavy bedded rock in which chert and iron minerals alternately dominate in layers 2–30 cm thick. Granules and oolites are present. In the Gogebic district the iron formation is 150–310 m thick and consists of an alternation of wavy to irregularly bedded rocks characterized by granule and oolitic textures. The iron in the irregularly bedded rocks is principally in the form of magnetite and iron silicates, and granule textures are common. The evenly bedded iron formation is mineralogically complex, consisting of chert, siderite, iron silicates and magnetite. Each mineral may dominate a given layer and may be accompanied by one or more of the other minerals.

PHANEROZOIC IRONSTONES

These are usually classified into two types, Clinton and Minette, but both are now of very diminished economic importance as they are of low grade and impossible to beneficiate economically on account of their silicate mineralogy. Mining of ironstones in the UK, once very important, has now ceased, and exploitation of these ores within the EEC will probably come to an end in a few years. They are moving, as it were, from the category of reserve to that of resource even though there are still many megatonnes in the ground.

(a) *Clinton type*. This forms massive beds of oolitic hematite-chamosite-siderite rock. The iron content is about 40–50% and they are higher in Al and P than BIF. They also differ from BIF in the absence of chert bands, the silica being mainly present in iron silicate minerals with small amounts as clastic quartz grains. Clinton ironstones form lenticular beds usually 2–3 m thick and never greater than 13 m. This type of ironstone appears to have formed in shallow water along the margins of continents, on continental shelves or in shallow parts of miogeosynclines. It is common in rocks of Cambrian to Devonian age. One of the best examples is the Ordovician Wabana ore of Newfoundland (Gross 1970).

(b) *Minette type*. These are the most common and widespread ironstones. The principal minerals are siderite and chamosite or another iron chlorite, the chamosite often being oolitic; the iron content is around 30% while lime runs 5–20% and silica is usually above 20%. The high lime content forms one contrast with BIF and often results in these ironstones being self-fluxing ores.

Minette ironstones are particularly widespread and important in the Mesozoic of Europe, examples being the ironstones of the English Midlands, the minette ores of Lorraine and Luxembourg, the Salzgitter ores of Saxony and the iron ores of the Peace River area, Alberta. Unlike the BIF, neither the Minette nor the Clinton ironstones show a separation into oxide, carbonate and silicate facies. Instead the minerals are intimately mixed, often in the same oolite.

SEDIMENTARY MANGANESE DEPOSITS

Sedimentary manganese deposits and their metamorphosed equivalents produce the bulk of the world's output of manganese. Residual deposits are the other main source. The USSR is the world leader in manganese ore production, and in 1984 produced 11 600 000 t, i.e. about 45% of world production. Approximately 75% of this came from the Nikopol Basin in the Ukraine and much of the remainder from the Chiatura Basin in Georgia. The other important producers are the Republic of South Africa, Gabon, India, Australia and Brazil, though the latter has been cutting down production to conserve its resources for its own steel industry.

The geochemistry of iron and manganese is very similar and the two elements might be expected to move and be precipitated together. This is indeed the case in *some* Precambrian deposits; for example, in the Cuyuna District (Fig. 19.9), manganese is abundant in some of the iron formations and forms over 20% of some of the ores. In other areas there seems to have been a complete separation of manganese and iron during weathering, transportation and deposition so that many iron ores are virtually free of manganese and many manganese ores contain no more than a trace of iron. The mechanism of this separation is still unknown. Stanton (1972) and Roy (1976) discuss a number of possibilities. Firstly, there is the possibility of segregation at source due to manganese being leached more readily from source rocks because of its relatively low ionic potential. This is, however, unlikely to produce more than a few per cent difference in the Mn/Fe of the extracted material compared with the ratio in the source rocks. A second possibility arises from the observation that many hot springs produce more manganese than iron, suggesting that iron has been precipitated preferentially from these hydrothermal solutions before they reached the surface. The third possibility is segregation by differential precipitation. Chemical considerations suggest that a limited increase in pH in *some* natural situations may lead to the selective elimination of iron from iron-manganese solutions. A fourth possibility is that separation occurs during diagenesis. The development of reducing conditions will cause both the iron and the manganese to be reduced and to go into solution and move laterally and upward. When the solutions reach an oxidizing environment the two are precipitated. However, since iron will always be the last to be reduced and hence mobilized, and the first to be oxidized and hence immobilized, manganese will tend to be progressively separated in diagenetic solutions.

There are two main classes of ancient sedimentary deposits: nonvolcanogenic and volcanogenic-sedimentary (Roy 1976). Among the nonvolcanogenic, the quartz-glauconite sand-clay association, which includes the chief USSR deposits, and the manganiferous carbonate association are the most important (Varentsov & Rakhmanov 1977).

(a) *The quartz-glauconite sand-clay association*. This formed in a shelf environment under estuarine and shallow marine conditions. On one side it passes into a non-ore-bearing coarse clastic succession, sometimes with coal seams, that lies between the manganese orefields and the source area for the sediments. On the other side it passes into an argillaceous sequence that marks deeper water deposition.

The largest manganese ore basin of this type is the South Ukrainian Oligocene Basin and its deposits include about 70% of the world's reserves of manganese ores. It forms a part of the vast South European Oligocene Basin which also contains the deposits of Chiatura and Mangyshlak in the USSR and Varnentsi in Bulgaria. The distribution of ore deposits is shown in Fig. 19.10.

The manganese ore forms a layer interstratified with sands, silts and clays. It is 0–4.5 m thick, averages 2–3.5 m and extends for over 250 km. There are intermittent breaks due to post-Oligocene erosion. A glauconitic sand is frequently present at the base of the ore layer which consists of irregular concretions, nodules and rounded earthy masses of manganese oxide and/or

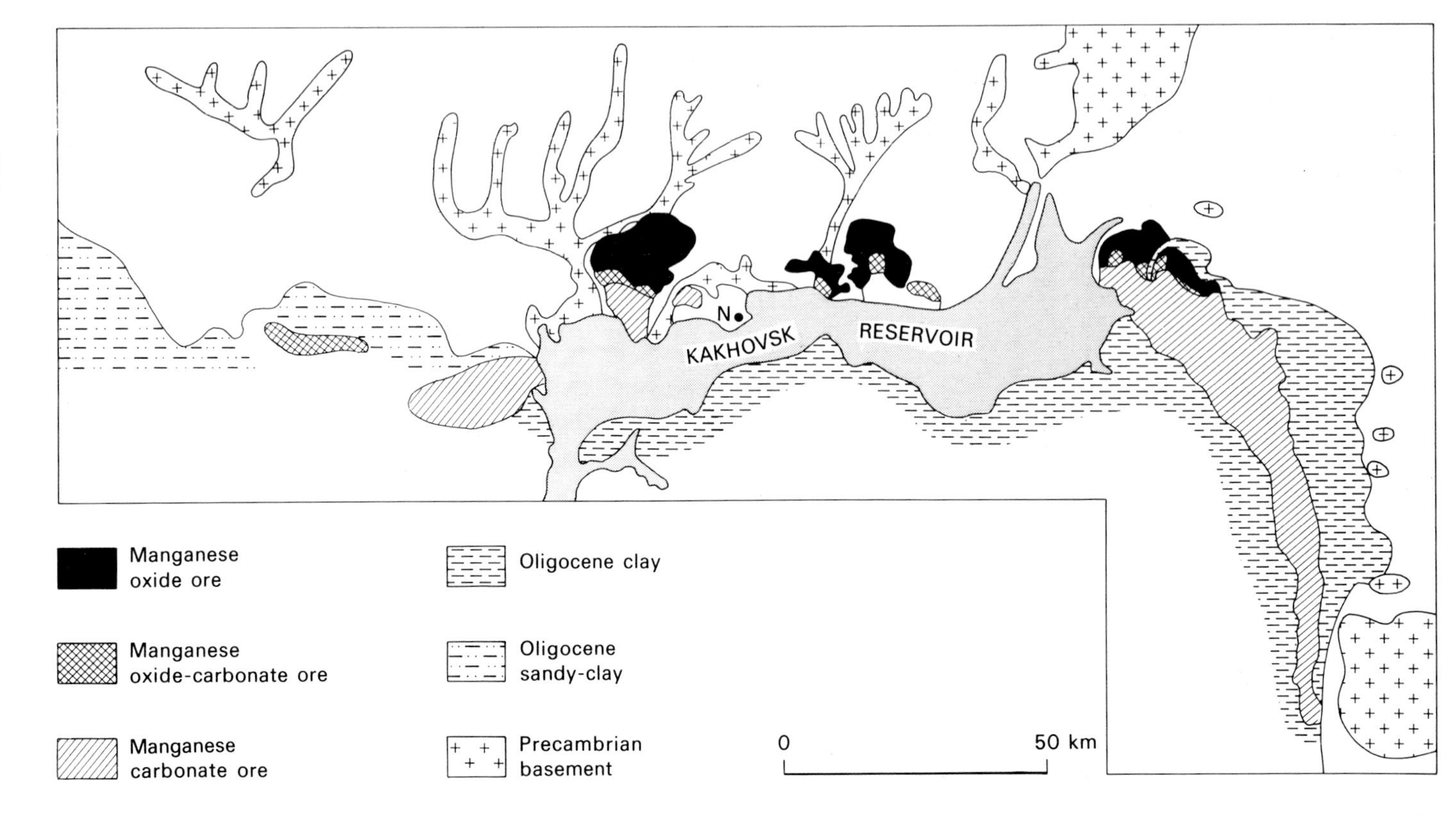

Fig. 19.10. Distribution of manganese ore in the South Ukranian Basin. (Modified from Varentsov & Rakhmanov 1977). The northern and eastern parts of the map area with outcrops of Precambrian basement are largely covered by Quaternary sediments. N = Nikopol.

carbonate in a silty or clayey matrix. A shoreward to deeper water zoning is present (Figs. 19.10 and 11). The dominant minerals of the oxide zone are pyrolusite and psilomelane. The principal carbonate minerals are manganocalcite and rhodochrosite. Progressing into a deeper water environment, the ore layer in the carbonate zone grades into green-blue clays with occasional manganese nodules. The Chiatura deposit of Georgia shows a similar zonal pattern. The average ore grade in the Nikopol deposits is 15–25% Mn and at Chiatura it varies from a few to 35%.

The lack of associated volcanic rocks suggests that these deposits were formed by weathering and erosion from the nearby Precambrian shield which contains a number of rock-types (spilites, etc) that could have supplied an abundance of manganese. Palaeobotanical research has shown that the deposition of the ores coincided with a marked climatic change from humid subtropical to cold temperate. This change could have affected the pattern of weathering and transportation of the manganese. An older but similar deposit is the huge Groote Eylandt Cretaceous deposit, Gulf of Carpentaria, Australia, which was also formed in shallow marine conditions just above a basal unconformity during a marine transgression (Ostwald 1981).

(b) *Manganiferous carbonate association*. Varentsov & Rakhmanov (1977) divide this association into two types:

(1) a geosynclinal manganiferous dolomite-limestone formation;

(2) a manganiferous limestone-dolomite formation developed on rigid cratonic blocks.

These deposits have many features in common and both are basically

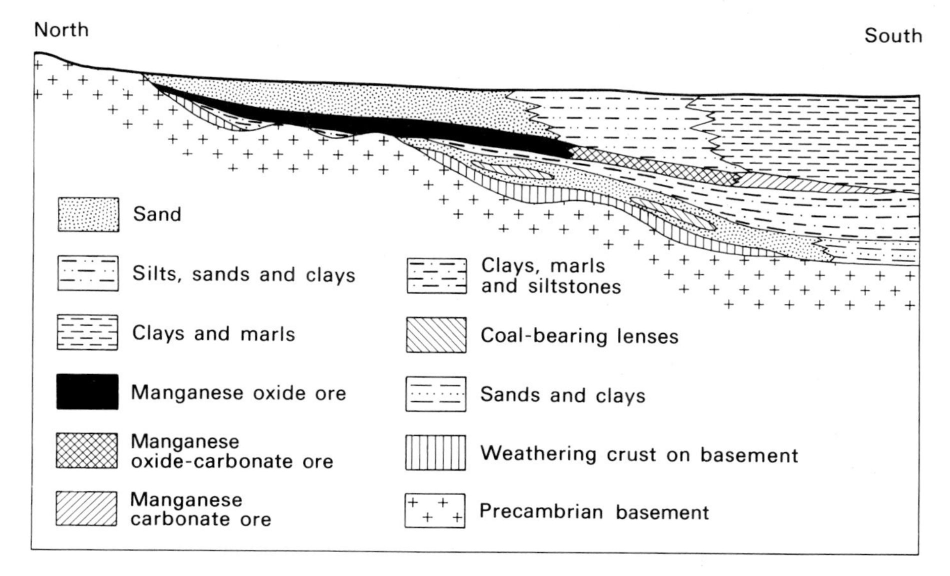

Fig. 19.11. Diagrammatic cross section through the Nikopol manganese deposits showing the zonation of the manganese ores and the transgressive nature of the sedimentary sequence with its overlap on to the Precambrian basement of the Ukranian Platform. (After Varentsov 1964).

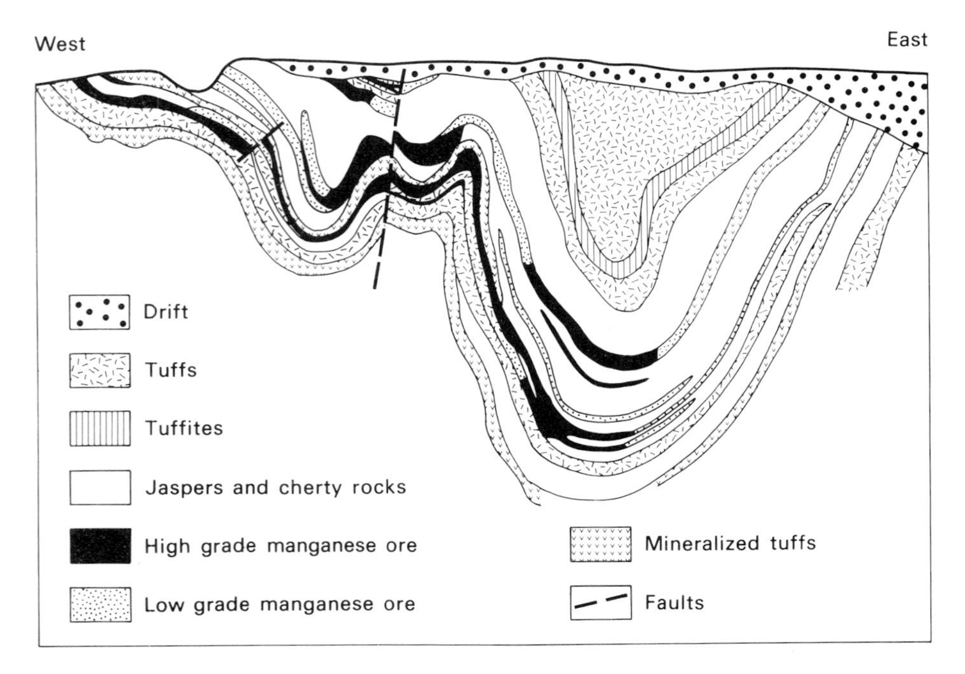

Fig. 19.12. Geological section through the Kusimovo deposits, Southern Urals. (After Varentsov & Rakhmanov 1977).

minor manganese-rich divisions of larger non-ore carbonate formations. Like the quartz-glauconite sand-clay association, deposits of significant size are not common. The principal deposits are in the USSR and north Africa, in particular Morocco. They vary from small lenses to quite extensive, continuous beds of manganese limestone with the lenses often occurring in swarms within a manganiferous section of the general carbonate succession.

The geosynclinal type includes the Usa deposit in the Lower Cambrian of south-western Siberia where 1000 m of limestones and dolomites underlie the ore horizon. This is succeeded conformably by interstratified manganese limestones, weakly manganous limestones and slates. These carry manganocalcite and calcic rhodochrosite ores. The deposit consists of three lenses elongated parallel to the strike, the largest of these being 370 m thick. The others are 215 and 170 m thick. Most of the manganese is present as carbonates with some in manganostilpnomelane, and ore grades vary from 5 to over 30%.

The cratonic type occurs on the eroded surfaces of cratons close to uplifted areas that formed land at the time of sedimentation. Usually the succession contains three distinct units: (1) a lower terrigenous, often red bed, unit; (2) the ore-bearing unit, dominantly a carbonate formation with subordinate red clay and sometimes gypsum; (3) an upper terrigenous unit, again often red. The manganese ore occurs as layers and lenses in the central unit. The Lower Permian Ulutelyak deposit of the USSR occurs on the gently dipping slope of the south-eastern portion of the Russian Platform where it merges with the

Pre-Urals Marginal Foredeep. Here the seams of carbonate manganese ore are up to 1 m thick with a grade of 12–18% Mn.

VOLCANOGENIC — SEDIMENTARY DEPOSITS

These are geologically the most widespread manganese deposits, but economically they are of minor importance compared with the nonvolcanogenic deposits described above. They were divided by Shatsky (see Varentsov & Rakhmanov 1977) into a greenstone association with spilite-keratophyre volcanism and a porphyry association with trachy-rhyolite volcanism. The greenstone association is the important one. It occurs in geosynclinal-type terrane and consists mainly of a silicate facies; sometimes carbonates are dominant but rarely oxides. A good example is the Kusimovo deposit (Fig. 19.12) which occurs in a group of deposits near Magnetogorsk in the Southern Urals. There, an intimate association exists in strongly folded rocks with basic and intermediate volcanics and the orebodies are layers or lenses which rest conformably on the underlying pyroclastics. They range from a few centimetres to five metres in thickness and are bedded, with bands of braunite alternating with cherty jasperoid layers. The manganese mineralogy of these deposits varies with the grade of metamorphism in a complicated manner and the tenor of the ores is commonly 15–25% Mn.

The evidence for a volcanic-exhalative origin of manganese ores of this class is fairly compelling. Not only are they intimately associated with volcanic activity in many different settings including modern island arcs (Stanton 1972), but they frequently carry volcanic fragments including shards and they are enriched in elements such as As, Ba, Cu, Mo, Pb, Sr and Zn that appear to be volcanic contributions (Zantop 1981).

20

Residual Deposits and Supergene Enrichment

In the previous chapter we considered the concentration into orebodies of sedimentary material removed by mechanical or chemical processes and redeposited elsewhere. Sometimes the material left behind has been sufficiently concentrated by weathering processes and ground water action to form residual ore deposits. For the formation of extensive deposits, intense chemical weathering, such as in tropical climates having a high rainfall, is necessary. In such situations, most rocks yield a soil from which all soluble material has been dissolved and these soils are called laterites. As iron and aluminium hydroxides are amongst the most insoluble of natural substances, laterites are mainly composed of these materials and are, therefore, of no value as a source of either metal. Sometimes, however, residual deposits can be high grade deposits of one metal only.

Residual deposits of aluminium

When laterite consists of almost pure aluminium hydroxide, it is called bauxite and is the chief ore of aluminium — the metal used in everything from beer cans to jumbo jets. Bauxite will develop on any rock with a low iron content or one from which the iron is removed during weathering. As with placer deposits, bauxites are vulnerable to erosion and most deposits are therefore post-Mesozoic. Older deposits, however, are known, for example those in the Palaeozoic of the USSR. Some eroded bauxite has been redeposited to form what are called transported or sedimentary bauxites.

After oxygen and silicon, aluminium is the third most common element in the earth's crust, of which it forms 8.1%. Aluminium displays a marked affinity for oxygen and is not found in the native state. In weathered materials it accumulates in clay minerals or in purely aluminous ones such as gibbsite, boehmite and diaspore which are the principal minerals of bauxite. The mineralogy of bauxites depends on their age. Young bauxites are gibbsitic. With age, gibbsite gives way to boehmite and diaspore. There are certain chemical requirements that bauxites must meet to be economic (Table 20.1). Bauxite deposits are usually large deposits worked in open pits. The largest deposit is that of Sangaredi in Guinea, where there is at least 180 Mt forming a plateau up to 30 m thick and averaging 60% alumina, and this country is, after Australia (25 000 Mt), the western world's largest producer — 13 000 Mt

Table 20.1. Bauxite grades by end use. (From Edwards & Atkinson 1986).

		Metallurgical	Chemical	Cement	Refractory (calcined)	Abrasive (calcined)	Proppants
Requirements		High alumina, low iron, low silica, low titania	Must be gibbsitic, very low iron, silica low but not critical	Moderately high alumina, low silica, iron may be high, diaspore preferred	High alumina, low iron, low silica, very low alkalies	High alumina, moderately low iron, low silica, low titania	High alumina, low silica, low clays, iron content not critical
Typical analyses	Al_2O_3	50–55%	>55%	45–55%	>84.5%	80–88%	None available
	SiO_2	0–15%	5–18%	<6%	<7.5%	4–8%	
	Fe_2O_3	5–30%	<2%	20–30%	<2.5%	2–5%	
	TiO_2	0–6%	0–6%	3%	<4%	2–5%	

in 1984. Next in line in that year were Jamaica (8570 Mt), Brazil (5200 Mt) and Surinam and Yugoslavia, both with 3000 Mt.

CLASSIFICATION OF BAUXITES

Bauxite deposits are extremely variable in their nature and geological situations. As a result, many different classifications have been put forward although here we have space to look at only four, those of Harder, Hose, Grubb and Hutchison (Table 20.2). Grubb's simple scheme is based on the topographical levels at which these deposits were formed and it involved assigning some karst bauxites to the high level class and some to the low level — an opinion not supported by Bardossy (1982). Hutchison (1983) combined Grubb's two classes into one, which he terms lateritic crusts, whilst recognizing karst and sedimentary bauxites as separate entities — a scheme adopted by Evans (1980). A useful discussion of the mineralogy and geochemistry of bauxites can be found in Maynard (1983).

(a) *High level or upland bauxites.* These generally occur on volcanic or igneous source rocks forming thick blankets of up to 30 m which cap plateaux in tropical to sub-tropical climates. Examples occur in the Deccan Traps of India, southern Queensland, Ghana and Guinea. These bauxites are porous and friable, show a remarkable retention of parent rock textures, and are dominantly gibbsitic. They rest directly on the parent rock with little or no intervening underclay. Bauxitization is largely controlled by joint patterns in the parent rock, with the result that chimneys and walls of bauxite often extend deep into the footwall.

(b) *Low level peneplain-type bauxites.* These occur at low levels along tropical coastlines such as those of South America, Australia and Malaysia. They are distinguished by the development of pisolitic textures and are often boehmitic in composition. Peneplain deposits are generally less than 9 m thick and are usually separated from their parent rock by a kaolinitic underclay. They are frequently associated with detrital bauxite horizons produced by fluvial or marine activity.

Table 20.2. Classification of bauxite deposits.

Harder & Greig (1960)	Hose (1960)	Grubb (1973)	Hutchison (1983)
Surface blanket deposits	Bauxites formed on peneplains	High level or upland bauxites	Lateritic crusts
Interlayered beds or lenses in stratigraphic sequences	Bauxites formed on volcanic domes or plateaux		
Pocket deposits in limestones, clays or igneous rocks	Bauxites formed on limestones or karstic plateaux	Low-level peneplain-type bauxites	Karst bauxites
Detrital bauxites	Sedimentary reworked bauxites	Sediment bauxites	

(c) *Karst bauxites*. These include the oldest known bauxites — those in the lands just north of the Mediterranean which range from Devonian to mid-Miocene. Other major deposits are the Tertiary ones of Jamaica and Hispaniola. These bauxites overlie a highly irregular karstified limestone or dolomite surface. Texturally, karst bauxites are quite variable. The West Indian examples are gibbsitic ores with a structureless earthy, sparsely concretionary texture. European karst bauxites, on the other hand, are generally lithified and texturally pisolitic, oolitic, fragmental or even bedded. Mineralogically they are predominantly boehmitic ores. From these and other facts Grubb contended that the West Indian bauxites have strong affinities with upland deposits, whilst the European karst bauxites are more reminiscent of peneplain deposits.
(d) *Transported or sedimentary bauxites*. This is a small class of non-residual bauxites formed by the erosion and redeposition of bauxitic materials.

Iron-rich laterites

Most iron-bearing laterites are too low in iron to be of economic interest. Occasionally, however, laterites derived from basic or ultrabasic rock may be sufficiently rich in iron to be workable, though in some cases other metals such as cobalt and nickel may also have been enriched to such an extent as to poison the ore. These deposits, which may be as much as 20 m thick but are usually less than 6 m, consist of nodular red, yellow or brown hematite and goethite which may carry up to 20% alumina. Deposits of this type form mantles on plateaux and are worked in Guyana, Indonesia, Cuba, the Philippines, etc.

A good example is the Conakry deposit in Guinea which is developed on dunite, the change from laterite to dunite being sharp. Most of the laterite consists of a hard crust, usually about 6 m thick. The ore as shipped contains 52% Fe, 12% alumina, 1.8% silica, 0.25% phosphorus, 0.14% S, 1.8% Cr, 0.15% Ni, 0.5% TiO_2 and 11% combined water. This last figure illustrates one of the drawbacks of these ores — their high water content, which may range up to 30%. This has to be transported and then removed during smelting.

Auriferous bauxites and laterites

An exciting development in the exploration of bauxites and laterites has been the discovery of a large tonnage (50–100 Mt) of bauxite at Boddington in Western Australia (Anon. 1985d) that runs 3–3.5 g t^{-1} gold. A project costing 112 million Australian dollars will turn this into a bauxite-gold producer supplying the market with over 5 t (165 000 troy oz) gold p.a., probably making it Australia's largest gold operation and boosting Australian gold production by 15%. Naturally adjacent lateritic ground lying over the same greenstone belt is being actively explored for further deposits.

Secondary enrichment of gold is well known in Australian gold deposits and many 5–20 oz nuggets have recently been found, using metal detectors, in laterites of the Coolgardie district, Western Australia. Secondary gold concentrations in Australia, like the nugget-bearing bauxite of the Cloncurry region of Queensland are often found where little or no basement gold mineralization is known (Wilson 1983). Clearly we have here a new deposit type of world-wide significance, particularly for the shield areas of Gondwanaland, and residual deposits developed over greenstone belts will be the first exploration target in the search for further deposits. It is possible that economic platiniferous laterites may be found in the future (Bowles 1985).

Residual deposits of nickel

The first major nickel production in the world came from nickeliferous laterites in New Caledonia where mining commenced in about 1876. It has been calculated that there are about 64 Mt of economically recoverable nickel in land-based deposits. Of this, about 70% occurs in lateritic deposits, although less than a half of current nickel production comes from these ores. Golightly (1981) has given a good review of this deposit type.

Residual nickel deposits are formed by the intense tropical weathering of rocks rich in trace amounts of nickel such as peridotites and serpentinites, running about 0.25% Ni. During the lateritization of such rocks, nickel passes (temporarily) into solution but is generally quickly reprecipitated either on to iron oxide minerals in the laterite or as garnierite and other nickeliferous phyllosilicates in the weathered rock below the laterite. Cobalt too may be concentrated, but it is usually fixed in wad. Grades of economic deposits range from 1.5–3% Ni.

(a) *Nickel deposits of New Caledonia*. Much of New Caledonia is underlain by ultrabasic rocks, many of which are strongly serpentinized. A typical environment of nickel mineralization is shown in Fig. 20.1 and a more detailed profile in Fig. 20.2. The nickel occurs in both the laterite and the weathered rock zone. In the latter it forms distinct masses, veins, veinlets or pockets rich in garnierite which occur around residual blocks of unweathered ultrabasic rock and in fissures running down into the underlying rock. The material mined is generally a mixture of the lower parts of the laterite and the weathered rock zone. Above the nickel-rich zone there are pockets of wad containing significant quantities of cobalt. Grades of up to 10% Ni were worked in the past, but today the grade is around 3% Ni. It has been estimated that there are 1.5 Gt of material on the island assaying a little over 1% Ni. All laterites take time to develop and it is thought that those on New Caledonia began to form in the Miocene.

(b) *The Greenvale Nickel Laterite, north Queensland*. This deposit was discovered in 1966 as a result of the comparison of the geological environment with that of New Caledonia (Fletcher & Couper 1975). The section above the fresh serpentinite (which runs 0.28% NiO) is similar to that of the New Caledonian occurrences (Fig. 20.3). Nickel and cobalt are concentrated to ore grade in a laterite mantle covering about two thirds of the serpentinite.

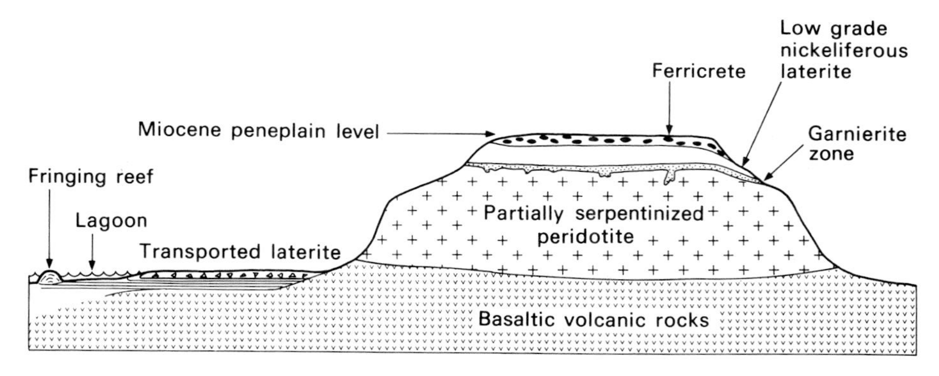

Fig. 20.1. Diagrammatic profile of a peridotite occurrence in New Caledonia showing the development of a residual nickel deposit. (After Dixon 1979).

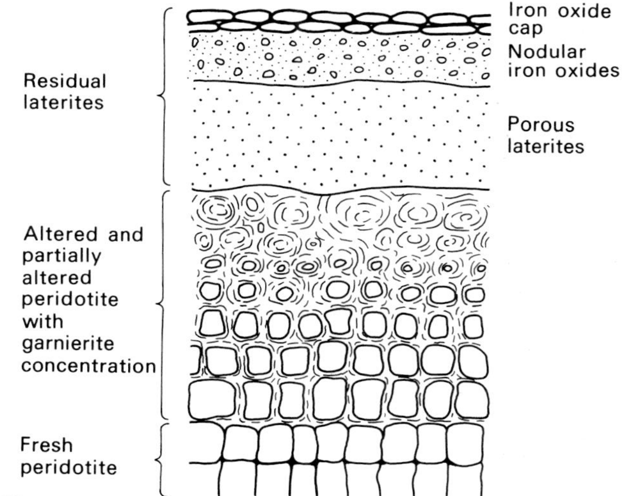

Fig. 20.2. Section through nickeliferous laterite deposits, New Caledonia. (After Chételat 1947).

Erosion has removed the ore zone from the rest of the peridotite. Ore reserves run to 40 Mt averaging 1.57% Ni and 0.12% Co. The ore zone occurs mainly in the weathered serpentinite, often towards the top, and partially in the overlying limonitic laterite.

Supergene enrichment

Although it is more commonly applied to the enrichment of sulphide deposits, the term supergene enrichment has been extended by many workers to include similar processes affecting oxide or carbonate ores and rocks such as those of iron and manganese. In supergene sulphide enrichment the minerals of economic interest are carried down into hypogene (primary) ore where they are

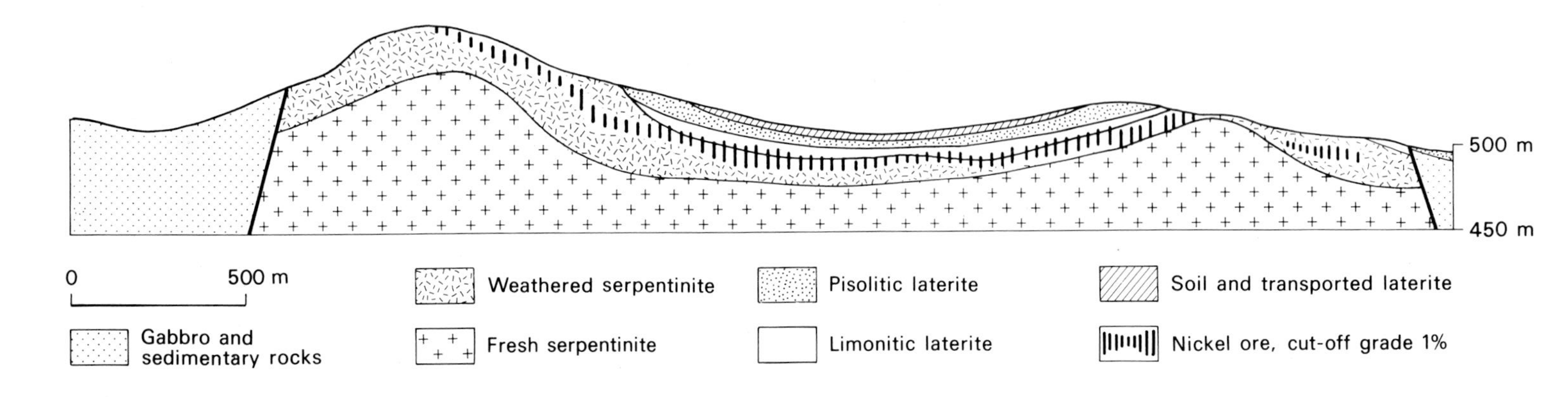

Fig. 20.3. Diagrammatic section of the Greenvale Serpentinite, Queensland, Australia, showing the position of the nickel orebody. (After Fletcher & Couper 1975).

precipitated with a resultant increase in metal content, whereas in the case of iron and manganese ores it is chiefly the gangue material that is mobilized and carried away to leave behind a purer metal deposit.

SUPERGENE SULPHIDE ENRICHMENT

Surface waters percolating down the outcrops of sulphide orebodies oxidize many ore minerals and yield solvents that dissolve other minerals. Pyrite is almost ubiquitous in sulphide deposits and this breaks down to produce insoluble iron hydroxides (limonite) and sulphuric acid:

$$2FeS_2 + 15O + 8H_2O + CO_2 \rightarrow 2Fe(OH)_3 + 4H_2SO_4 + H_2CO_3 \text{ and}$$
$$2CuFeS_2 + 17O + 6H_2O + CO_2 \rightarrow 2Fe(OH)_3 + 2CuSO_4 + H_2CO_3.$$

Copper, zinc and silver sulphides are soluble and thus the upper part of a sulphide orebody may be oxidized and generally leached of many of its valuable elements right down to the water table. This is called the zone of oxidation. The ferric hydroxide is left behind to form a residual deposit at the surface and this is known as a gossan or iron hat — such features are eagerly sought by prospectors. As the water percolates downwards through the zone of oxidation, it may, because it is still carbonated and still has oxidizing properties, precipitate secondary minerals such as malachite and azurite (Fig. 20.4).

Often, however, the bulk of the dissolved metals stays in solution until it reaches the water table below which conditions are usually reducing. This leads to various reactions which precipitate the dissolved metals and result in the replacement of primary by secondary sulphides. At the same time, the grade is increased and in this way spectacularly rich bonanzas can be formed. Typical reactions are as follows:

$$PbS + CuSO_4 \rightarrow CuS + PbSO_4 \text{ (Covellite + anglesite)},$$
$$5FeS_2 + 14CuSO_4 + 12H_2O \rightarrow 7Cu_2S + 5FeSO_4 + 12H_2SO_4 \text{ (Chalcocite)},$$
$$CuFeS_2 + CuSO_4 \rightarrow 2CuS + FeSO_4 \text{ (Covellite)}.$$

This zone of supergene enrichment usually overlies primary mineralization which may or may not be of ore grade. It is thus imperative to ascertain whether newly discovered near surface mineralization has undergone supergene enrichment, for, if this is the case, a drastic reduction in grade may be encountered when the supergene enrichment zone is bottomed. For this purpose a careful polished section study is often necessary.

Clearly, such processes require a considerable time for the evolution of significant secondary mineralization. They also require that the water table be fairly deep and that ground level is slowly lowered by erosion — this usually means that such phenomena are restricted to non-glaciated land areas.

Supergene enrichment has been important in the development of many porphyry copper deposits and a good example occurs in the Inspiration orebody of the Miami district, Arizona (Fig. 20.5). Primary ores of this district

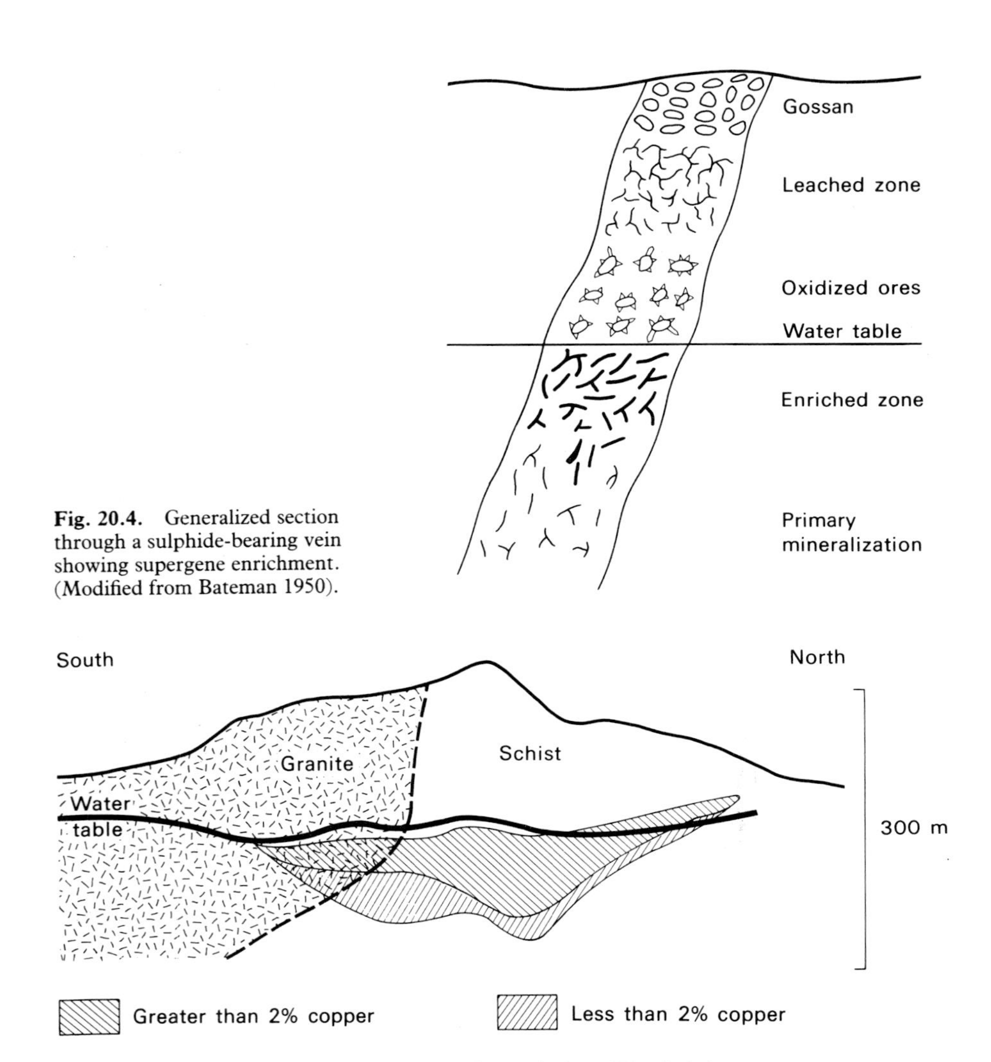

Fig. 20.4. Generalized section through a sulphide-bearing vein showing supergene enrichment. (Modified from Bateman 1950).

Fig. 20.5. Cross section through the Inspiration orebody at Miami, Arizona, showing the relationship between the high and low grade ores and the position of the supergene enriched zone. (Cu over 2%) relative to the water table. (Modified from Ransome 1919).

are developed along a granite-schist contact with most of the ore being developed in the schist. The unenriched ore averages about 1% Cu (Ransome 1919) and consists of pyrite, chalcopyrite and molybdenite. Supergene enrichment increased the grade up to as much as 5% in some places. The schist is more permeable than the granite and more supergene enrichment occurred within it. The enrichment shows a marked correlation with the water table (Fig. 20.5) where it starts abruptly. Downwards, it tapers off in intensity and dies out in primary mineralization. Chalcocite is the main secondary sulphide, having replaced both pyrite and chalcopyrite.

The leached and oxidized zones may not be without economic importance as ores. One of the world's largest open pit gold-silver mines, Pueblo Viejo, Dominican Republic (27 Mt grading 4.23 g t^{-1} Au and 21.6 g t^{-1} Ag plus another body of 14 Mt) is developed in the oxidized zone of sulphide protore (Russell *et al.* 1981). Another and fascinating example is the world's only germanium-gallium mine, the Apex Mine, Utah (Bernstein 1986). At this locality germanium and gallium have been concentrated in the secondary iron oxides — material which is often only examined for the exploration data it may yield. It is very probable that there are more such orebodies awaiting discovery by mineral sleuths who have in mind the needs of the high tech. industries.

SUPERGENE ENRICHMENT OF LOW GRADE IRON FORMATION

Most of the world's iron ore is won from orebodies formed by the natural enrichment of banded iron formation (BIF). Through the removal of silica from the BIF the grade of iron may be increased by a factor of two to three times. Thus, for example, the Brookman Iron Formation of the Hamersley Basin in Western Australia averages 20–35% Fe but in the orebodies of Mount Tom Price it has been upgraded to dark blue hematite ore running 64–66% Fe.

The agent of this leaching is generally considered to be descending ground water, though a minority school in the past has argued the case for leaching by ascending hydrothermal water. In general, these orebodies show such a marked relationship to the present (or a past) land surface that there is little doubt that we are dealing with a process akin to lateritization. This relationship is clearly exemplified by the orebodies of Cerro Bolivar, Venezuela (Ruckmick 1963) which are shown in Fig. 20.6. These orebodies are developed in a tropical area having considerable relief and thus the ground water passing through them can be sampled in springs emerging from their flanks. This

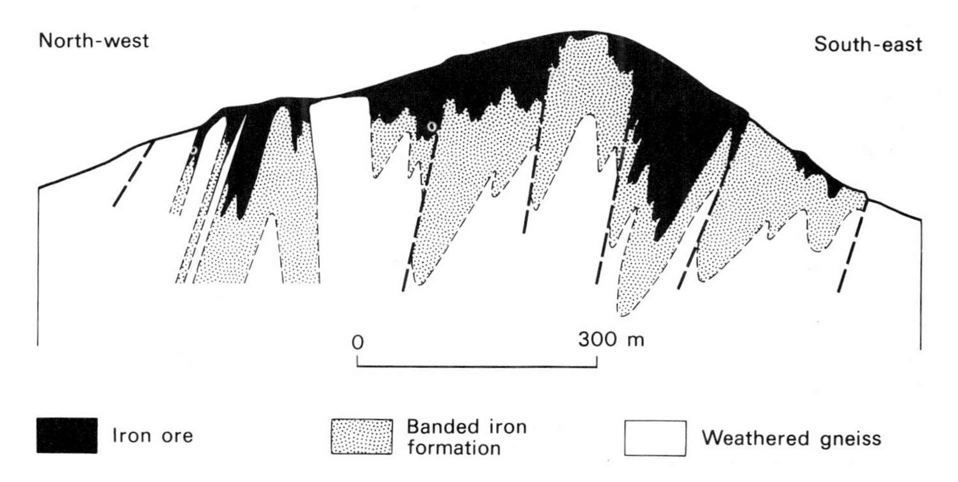

Fig. 20.6. Cross section through the iron orebodies of Cerro Bolivar, Venezuela. (After Ruckmick 1963).

water carries 10.5 ppm silica and 0.05 ppm iron. Its pH averages 6.1. Clearly, the rate of removal of silica is about two hundred times that of iron so that the iron tends to be left behind whilst the silica is removed. The iron is not entirely immobile and this probably accounts for the fact that many orebodies of this type consist of compact high grade material of low porosity. Obviously, the voids created by the removal of silica have been filled by iron minerals, normally hematite. This process seems to be taking place at Cerro Bolivar at the present time and Ruckmick has calculated that if the leaching process occurred in the past at the same rate as it is proceeding today then the present orebodies would have required 24 Ma for this development.

That downward-moving waters were the agent of leaching is attested to by the frequency with which enriched zones of BIF occur in synclinal structures into which downward-moving ground water has been concentrated. The leaching is, of course, intensified if the BIF is underlain by impervious formations such as slate as in the Western Menominee District of Michigan and schist in the Middleback Ranges of South Australia. The BIF of the Middleback Ranges occurs in the Middleback Group. The rocks of this group have suffered folding along northerly axes followed by easterly cross-folding. This produced a number of domes and basins. The domes have largely been removed by erosion leaving isolated cross-folded synclinal areas scattered across an older basement. Downward-moving ground waters have produced orebodies by leaching the BIF in the keels of these synclines. The evidence for

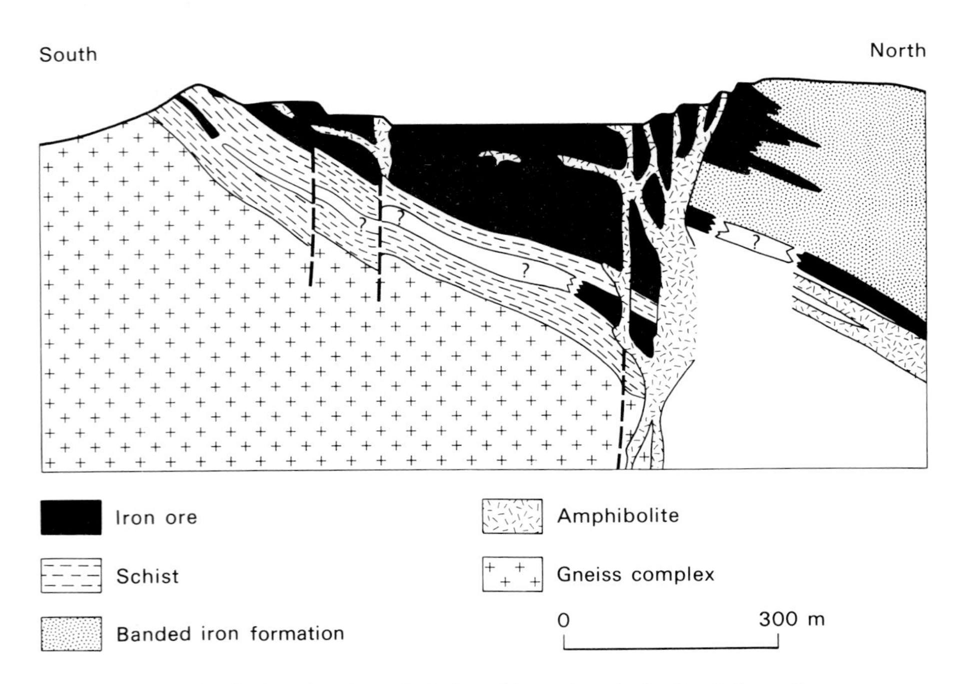

Fig. 20.7. Longitudinal section through the Iron Monarch orebody, South Australia. (Modified from Owen & Whitehead 1965).

this mechanism of formation is particularly good at the Iron Monarch body. This orebody occupies a northward, plunging syncline. The main northern down plunge termination of the orebody is a north-west trending dyke complex which appears to have acted as a dam to the downward-moving water preventing extensive leaching from proceeding further down the keel of the syncline (Fig. 20.7). For this reason, very little enrichment has occurred on the down plunge side.

SUPERGENE ENRICHMENT OF MANGANESE DEPOSITS

Deep weathering processes akin to lateritization can also give rise to the formation of high grade manganese deposits. Although not comparable in size with the previously described sedimentary deposits (Chapter 19), they nevertheless form important accumulations of manganese. Large deposits of this type occur at Postmasburg in South Africa, and in Gabon, India, Brazil and Ghana.

Since manganese is more mobile than iron, the problem arises as to why manganese is retained in the weathering profile. The reason is probably because in most cases the residual manganese deposits are formed on low grade manganiferous limestones and dolomites. These rocks are low in iron and silica, and manganese takes the place of iron in the laterite profile. The carbonates are easily dissolved (by comparison with silica) and the higher pH which probably prevailed in these environments would have tended to immobilize the manganese.

Residual and other manganese deposits are described in detail by Roy (1981). In view of the great economic importance of the vast deposits of the Postmasburg Field it is pertinent to point out that not all workers accept a supergene origin, and De Villiers (1983) has detailed evidence for a hydrothermal origin.

SUPERGENE ENRICHMENT OF URANIUM DEPOSITS

An important example of this enrichment is the Rössing Mine (Chapter 10), where 40% of the uranium is present in secondary minerals.

21

The Metamorphism of Ore Deposits

It is all too often forgotten that the majority of ore deposits occur in metamorphosed host rocks. Among these deposits, the substantial proportion that are syndepositional or diagenetic in age must have been involved in the metamorphic episode(s) and consequently their textures and mineralogy may be considerably modified, often to the advantage of mineral separation processes. In addition, parts of the orebody may be mobilized and epigenetic features, such as cross-cutting veins, may be imposed on a syngenetic deposit. Ores in metamorphic rocks may be metamorph*ic* in the sense that their economic minerals have been largely concentrated by metamorphic processes; these include the pyrometasomatic deposits and some deposits of lateral segregation origin. On the other hand they may be metamorph*osed*, in the sense that they predate the metamorphism and have therefore been affected by a considerable change in pressure-temperature conditions which may have modified their texture, mineralogy, grade, shape and size — structural deformation often accompanies metamorphism. This chapter will be concerned entirely with metamorphosed ores.

It is important from the economic as well as the academic viewpoint to be able to recognize when an ore has been metamorphosed. Strongly metamorphosed ores may develop many similar features to high temperature epigenetic ores for which they may be mistaken. This can be very important to the exploration geologist, for in the case of an epigenetic ore, particularly one localized by an obvious structural control, the search for further orebodies may well be concentrated on a search for similar structures anywhere in the stratigraphic column, whereas repetitions of a metamorphosed syngenetic ore should be first sought for along the same or similar stratigraphic horizons. Notable examples of orebodies long thought to be high temperature epigenetic deposits, but now considered to be metamorphosed syndepositional ores, are those of Broken Hill, New South Wales and the Horne Orebody, Quebec. In both areas recent exploration having a stratigraphic basis has been successful in locating further mineralization.

Certain types of deposit, by virtue of their structural level or development in geological environment, are rarely, if ever, seen in a metamorphosed state. Some examples are porphyry copper, molybdenum and tin deposits, ores of the carbonatite association and placer deposits. On the other hand, certain deposits such as the volcanic massive sulphide class have generally suffered some degree or other of metamorphism such that their original textures are commonly much modified.

Three types of metamorphism are normally recognized on the basis of field occurrence. These are usually referred to as contact, dynamic and regional metamorphism. Contact metamorphic rocks crop out at or near the contacts of igneous intrusions and in some cases the degree of metamorphic change can be seen to increase as the contact is approached. This suggests that the main agent of metamorphism in these rocks is the heat supplied by the intrusion. As a result, this type is sometimes referred to as thermal metamorphism. It may take the form of an entirely static heating of the host rocks without the development of any secondary structures such as schistosity or foliation, though these are developed in some metamorphic aureoles.

Dynamically metamorphosed rocks are typically developed in narrow zones such as major faults and thrusts where particularly strong deformation has occurred. Epigenetic ores, developed in dilatant zones along faults, often show signs of dynamic effects (brecciation, plastic flowage, etc.) due to fault movements during and after mineralization.

Regionally metamorphosed rocks occur over large tracts of the earth's surface. They are not necessarily associated with either igneous intrusions or thrust belts, but these features may be present. Research has shown that regionally metamorphosed rocks generally suffered metamorphism at about the time they were intensely deformed. Consequently they contain characteristic structures such as cleavage, schistosity, foliation or lineation which can be seen on both the macroscopic and microscopic scales, producing a distinctive fabric in rocks so affected.

How can we recognize when ores have been metamorphosed? One way is to compare their general behaviour with that of carbonate and silicate rocks which have undergone metamorphism. In contact and regionally metamorphosed areas the rocks generally show:
(1) the development of metamorphic textures;
(2) a change of grain size — usually an increase;
(3) the progressive development of new minerals. In addition, as noted above, regionally metamorphosed rocks usually develop certain secondary structures. Let us examine each of these effects in turn.

Development of metamorphic textures

DEFORMATION

All three types of deformation, elastic, plastic and brittle, are important — elastic deformation largely because it raises the internal free energy of the grains and renders them more susceptible to recrystallization and grain growth. Plastic deformation occurs by primary (translation) gliding or by secondary (twin) gliding. In translation gliding, movement occurs along glide planes inside a grain without any rotation of the crystal lattice. Translation gliding can be readily induced in the laboratory in galena which has glide planes parallel to {100}. In twin gliding, rotation of part of the crystal lattice occurs so that it takes up a twinned position. Rotation is initially on the molecular scale, but with continued applied stress it spreads across the whole grain by rotation of one layer after another. The normal result is the development of polysynthetic twins. One example familiar to geologists is the secondary twinning of carbonates in deformed marbles. Most of the soft opaque minerals contain many potential glide planes, there being eight in galena, and as a result, the stresses which may affect these minerals during dynamic and regional metamorphism can lead to plastic flowage. Elevated temperatures facilitate such flowage, but flow can be induced at lower temperatures by increasing the applied pressure.

Brittle deformation takes the form of rupturing or shearing and both generally follow lines of weakness in grains such as cleavages, twin planes, etc. Ruptures are often healed by new growth of the same or another mineral in the space created, but on the other hand, softer minerals may flow into this space thus isolating the fragments of the ruptured mineral. This is particularly the case in polyphase grain aggregates made up of strong and weak minerals. At a given degree of deformation the stronger minerals such as pyrite and arsenopyrite may fail by rupturing whilst the softer minerals such as galena and the sulphosalts may flow. These processes can give rise to grain elongation and the development of schistose textures.

RECRYSTALLIZATION

The deformed state is one of high potential energy and, if the temperature is high enough, annealing will occur leading to a reduction of this high energy level. Of the various processes that take place during annealing, recrystallization is the most important. It consists of the replacement of strained grains by strain-free grains followed by grain growth, the new grains meeting in growth impingement boundaries which represent an unstable configuration. The presence of these grain boundaries leads to an increase in the free energy of the system over that which it would possess if no grain boundaries were present. This extra free energy is that of unsatisfied bonds at or near the surface of the grains; it is called the interfacial free energy.

The shapes of grains in a polycrystalline aggregate are governed by two main factors; firstly, the need to reduce the overall free energy level to a minimum, and secondly, the requirement to fill space. The first requirement would be fulfilled by spherical grains but these would clearly not fill space. The microscopic examination of polished sections of artificially annealed metals and minerals sulphides shows that most grains meet three at a time at a point. Separation of these grains shows that they are bounded by a number of flat surfaces identical with those in a soap froth (Smith 1964). Inspection of such froths (which can be done by shaking up some soap solution in a plain tumbler) shows that in three dimensions the great majority of bubbles (and, by inference, grains too) meet in threes along lines. These lines are called triple junctions. When intersected by a surface (e.g. that of a polished section) these junctions appear as points (Fig. 21.1A). In a monominerallic aggregate the angles between the grain boundaries around such points are equal to or close to 120°, provided the section is normal to the triple junction. Such equilibrium grain configurations will normally be present in annealed (i.e. metamorphosed) ores. They are also present in some autoannealed ores such as chromite deposits in large lopoliths and at least one case of their occurrence in an epigenetic ore has been recorded (Burn 1971). In investigating such grain configurations in ores, it is usual to measure a number of these angles and to employ a frequency plot to determine whether they peak at 120° (Stanton 1972). Calculation of the standard deviation will produce a measure of the degree of perfection of the annealing. Very beautifully annealed mineral aggregates can be seen in the gold-copper ores of Mount Morgan (Lawrence 1972).

So far we have been concerned with monominerallic aggregates. In the case where a phase is in contact with two grains of a different phase the dihedral angel (θ) is no longer 120° (Fig. 21.1B) but some other constant angle which varies according to the composition of the phases. The values of a number of these dihedral angles are known, mainly from the work of Stanton, and again measurements can be made to see if the grain aggregate has been annealed.

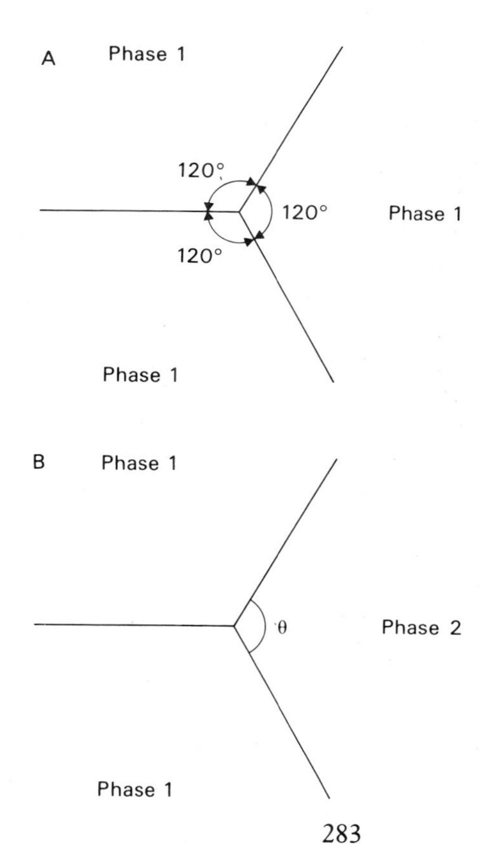

Fig. 21.1. A — Grain boundary configuration about a triple junction point in a monominerallic aggregate. B — The same configuration when a second phase is present. θ is the dihedral angle.

Increase in grain size

In an annealed aggregate, grain boundary energy is the dominant force leading to structural modification. An influx of heat will raise the energy level and if the temperature is high enough to permit diffusion, grain boundary adjustments will occur. Since the total grain boundary free energy is proportional to the grain boundary area, grain growth will occur in order to reduce the number of grains per unit volume. Grain growth does not go on indefinitely and it has been shown empirically by metallurgists that the stable average grain diameter is given by:

$$De^2 = C(T - T_0)$$

where De is the mean grain diameter, C and T_0 are constants for a given metal, and T is the absolute temperature. Thus, with increasing grades of metamorphism, monominerallic aggregates in particular, and indeed all grain aggregates should show an increase in grain size. Vokes (1968) has demonstrated that this is the case for metamorphosed volcanic massive sulphide deposits in Norway.

When recrystallization occurs under a directed pressure then the effects known as Riecke's Principle may become important. This is the phenomenon of pressure solution causing parts of grains to go into solution at points of contact, with the dissolved material being redeposited on those parts of grains which are not under such a high stress. This produces flat elongate grains which give the ore a schistose or gneissose appearance.

Textural adjustments and the development of new minerals

As Vokes (1968) has pointed out, because of the considerable ranges of pressure and temperature over which sulphide minerals are stable, we do not find the progressive development of new minerals with increasing grade of metamorphism that are present in many classes of silicate rocks. Banded iron formations on the other hand do show the abundant development of new mineral phases (James 1955, Gole 1981, Haase 1982), as do many manganese deposits. Sulphide assemblages by comparison only show some minor mineralogical adjustments.

TEXTURAL ADJUSTMENTS

Two commonly noted phenomena are the reorganization of exsolution bodies and the destruction of colloform textures. Though solid solutions may be created during metamorphism by the diffusion of one or more elements into a mineral, the general tendency is for exsolution bodies to migrate to grain boundaries to form intergranular films and grains during the very slow cooling period which follows regional metamorphism.

A general decrease in the incidence of colloform textures with an increasing grade of metamorphism has been reported from pyritic deposits in the Urals, Japanese sulphide deposits and Canadian Precambrian volcanic massive sulphide deposits. Both these adjustments generally improve the ores from the point of view of the mineral processor. Exsolution bodies do not pass with their host grains into the wrong concentrate, and the replacement of colloform textures by granular intergrowths leads to better mineral separation during grinding.

MINERALOGICAL ADJUSTMENTS

In contact metamorphism the main effects so far studied are those seen next to basic dykes cutting sulphide orebodies. These include the following reactions and changes:

(1) pyrite + chalcocite → chalcopyrite + bornite + S;
(2) pyrite + chalcocite + enargite → chalcopyrite + bornite + tennantite + 4S;
(3) the iron content of sphalerite increases;
(4) copper diffuses into sphalerite;
(5) marginal alteration of pyrite to produce pyrrhotite + magnetite.

The principal mineralogical change reported from regionally metamorphosed sulphide ores is an increase in the pyrrhotite:pyrite ratio with the appearance of magnetite at higher grades. The sulphur given off by the reactions involved in pyrrhotite and magnetite production may diffuse into the wall rocks, promoting the extraction of metals from them and the formation of additional sulphide minerals.

Some effects on orebodies and implications for exploration and exploitation

Some results of fold deformation of orebodies have been touched on in various places in earlier chapters. For example, in Fig. 16.5 the effective thickening of the orebody in part of the Luanshya deposit is evident, and in Fig. 17.9, the migration of ore components into fold hinges during deformation and metamorphism is shown. These orebodies at the Homestake Mine were further deformed during later phases of folding (Fig. 17.8), and became spindle-shaped.

Many massive sulphide deposits occur in regionally metamorphosed terrane of all grades of metamorphism; some are very severely deformed and there are many excellent studies of their metamorphism, e.g. Selkman (1984) and Frater (1985) as well as of their deformation (van Staal & Williams 1984) and of the reconstruction of original features by employing 'unfolding techniques' (Zachrisson 1984). Sangster & Scott (1976) discussed some of the implications for exploration arising from the metamorphism of these bodies and their country rocks. They pointed out that a massive sulphide deposit after undergoing even mild metamorphism is often radically different in many ways from its original form, and that metamorphic changes can directly influence the exploration for, and the development of, these orebodies. Metamorphism of

Table 21.1. Some metamorphic equivalents of primary rock types accompanying volcanic-associated massive sulphide deposits.

Primary rock type	Medium grade metamorphism	High grade metamorphism
Chert	Quartzite	Quartzite
Pyritic BIF	Pyrite-pyrrhotite-mica schist	Pyrite-pyrrhotite-mica gneiss
Rhyolite Rhyolite-tuff Rhyolite-breccia Rhyolite-agglomerate	Muscovite-feldspar-quartz schist	Granitic gneiss
Andesite Andesitic volcaniclastics	(Biotite)-hornblende-plagioclase schist	Amphibolite with clinopyroxene ± biotite ± garnet
Basalt	Oligoclase or andesine amphibolites ± epidote ± garnet	Andesine or labradorite amphibolite ± garnet

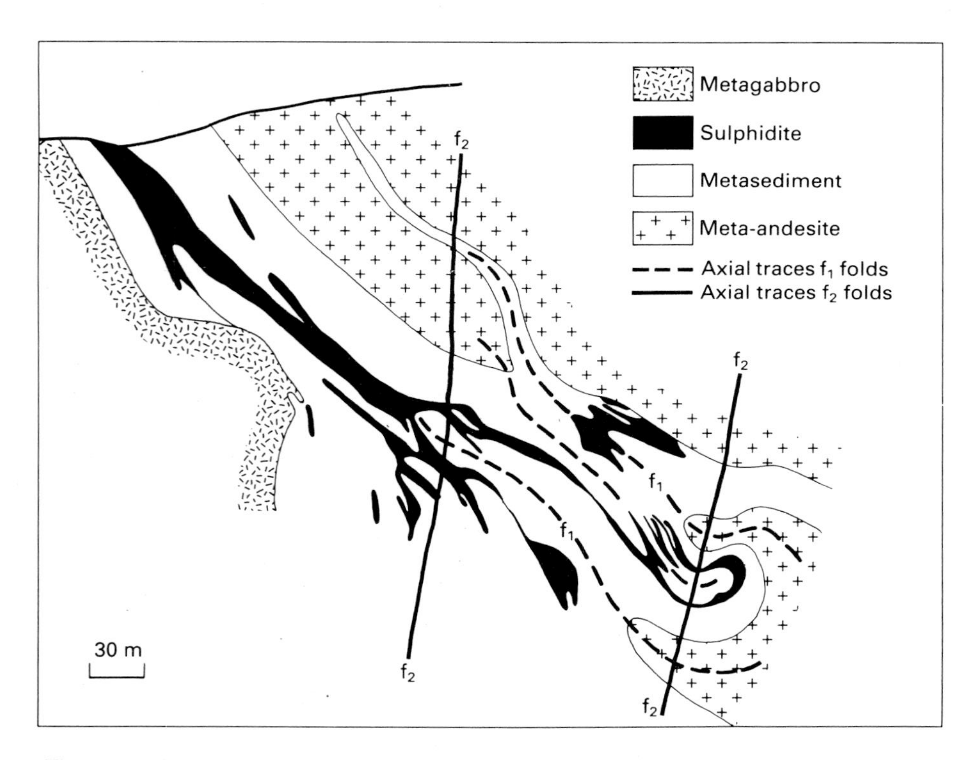

Fig. 21.2. Cross section through the Chisel Lake ore zone, Manitoba. (After Martin 1966).

the host rocks can so change them that it is difficult for the geologist to recognize and trace favourable rock environments. Because many of these orebodies are associated with acid volcanics, it is imperative that the distinction be made between metamorphosed silicic volcanics and metaquartzites,

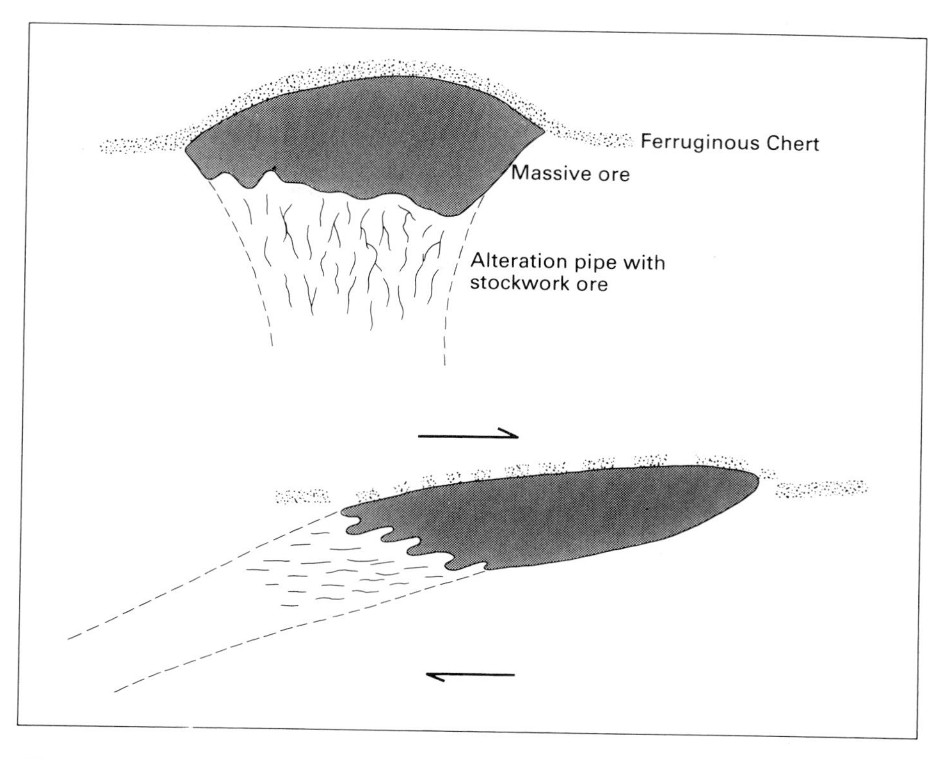

Fig. 21.3. Schema showing the effect of shearing (below) on an undeformed massive sulphide deposit (above). (After Sangster & Scott 1976).

meta-arkoses or granite-gneisses. Similarly, the recognition of a metamorphosed cherty tuff layer can be of great assistance as a marker horizon and in structural interpretation. The explorationist has to 'see through' metamorphosed suites of rocks to reconstruct the original stratigraphy and environment in the search for favourable areas and horizons. Some of these problems will be clearly grasped from a perusal of Table 21.1.

Metamorphism and deformation of a volcanic-associated massive sulphide orebody may also give rise to a different mineralogy, grain size and shape of the deposit, which can affect its geophysical and even its geochemical response. Such deposits were probably originally roughly circular or oval in plan and lenticular in cross section. Deformation may change them to blade-like, rod-shaped or amoeba-like orebodies, which may even be wrapped up and stood on end, and then be mistaken for replacement pipes, e.g. the Horne Orebody of Noranda, Quebec! A good example of what can happen to what was possibly originally one single massive sulphide body is shown in Fig. 21.2. Martin's paper is well worth studying as it is a good, but succinctly described, example of the problems that such deformation can pose to the mine geologists and engineers.

When considerable shearing is involved in the deformation, a flattening effect by transposition along shear or schistosity planes may transpose the alteration pipe and the stockwork beneath the massive sulphide lens into a position nearly parallel to the schistosity and the lens (Fig. 21.3). In extreme cases, the sulphide lens may be sheared off the stockwork zone, separating the two. Metamorphism may also have a profound effect on the rocks of the alteration zone. Chloritized andesites in alteration pipes have frequently been metasomatized to a bulk composition that will produce cordierite-anthophyllite hornfelses — the 'spotted dog rock' or 'dalmatianite' of the Noranda district — in the thermal aureoles of intrusives.

Part III
Mineralization in Space and Time

22

Plate Tectonics and the Global Distribution of Ore Deposits

Introduction

Mining geologists have for many decades attempted to relate various types of mineralization to large scale crustal structures. Prior to the development of plate tectonic theories of crustal evolution and deformation, the Hall-Dana theory (of geosynclines and their deformation and uplift to form mountain chains) dominated geological thought. Various schemes of pre-orogenic, syn-orogenic and post-orogenic stages of magmatism were proposed as integral parts of this orogenic cycle. It was noted by mining geologists that many ore deposits occur in geosynclinal regions, and the hypothesis of distinct stages of magmatism linked to the evolution of geosynclines naturally led to the concept of accompanying stages of metallogenesis, since in the forties and fifties much emphasis was still placed on the magmatic-hydrothermal theory for the origin of the majority of metallic deposits.

When relating the genesis of orebodies to major tectonic features such as island arcs and continental mountain chains it is essential to know the age relations between the mineral deposits and their host rocks; that is, whether the ores are syngenetic and thus part of the stratigraphic sequence, or whether they are epigenetic and therefore younger, perhaps much younger, than their host rocks. One of the reasons why a knowledge of age relations is important is that in the classical geosynclinal theory it was held that all the rocks of the geosyncline were formed within a subsiding linear trough. The plate tectonic theory tells us that some components of orogenic belts may have been transported for hundreds or even thousands of kilometres from their place of origin. If they contain syngenetic ore deposits then these must have evolved in a very different environment from that in which we now find them. The increasing trend towards syngenetic interpretations of many orebodies is therefore very important.

Since this chapter was originally written in 1979 three books have been published on this subject alone: Hutchison (1983), Mitchell & Garson (1981) and Sawkins (1984). In what follows I have drawn on all three for data and ideas but, for reasons of space, this chapter must be very selective and for broader treatments of this subject the reader is referred to these three works.

Plate tectonics

We now know a great deal concerning plate tectonic development during the

Phanerozoic; evidence for similar activity during the Proterozoic is also abundant and, for many, convincing (Windley 1984). A number of workers have recently applied plate tectonic principles to explain Archaean geology, particularly the genesis of greenstone belts. These belts are generally considered to be restricted to the Archaean and early Proterozoic, but Tarney *et al.* (1976) argued persuasively that the Rocas Verdes Marginal Basin in southern Chile represents a young example of greenstone belt formation.

Plate convergence and spreading centres are among the important features which control the global location of mineral deposits. These features are depicted in Fig. 22.1 Plate convergence due to subduction can occur entirely within oceanic areas, adjacent to continental margins or within such margins, and this process is accompanied by extensive activity involving materials derived from the mantle wedge above the Benioff Zone, and also from the subducted plate. Other important features to which mineralization appears to be related include hot spots (mantle plumes), rifting and other extensional tectonics, and collision tectonics. The various rock and tectonic environments involved are reviewed in the above books, and a good discussion of the sedimentation in these different tectonic environments was given by Mitchell and Reading (1986). In this chapter I will adopt the six tectonic settings discussed by Mitchell and Reading, which are:

(1) Interior basins, intracontinental rifts and aulacogens;
(2) Oceanic basins and rises;
(3) Passive continental margins;
(4) Subduction-related settings;
(5) Strike-slip settings;
(6) Collision-related settings.

Interior basins, intracontinental rifts and aulacogens

There are two types of sedimentary basins within continental interiors: large basins often over 1000 km across, and relatively narrow, fault-bounded rift valleys.

CONTINENTAL INTERIOR BASINS (INTRACRATONIC BASINS)

These may contain entirely continental sediments, much of which may have been deposited in large lakes, e.g. the Chad Basin in Affica in which the palaeo-lake Mega Chad extended over at least 300 000 km^2. The Eyre Basin, Australia, is another good example. The Chad Basin has an area of about 600 000 km^2 and contains up to 2 km of Mesozoic and Tertiary sediments. Some interior basins, e.g. Hudson Bay, have been inundated by the sea and may contain mainly marine sediments. In some of these the marine transgression is accompanied by mineralization e.g. the Permian Kupferschiefer of northern Europe and the late Proterozoic Central African Copperbelt (see Chapter 16), although it must be mentioned that recent workers have sugges-

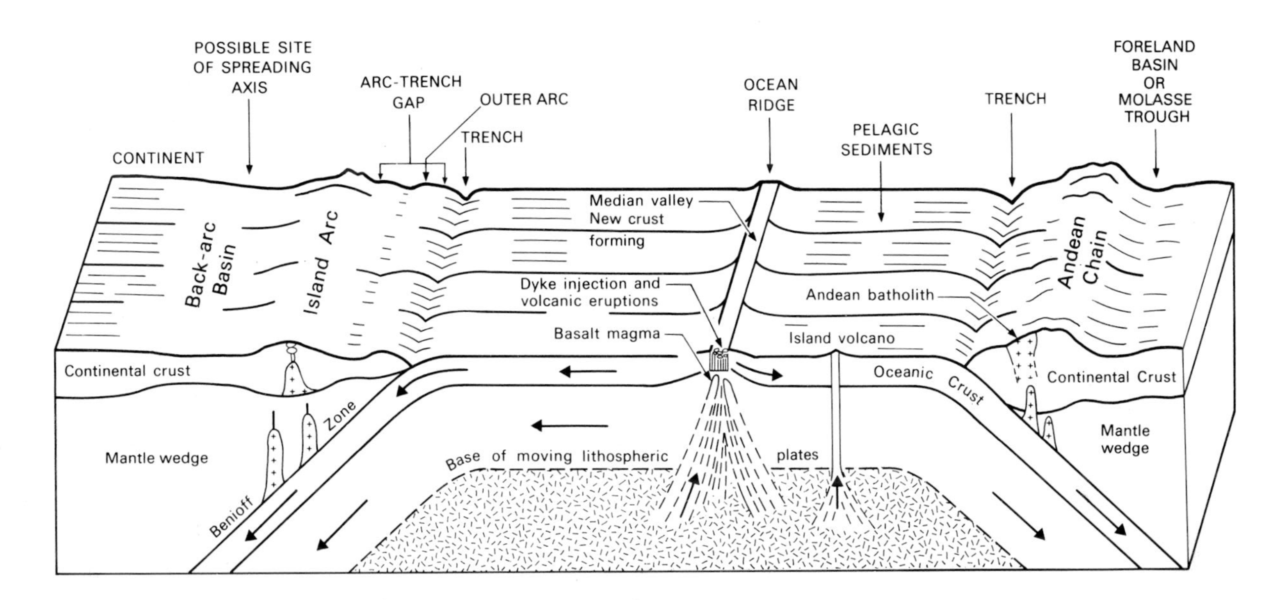

Fig. 22.1. Block diagram of the basic elements of oceanic crust formation and plate movements. On the right is the site of the Andean-type arc with a molasse trough on the continental side, its deposits overlapping on to Precambrian crust, and an arc-trench gap and trench on the oceanic side. On the left is the island arc-setting with its trench, troughs of deposition in the arc-trench gap, volcanic arc and (not shown) the back-arc wedges of clastic sedimentation. The back-arc basin represents the site of the Japan Sea-type setting which may or may not have its own spreading axis.

ted a rift setting for the latter. The first continental basins were developed as soon as sufficient craton was present, and this seems to have been about 3000 Ma ago in southern Africa (Windley 1984). Here the Dominion Reef and Witwatersrand Supergroups with their important gold mineralization (Chapter 19) were laid down. Other Proterozoic basins with associated mineralization include the one containing the Huronian Supergroup with the uraniferous conglomerates of Blind River (Elliot Lake), and those of the Athabaska (Canada) and Alligator River (Australia) successions with their unconformity-associated uranium deposits.

Phanerozoic intracontinental basins are often important for their evaporite deposits such as the Permian Zechstein evaporites of Europe, the Devonian evaporites of the Elk Point Basin of western Canada and the Silurian evaporites of the Michigan Basin. These and similar evaporites are important for potash and soda production. In some basins, e.g. the Paris Basin (France) and the Triassic basins of the British Isles, important gypsum deposits occur. Platform carbonates are present around some basins and these may host lead-zinc mineralization e.g. the coarse, dolomitic Presqu'ile Formation that contains many of the orebodies of the important Pine Point Field on the margin of the Elk Point Basin. This plate tectonic setting is of course transitional to the shelf environment of passive continental margins, and carbonate-hosted base metal deposits occur in both settings.

In this section, mention must be made of the sandstone uranium-type deposits developed in the Wyoming and Colorado Basins (Fig. 18.4), the Mississippi Embayment, and clastic basins in other continents (Chapter 18). Lastly, though not strictly pertinent to the theme of this book, the importance of these basins as gas, oil and coal producers should not be forgotten!

DOMES, RIFTS (GRABEN) AND AULACOGENS

These are initiated by the doming of continental areas which, due to stretching, develop three rift valleys that meet at a 120° triple junction (Burke & Dewey 1973). As is shown in Fig. 22.2, two of the rift valleys may combine to form a divergent plate boundary leading through graben to ocean spreading, whilst the third arm may only show partial development. This third arm may develop a considerable thickness of sediments, with some volcanics and igneous intrusions. All these may be structurally deformed but the geological history of such zones is relatively simple. They do not often progress much beyond the graben stage and are called failed arms or aulacogens. The associated domes are believed to have formed over mantle plumes (hot spots).

Hot spot-associated mineralization

The triple junction by the shoulder of Brazil is thought to have been generated by the Niger Mantle Plume. Plumes may cause melting of the continental crust forming granite intrusions, e.g. the Cabo Granite of Brazil. Now the big

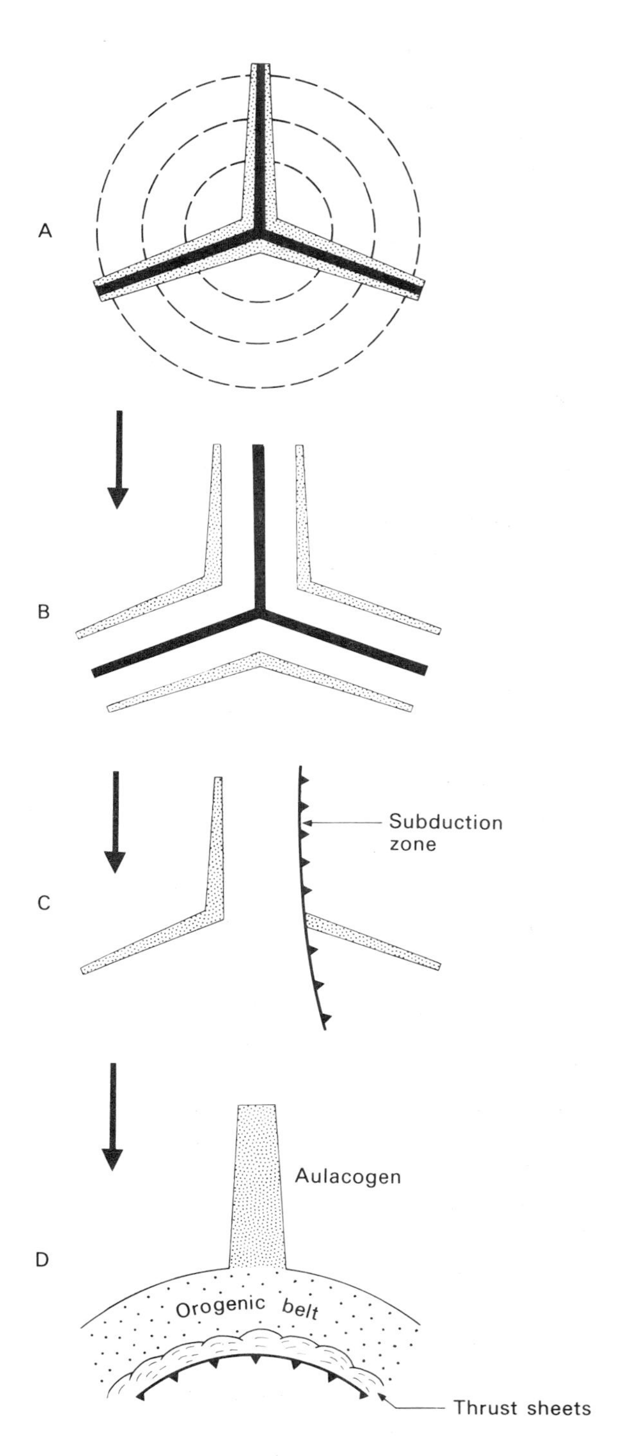

Fig. 22.2. Schematic diagrams showing the development of an aulacogen. A — Development of three rift valleys with axial dykes in a regional dome. B — The three rifts develop into accreting plate margins (e.g. early Cretaceous history of Atlantic Ocean-Benue Trough relationship). C — One arm of system begins to close by marginal subduction causing deformation of sediments and volcanics. D — continental margin collides with subduction zone forming orogenic belt, and the failed arm is preserved as an aulacogen. (After Burke & Dewey 1973).

tin-tungsten provinces are associated with subduction zones throughout the world, but there are some tin provinces which do not fit this pattern, e.g. those of Nigeria, Rondônia (Brazil) and the Sudan. Tin in these provinces is associated with sodic granites in ring complexes and the specific association in Nigeria is with A-type hypersolvus biotite granites (Bowden 1982). Mitchell & Garson (1976) suggested that mantle plumes were responsible for the generation of these granites and their assoicated mineralization. Both the granitic magma and the associated metals were probably derived from the crust.

Tin deposits are also associated with the granites of the Bushveld Complex, RSA. This huge igneous intrusion and that of the Stillwater Complex, USA (Chapter 11) have been attributed to hot spot activity in an anorogenic setting. The huge Dufek Complex in Antarctica may also be related to a mantle hot spot, and in the future it could become a hunting ground for platinum (de Wit 1985).

The Sudbury Structure, Canada, with its copper-nickel ores, could also have been developed over a hot spot, but much current opinion favours a meteorite impact origin (Chapter 12), and this has also been suggested as the origin of the Bushveld Complex. The Palabora Complex with its copper-bearing carbonatites is about the same age as the Bushveld Complex and may be related to the same hot spot activity. This and other hot spot activity, rifting and associated mineralization in Africa have been reviewed by Olade (1980). The alkaline igneous activity and the associated carbonatites of the Kola Peninsula (Chapter 9) may also be related to hot spot activity.

The mid Proterozoic (1700–1000 Ma) bimodal association of anorthosite and granite, which gave rise to the enormous anorthositic massifs with their associated titanium mineralization (p. 90 and Chapter 11), probably represents a major but abortive attempt world-wide to develop continental fragmentation. A possible modern analogue is the line of Phanerozoic intrusions in a belt 200 km wide and 1600 km long stretching from Hoggar in Niger down to the Benue Trough (Windley 1984).

Mineralization associated with continental rifting

The initial stage of graben development is marked by deeply penetrating faults forming pathways to the mantle and giving rise to volcanism. This is usually of alkaline type, sometimes with the development of carbonate lavas and intrusives and occasionally kimberlites (Fig. 22.3). Erosion of these may lead to the formation of soda deposits (e.g. Lakes Natron and Magadi in East Africa) and the intrusive carbonates may carry a number of metals of economic interest (Chapter 9) as well as being a source of phosphorus and lime.

In the graben themselves, sediments with or without volcanics may accumulate. In the East African rift valleys, red bed type deposits are common, with conglomerate fans along the escarpments, and playa-lakes have produced evaporites. The water of Lake Kivu is rich in zinc and precipitates sphalerite. The deposition of this and other metals is believed to be due to hydrothermal

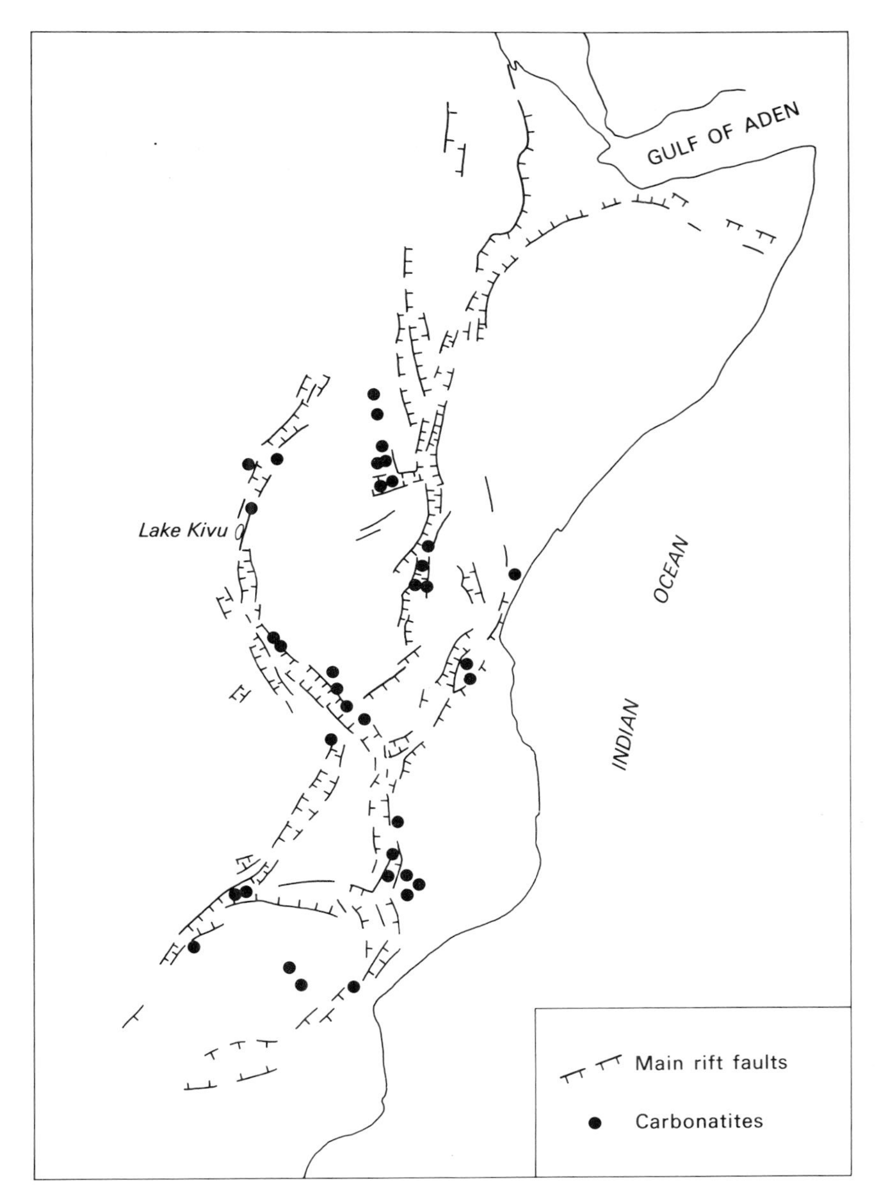

Fig. 22.3. Distribution of carbonatite intrusions relative to the East African Rift Valleys. (After Mitchell & Garson 1976).

solutions which represent ground water that has been cycled through volcanics and sediments at times of high rainfall (Degens & Ross 1976).

Recently-formed aulacogens occur at both ends of the Red Sea and the Gulf of Suez is one of these. It contains a 4 km thick succession of Neogene salt, limestone and clastic sediments, and in it lie the Ras Morgan Oilfields. Other young aulacogens also contain salt deposits and oilfields, e.g. the North

Sea. Slightly older aulacogens are present on both sides of the Atlantic (Fig. 22.4), and one occurs where the shoulder of Brazil fits into Africa. At this point, spreading occurred on all three arms of a triple junction 120–80 Ma ago; then the Benue Trough closed with, Burke & Dewey suggested, a short-lived period of subduction, whilst the South Atlantic and Gulf of Guinea arms continued to spread. The Benue Trough is an aulacogen about 560 km long having a central zone of high Bouguer anomalies flanked on either side by elongate negative anomalies. This pattern is believed to be due to the presence of near surface masses of crystalline basement and intermediate igneous rocks below the central zone. Lead-zinc-fluorite-baryte mineralization occurs in fractures in Lower Cretaceous limestone, and similar lead mineralization is present

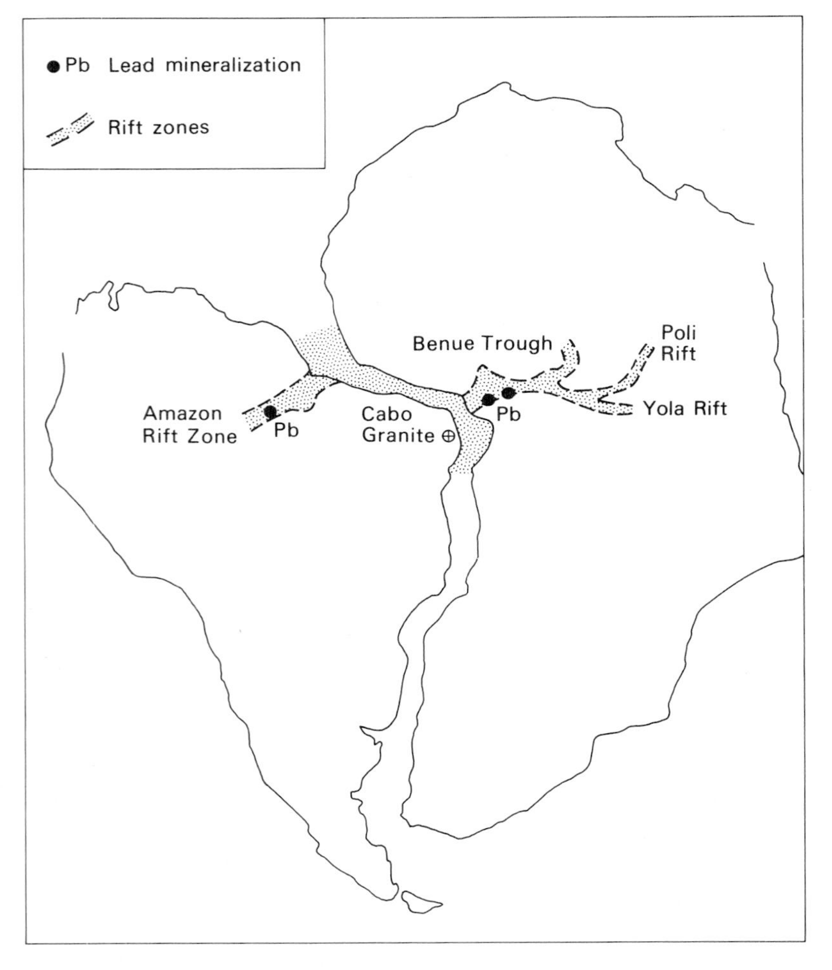

Fig. 22.4. Sketch map showing the locations of the Benue Trough, the Amazon Rift Zone and the lead mineralization within these aulacogens. (After Burke & Dewey 1973 and Mitchell & Garson 1976).

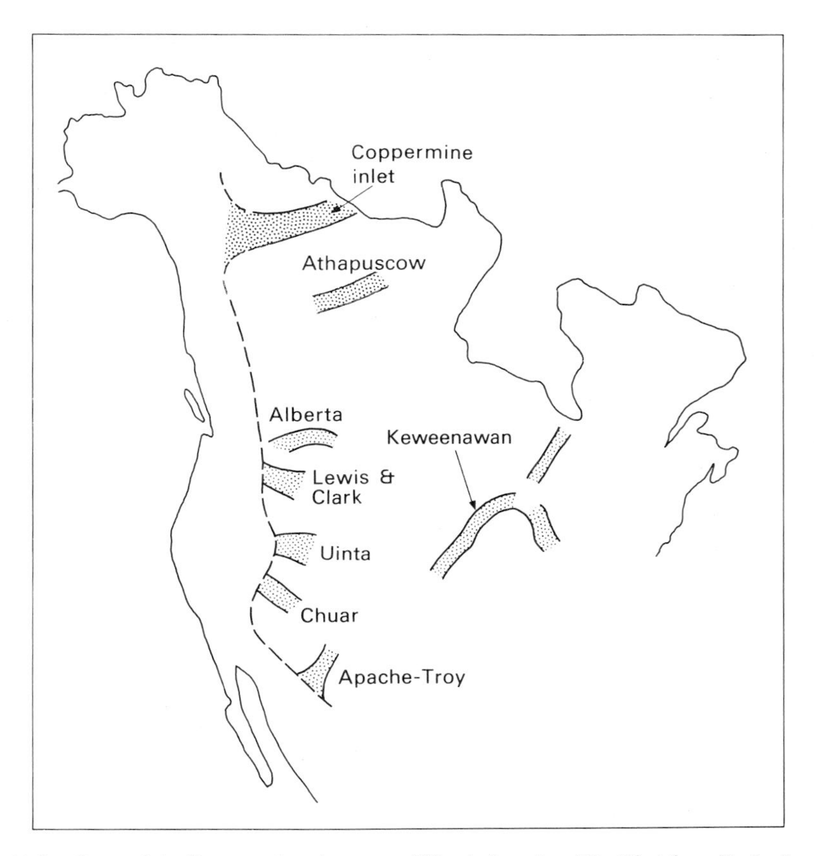

Fig. 22.5. Some of the Proterozoic aulacogens of North America. (Modified from Burke & Dewey 1973).

in the Amazon Rift Zone. Tholeiitic igneous activity also took place and intrusions of diorite, gabbro and pyroxenite are present. The origin of the mineralization in both aulacogens is controversial as it may be associated with basic magmatism or with circulating brines which have passed through evaporite deposits.

A number of Proterozoic aulacogens that carry more important mineralization than that in the above examples have been recognized and Burke & Dewey (1973) have identified a number of these in North America (Fig. 22.5). Those running perpendicular to the western margin of the continent carry great thicknesses of rocks of the Belt Series. The northernmost trough contains the Coppermine River Group consisting of more than 3 km of basalt flows with native copper mineralization overlain by greater than 4 km of sediments with evaporites near the top. Epigenetic deposits associated with the rifting include the Great Bear Lake Uranium Field. To the south is the Athapuscow Aulacogen with red beds, sedimentary uranium deposits and evidence of the former presence of evaporites (Stanworth & Badham 1984).

Further south is the Alberta Rift which passes into British Columbia and contains about 11 km of late Precambrian sediments in which important stratiform and epigenetic lead-zinc mineralization occur, including the famous Sullivan Mine at Kimberly, British Columbia. Phosphorite deposits are also present. To the east, the Keweenawan Aulacogen, which is the same age as the Coppermine Aulacogen, also carries a considerable thickness of basalt and clastic sediments, and again native copper mineralization is present (Elmore 1984). Other mining camps which are believed to lie within aulacogens include the copper and lead-zinc ores of Mount Isa, Queensland, and the McArthur River Pb-Zn deposit (Dunnet 1971). Red bed coppers are well developed in upper Palaeozoic aulacogens of the USSR, and uranium mineralization is also present in some of these.

Among orthomagmatic ore deposits in aulacogens, mention must be made of the Great Dyke of Zimbabwe (Fig. 11.3), the copper-nickel ores of the Noril'sk-Talnakh region of the USSR (Fig. 12.11), and the Duluth Complex, USA.

A number of porphyry molybdenum deposits have recently been discovered in aulacogens, e.g. in the felsic rocks of the Oslo Graben (Ihlen *et al.* 1982) and the Malmbjerg deposit of Greenland (Sawkins 1984).

Ocean basins and rises

We must now pass to the later stages of rifting that lead on to the development of embryonic oceans like the Red Sea. With further extension of the crust and the commencement of continental drift, deep crustal flowage and tensional faulting will combine to thin the crust along the graben. At some stage during this process an opening to the sea may initiate marine conditions. Observations from the Rhine Graben, Mesozoic deposits along the Atlantic coastlines and the Miocene of the Red Sea region indicate that evaporite series of great thickness may form at this time. These evaporites contain halite as well as gypsum and therefore have a double economic importance.

The Red Sea is known to be floored by basalt and there is evidence that new oceanic crust is being formed along its median zone. On either side, pelagic sediments are forming, but the depths are not yet sufficiently great nor the other factors present which lead to phosphorite development as in areas of upwelling oceanic currents along some Atlantic and Pacific coasts where phosphorite deposits are forming at the present day. The recently formed carbonate miogeosynclinal area of Florida is well known for its phosphorites which yield by-product uranium. Whether such deposits can occur in Red Sea basins depends on where one draws the boundary between these and the Atlantic-type situation into which they evolve.

The possible mode of development of oceanic crust along the median zone of the Red Sea is shown in Fig. 22.6. As this new crustal material moves away from the median zone, layer 1 of the oceanic crust, in the form of pelagic sediment, is added to it. As Sillitoe (1972a) has suggested, there is strong

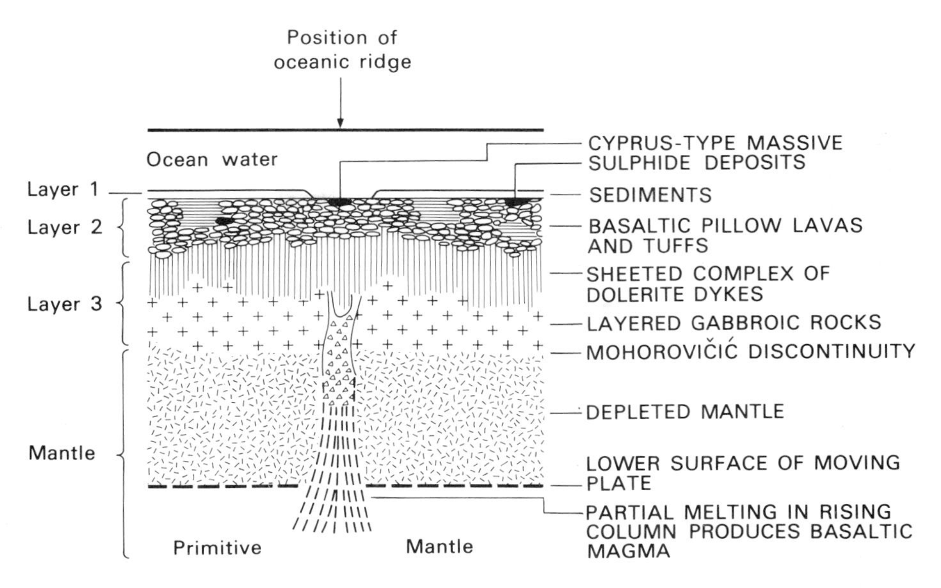

Fig. 22.6. Schematic representation of the development of oceanic crust along a spreading axis. The crustal layering and the possible locations of Cyprus-type massive sulphide deposits are shown. (After Cann 1970 and Sillitoe 1972a).

evidence that many Cyprus-type sulphide deposits are formed during this process of crustal birth. If slices of the oceanic crust are thrust into melanges at convergent junctures or preserved in some other way then sections similar to that shown in Fig. 22.6 may be expected. This is the case where these cupriferous pyrite orebodies are found in Cyprus, Newfoundland, Turkey, Oman, etc. Francheteau *et al.* (1979) have recently described deposits of massive sulphides from the East Pacific Rise which occur on a spreading ridge away from transform faults and which show a number of similarities to the Cyprus deposits. Similar situations to the oceanic spreading ridges are found in back-arc basins and indeed many workers hold that the Troodos Massif of Cyprus was developed in such a milieu. In this massif, non-economic nickel-copper mineralization and economic chromite deposits occur in the basic plutonic rocks.

Hydrothermal mineralization with the development of copper, zinc, silver and mercury has been reported from oceanic ridges in the Atlantic and Indian Oceans by Dmitriev *et al* (1971). Surprisingly, tin mineralization was found in both ridges, accompanied in one case by typical hydrothermal minerals normally associated with granitic environments such as tourmaline, topaz, fluorite and baryte. Cupriferous pyrite mineralization of stockwork-type which occurs in metabasalts has been reported by Bonatti *et al.* (1976) from the Mid-Atlantic Ridge. They put forward evidence that it was generated by sea water solutions circulating through the oceanic crust in a manner similar to that described for the Cyprus deposits at the end of Chapter 4. As is the case with a number of ridges in the Pacific Ocean, black smokers have now been found on the Mid-Atlantic Ridge (Rona *et al.* 1986), indicating that the hydrothermal processes which produce them probably operate along most active sea floor spreading centres.

The median zone of the Red Sea is notable for the occurrence of hot brines and metal-rich muds in some of the deep basins. Those studied by Degens & Ross (1969, 1976) can be defined by the 2000 m contour. The largest of these deeps is the Atlantis II which is 14 × 5 km and 170 m deep. At the base, multicoloured sediments usually 20 m thick, but up to 100 m, rest on basalt and piston coring has demonstrated the presence of stratification and various sedimentary facies high in manganites, iron oxides and iron, copper and zinc sulphides. These sediments can run up to 20% Zn but the top 10 m in the Atlantis II deep run 1.3% Cu and 3.4% Zn, and feasibility studies are under way for their recovery. Above the sediments, the deeps contain brine solutions; these and the sediments run up to about 60°C and 30% sodium chloride. It has been suggested that the sediments and brines have originated either from the ascent of juvenile solutions, or from the recirculation of sea water which has leached salt from the evaporites flanking the Red Sea and passed through hot basalt from which it has leached metals. If this second mechanism is the sole or principal one, then metal-rich muds of this origin will only form when evaporites are sufficiently near to supply the required salt. This being the case, it is possible to outline areas of the present oceans which can be said, from the distribution of known evaporites, to have high or low probabilities of containing these deposits. A study of this nature has been made for the North Atlantic by Blissenbach & Fellerer (1973). However, as similar metal-rich muds have now been found on both the Atlantic and East Pacific Rises, this latter hypothesis may be suspect.

Passive continental margins

As ocean spreading gradually forces two continents apart, both sides of the original rift become passive margins (also termed trailing, inactive or Atlantic-type), with the development of a continental shelf, bounded oceanwards by a slope and landwards by a shoreline or an epicontinental sea. The type of sediment forming on shelves today depends on latitude and climate, on the facing of a shelf relative to the major wind belts and on tidal range (Mitchell & Reading 1986). In the past, shelves in low latitudes were often covered with substantial platform carbonate successions, and these can be hosts for base metal deposits of both epigenetic and syngenetic nature (so-called Mississippi Valley-type, Irish-type and Alpine-type deposits).

A number of small, stratiform, sandstone-hosted copper and lead-zinc deposits occur in the Cretaceous along the western edge of Africa from Nigeria to Namibia, and further north continue inland to the margin of the Ahaggar Massif. Similar deposits are known along the northern margin of the High Atlas Mountains on the margin of the Morocco Rift, and older ones, which are related to Infracambrian rifting and the development of a Palaeotethys Ocean, are found along the northern edge of Gondwanaland (Olade 1980, Sillitoe 1980). In many ways these copper deposits resemble those described in the first part of Chapter 16 and may be a variant of them. The lead ores of Laisvall in Sweden and Largentière in France appear to have been developed in a similar tectonic setting.

A number of the world's most important sedimentary manganese deposits occur just above an unconformity and were formed under shallow marine conditions on shelf areas e.g. Nikopol and Chiatura, USSR and Groote Eylandt in Australia (Chapter 19).

Passive continental margins that have suffered marine transgressions are also important for phosphorite deposits. There are a number of factors that control the development and distribution of phosphorites, including a low palaeolatitude and a broad shallow downwarp flanking a major seaway, such as may result from the complete rifting apart of a continent (Cook 1984). The Tertiary phosphorites of the Atlantic Coastal Plain of the USA appear to have formed under just these conditions. It should be noted that they are a source of by-product uranium.

Many workers now favour a continental shelf environment for the deposition of the Proterozoic Superior-type BIF (Windley 1984). McConchie (1984), in a comprehensive assessment of the evidence from the Hamersley Group, which contains the most extensive accumulation of sedimentary iron deposits known, makes a compelling case for a mid to outer shelf environment.

Finally, we must note the important placer deposits that are developed along the trailing edges of many continents, particularly those where trade winds blow in oblique to the shoreline combining with ocean currents to give rise to marked longshore drift. Among these we can list the diamond placers of the Namibian coast, the rutile-zircon-monazite-ilmenite deposits of the eastern and western coasts of Australia, and the similar deposits of Florida and the eastern coasts of Africa and South America.

Subduction-related settings

In these settings oceanic lithosphere is subducted beneath an arc system on the overriding plate. The subduction zone is usually marked by a deep sea *trench* and between this and the active *volcanic arc* is the *arc-trench gap*. The arc-trench gap in some arcs is comprised of an *outer arc* and a *fore-arc basin*. The outer arc, which may rise above sea-level, is usually built of oceanic floor or trench sediments scraped off the subducting plate above low angle thrusts and tectonically accreted to the overlying plate to form an *accretionary prism* or *tectonic mélange*. The fore-arc basin contains flat lying, undeformed sediments that may reach 5 km in thickness and that reflect progressive shallowing as the basin fills up: turbidites, shallow marine sediments and fluviatile-deltaic-shoreline complexes. The sediments are mainly derived from the volcanic arc or by erosion of uplifted basement (Windley 1984). Arc systems may be either *island arcs* or *continental margin arcs*. Island arcs are of several kinds: some are clearly intra-oceanic (Tonga, Scotia), some are separated from sialic continent by small semi-ocean basins (Japan, Kurile), some pass laterally into a continental margin fold belt (Aleutian) and some are built against continental crust (Sumatra-Java). In continental margin arcs, the volcanic arc is situated landward of the oceanic crust-continental crust boundary as in the Andes (Fig. 22.1). Behind volcanic arcs there is the back-arc area usually referred to when behind island arcs as the *back-arc basin* or *marginal basin* or when behind continental margin arcs as a *foreland basin* or *molasse trough*, e.g. the Amazon Basin behind the Andes.

MINERALIZATION IN ISLAND ARCS

It is convenient to divide the mineral deposits in these arcs according to whether they were formed outside the arc-trench environment and transported to it by plate motion (allochthonous deposits), or whether they originated within this geosynclincal complex (autochthonous deposits) (Evans 1976b).

Allochthonous deposits

Rocks and mineral deposits formed during the development of oceanic crust may by various trains of circumstances arrive at the trench at the top of a subduction zone (Fig. 22.7). This material is largely subducted, when some of it may be recycled, whilst some is thrust into mélanges. There is, however, only one area so far discovered with what appears to be Cyprus-type sulphide orebodies thrust into a mélange and that is north-western California, where we find the Island Mountain deposit and some smaller occurrences in the Franciscan mélange. Cyprus-type massive sulphides are also likely to be present at the base of island arc sequences, whether these are the initial succession or a second or later one formed by the migration of the Benioff Zone, because in most cases the arc basement will have originated at an oceanic ridge.

Clearly, any other deposits formed in new oceanic crust may also eventually be mechanically incorporated into island arcs and the most likely victims will be the chromite deposits of alpine-type peridotites and gabbros (Thayer 1964, 1967). Though some writers have proposed that these deposits were intruded partly as fluid magma and partly as crystal mush, Thayer (1969a, 1969b) has cited much evidence in favour of their being regarded as cumulates that originated in the upper mantle, often at mid-ocean ridges. Recent evidence from the Papuan ultramafic belt is particularly compelling in this respect (Thayer 1971). Here, in a block apparently composed of obducted oceanic crust and upper mantle, rock units of pillowed basalts, gabbro and peridotite rest upon each other in thicknesses exceeding 2000 m for each unit. The plutonic rocks all show well-developed cumulate textures. Podiform chromitites are the most important magmatic deposits in the allochthonous peridotites (Thayer 1971). Platinum metal deposits are known in some of these, but they are more important as the source rocks for placer deposits of these metals.

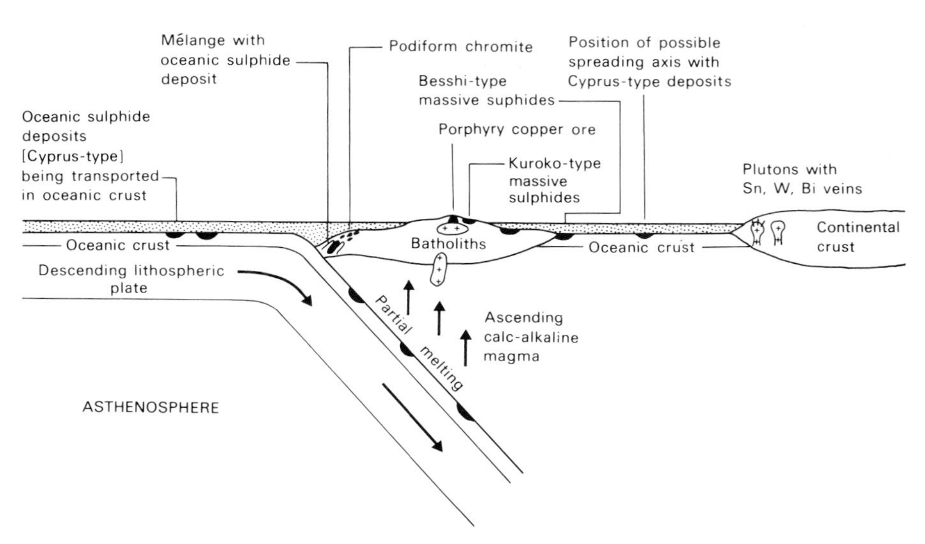

Fig. 22.7. Diagram showing the development and emplacement of some mineral deposits in an island arc and its adjacent regions. (After Sillitoe 1972a and 1972b with modifications).

Economic deposits of podiform chromitite occur in present day arcs in Cuba, where they are found in dunite pods surrounded by peridotite, and on Luzon in the Philippines, again in dunite in a layered ultramafic complex. Numerous occurrences are present in ancient island arc successions (cf., e.g. Thayer 1964). According to Thayer (1971), the allochthonous 'high temperature' lherzolitic peridotites and associated gabbroic 'granulites' of Green and the garnet peridotites of O'Hara contain no significant deposits of oxides or sulphides. A recent development of great importance in this context, and for exploration programmes, is the conclusion of Pearce *et al.* (1984) that all major podiform chromite deposits were formed in marginal basins and not at mid-ocean ridges.

If we follow Thayer (1971) in excluding the Precambrian ultramafic sheets from the class of alpine-type plutonic rocks of the ophiolite suite, then it must be recorded that at present remarkably few magmatic sulphide segregations are known in these rocks. Uneconomic pyrrhotite-pentlandite accumulations are known in Cyprus but the only deposits so far being exploited are those of the Acoje Mine in the Philippines, where nickel and platinum sulphides occur. The accelerated pace of mineral exploration in present day arcs will eventually indicate whether such deposits are truly rare or whether many are still awaiting discovery. At the moment, the latter possibility appears to be the more plausible.

Autochthonous deposits

In considering these deposits it is convenient to divide the geosynclinal development into three stages: initial or tholeiitic, main calc-alkaline stage and waning calc-alkaline stage.

(a) *Initial or tholeiitic stage.* The probable hydrous nature of such magmas and their derivation by partial melting of mantle material (Ringwood 1974) implies that cogenetic massive sulphide deposits may well be expected to form during this stage. However, no massive sulphide deposits have as yet been detected in the earliest tholeiitic sequences of island arcs, but such sequences must be considered as promising grounds for their discovery. Stanton (1972) has surmised that the early basic lavas of the Solomon Islands with sulphide amygdules and disseminations belong to this stage.

(2) *Main calc-alkaline stage.* This is the main period of arc development. Considerable sedimentation and subaerial volcanicity may have occurred during the tholeiitic stage, but the main island arc building and plutonic igneous activity belong to this stage. Baker (1968) showed that whilst the early stage of island arc volcanism is dominated by basalt and basaltic-andesite, the more evolved arcs have andesite as the dominant volcanic rock. Ringwood (1974) has summarized the investigations which have shown that the early tholeiitic stage is succeeded by magmas having a calc-alkaline trend and which are probably ultimately derived from two sources — subducted oceanic crust, and partial melting of the mantle wedge overlying the Benioff Zone. This will result in the diapiric uprise of wet pyroxenite from just above the Benioff Zone. The pyroxenite will undergo partial melting as it rises and the magmas so produced will fractionate as they rise to produce a wide range of magmas processing calc-alkaline characteristics.

The exposure of much more rock to subaerial erosion will greatly increase the volume of sediment which reaches inter-island gaps, the arc-trench gap and the back-arc area. Reef limestones will then be common. Rapid erosion on the flanks of sub-aerial volcanoes, ignimbritic activity (related to the increase in silica content of the magmas) giving rise to submarine lahars, submarine slumping and so on, would all contribute to produce a vast volume of material for local sedimentation and for transportation by turbidity currents to the arc-trench gap and along submarine canyons into the trench itself.

Volcanic massive sulphide, stockwork, pyrometasomatic and vein deposits are formed in this stage. Conformable sulphide orebodies of Besshi-type develop at this stage in back-arc basins (Fox 1984), accompanying the andesitic to dacitic volcanism. They occur in complex structural settings characterized by thick greywacke sequences as in the late Palaeozoic terrain of Honshu where the volcanism is basaltic to andesitic. The lack of ophiolitic components, the petrological association and the pyroclastic nature of the volcanics indicates a different environment from that of the Cyprus-type deposits. Sillitoe (1972a) suggested that the solutions which deposited these massive sulphide deposits might have come directly from subducted sulphide bodies (Fig. 22.7), with the metals in them being derived from the subducted bodies and layers 1 and 2 of the subducted ocean floor. Sillitoe (1972b) has further suggested that metal-bearing brines expelled from layers 1 and 2 would induce fusion in the wedge of mantle above the Benioff Zone, ascending with the resulting hydrous calc-alkaline magmas to form porphyry copper deposits when the volcanism is dominantly subaerial. Exploration in recent years has shown that many porphyry coppers are present in island arcs (Fig. 15.7) with, contrary to earlier ideas, both diorite and Lowell-Guilbert model types being present. There is no evidence that continental crust is required for their formation; therefore they can be sought for in any island arc with or without continental crust. The young age of the orebodies in some recent arcs indicates that they belong to the late stages of island arc evolution and perhaps overlap into the waning stage. Sillitoe's theory that the ultimate source of the metals in porphyry deposits is to be sought in subduction zones is attractive as it can account for the relatively uniform nature of porphyry copper deposits despite the great variety of their host rocks. His theory is attractive on other grounds. It can account for the regional differences seen in these deposits (Chapter 15), relating them to differences in the contents of copper, molybdenum and gold in the subducted oceanic crust.

Pyrometasomatic skarn deposits are common at igneous-carbonate country rock contacts in the older and more complex arcs such as Indonesia, the Philippines and Japan. Many of the granites and granodiorites with which these deposits are associated also possess spatially related pegmatitic and vein mineralization, and the resulting metallic association can be quite complex as in the Kitakami and Abukuma Highlands of Japan. Pyrometasomatic and vein deposits of tin are well known from parts of the Indonesian Arc. Less well known are the gold veins in the contact zones of diorites and granodiorites of the Solomon Islands. Other vein deposits are not common in the younger arcs, including those of mercury, which is surprising in view of their development in association with andesites in the Alpine orogenic belt, e.g. the eastern Carpathians. Cinnabar veins occur in the more complex arcs of the Philippines, Japan and New Zealand.

(c) *Waning calc-alkaline stage*. The massive sulphide deposits of Kuroko-type probably belong to the later stages of development of island arcs. They are associated with the more felsic stages of calc-alkaline magmatism and are marked by distinctive features that have been described in Chapter 16. Kuroko-type ores in ancient island arcs include the Palaeozoic deposits of Captain's Flat, New South Wales; Buchans, Newfoundland; and Avoca, Ireland.

As Stanton (1972) has pointed out, manganese deposits of volcanic affiliation are very common in eugeosynclinal successions of all ages from Archaean to Recent and island arcs are no exception. Large numbers of deposits of subeconomic size occur in the Tertiary volcanic areas of Japan, Indonesia and the West Indies and in the Recent volcanic sequences of the larger south-western Pacific islands. These deposits have been formed *in situ* and are not tectonic slices of pelagic manganese deposits of the oceanic crust. They belong to this and the preceding stage of island arc evolution.

The nature of back-arc basins and their sedimentary and volcanic contents are too variable and complex to be discussed here. A good summary can be found in Mitchell & Reading (1986). There is a scarcity of observed mineralization in modern back-arc basins, partly because of the thick blanket of sediment and volcanic ash present in most of them (Mitchell & Garson 1981), so we must turn to ancient marginal basins for evidence of the deposits that formed in them. According to Fox (1984) Besshi-type sulphide ores form in this environment (see above), and a number of ophiolites with their associated sulphide and chromite deposits have been assigned to this setting. The Taupo Volcanic Zone of North Island, NZ, with its geothermal systems, demonstrates the mineralizing processes that can take place in marginal basins (Chapter 4) and the similar but older Welsh Basin, UK, has Kuroko-style and Cu-Pb-Zn vein mineralization (Ball & Bland 1985, Reedman *et al.* 1985) and, in addition, sedimentary and volcanic manganese deposits and oolitic ironstones.

MINERALIZATION IN CONTINENTAL MARGIN ARCS

These arcs are also referred to as Andean or Cordilleran Arcs, but both terms are inappropriate because parts of some continental margin arcs, e.g. Sumatra and Java in the Sunda Arc, are separated from the continental interior by marine basins (Mitchell & Garson 1981). The principal features of these arcs include a trench with turbidites, high heat flow and regularly arranged metamorphic zones, as well as monzonitic-granodioritic plutons and batholiths, foreland basins that are often molasse-filled, and widespread sulphide and oxide mineralization (including the extensive development of porphyry copper and molybdenum deposits). Active volcanoes are frequently present in mountain chains underlain by crust of greater than normal thickness due either to crustal shortening or underplating. Although these are regions of plate convergence with its concomitant compression, many graben structures are developed running parallel or oblique to the arcs. Some magmatic activity and mineralization are related to this rifting.

The Andean Arcs lie along the western margin of South America where it abuts against the Nazca Plate which is moving down a Benioff Zone beneath the present mountain chain. This chain is immensely rich in orebodies of various types and metals, but on the whole these show a plutonic-epigenetic affiliation rather than a volcanic-syngenetic one. This is no doubt to some extent a function of erosion. The profusion of metals may be related to proximity to the East Pacific Rise and the abundance of metals that are being added to the oceanic crust at this spreading axis. Sawkins (1972) has listed the main features of Andean Cordilleran deposits, emphasizing their close relationship in time and space with calc-alkaline intrusives and their occurrence at high elevations (2000–4000 m above sea level) at relatively shallow levels in the upper crust. He reminded us that this implies that erosion of the Andes down to present sea level would remove this entire suite of ore deposits.

The Andes are characterized by linear belts of mineralization (Peterson

1970, Sillitoe 1976, Frutos 1982) that coincide with the morphotectonic belts which run down the mountain chain (Grant *et al.* 1980). The Coastal Belt, mainly consisting of Precambrian metamorphic rocks, is important for its iron-apatite deposits of skarn and other types. The Western Cordillera (made up mainly of Andean igneous rocks) is a copper province particularly important for porphyry-type deposits carrying copper, molybdenum and gold; and the Altiplano, a Cretaceous-Tertiary intermontane basin filled with molasse sediments, is part of a larger province of vein and replacement Cu-Pb-Zn-Ag deposits that extends along the Eastern Cordillera. This belt consists of Palaeozoic sedimentary rocks and Palaeozoic and Andean igneous rocks, with the tin (+ W-Ag-Bi) belt (Fig. 17.1) along its eastern margin.

This metal zonation has been related to the remobilization of metals, subducted in the oceanic crust, at different depths down the Benioff Zone (Wright & McCurry 1973), to migration of Benioff Zones, and to changes in their inclination (Mitchell & Garson 1972, Mitchell 1973, Sillitoe 1981b). The porphyry deposits are related to I-type granitoids probably generated along the Benioff Zone or in the mantle wedge above it, but the tin deposits are related to S-type granitoids, perhaps representing anatectic melts which derived their tin from continental crust. There is indeed good reason to believe that this is recycled tin (Dulski *et al.* 1982).

Metallogenic variations are known from other parts of the continental margin arcs of the Americas and they have been tabulated by Windley (1984).

The North American Cordillera is a vast storehouse of mineral treasures developed during a complicated tectonic history too involved for discussion here. A useful summary for the western USA can be found in Guild (1978). One point, however, should be made, and that is the important connexion between back-arc rifting and base and precious metal vein deposits (Ivosevic 1984). Extensional regimes in the back-arc region are also believed to have controlled the emplacement of Climax-type porphyry molybdenum deposits in the western States (Sawkins 1984).

Strike-slip settings

Two different types of fault are included here: the transform fault and the long known tear or wrench fault. Strike-slip faults vary in size from plate boundary faults such as the San Andreas Fault of California and the Alpine Fault of New Zealand, through microplate boundaries and intraplate faults, such as those of Asia north of the Himalayas, down to small scale fractures with only a few metres offset. Whether these structures occur as single faults or as fault zones (and where complex patterns with some extension are present in the resulting sedimentary basins), they may be important in locating economic mineral deposits. However, in this section there is only space to deal with structures having an obvious plate tectonic connexion, attractive though it might be to digress into the subject of lineaments and mineralization!

Well-recognized transform faults in continental crust have little or no associated mineralization, although some potential is perhaps indicated by the location of the Salton Sea geothermal system within the San Andreas Fault Zone, but it must be noted that this system is not underlain by normal continental crust. Transform faults are, however, loci of higher heat flow and could well be channel-ways for hydrothermal solutions, as certainly appears to be the case in the Red Sea. Here, the brine pools and metal-rich muds, described in *Ocean basins and rises* above, appear to be located above transform faults (Mitchell & Garson 1981). Other possible examples of their acting as structural controls of mineralization come from a study of the hypothetical continental continuations of transform faults. For example, many of the diamond-bearing kimberlites of West Africa lie along such lines (Williams & Williams 1977) and some onshore base metal deposits of the Red Sea region show a similar relationship.

Modern strike-slip basins occur in oceanic, continental and continental margin settings (Mitchell & Reading 1986). They may also occur in back-arc basins. The classic onshore strike-slip basin is the Dead Sea with its important salt production, and the best known offshore basins are those of the Californian Continental Borderland, some of which host commercial oil pools. Ancient strike-slip basins are difficult to identify. However, two examples of economic importance are the Tertiary Bovey Basin of England with its economic deposits of sedimentary kaolin and minor lignite, and the late Mesozoic, coal-bearing, lacustrine basins of north-eastern China.

Readers who wish to explore the possible connexion between mineralization and continental lineaments are recommended to read Heyl (1972) and Noble (1980) and to balance this pro-lineament reading with that of Gilluly (1976).

Collision-related settings

Continental collision results from the closing of an oceanic or marginal basin whose previous existence is revealed by the presence of lines of obducted ophiolites. In some orogens interpreted as collision belts, such as the Damarides of southern Africa, this evidence is missing, and palaeomagnetic studies have not yielded evidence for significant separation at any stage of the cratonic blocks on either side of the orogen. Now collision can occur between two active arc systems, between an arc and an oceanic island chain or between an arc and a microcontinent, but the most extreme tectonic effects are produced when a continent on the subducting plate meets either a continental margin or island arc on the overriding plate. Subduction can then lead to melting in the heated continental slab and the production of S-type granites (Pitcher 1983). In many continental collision belts, highly differentiated granites of this suite are accompanied by tin and tungsten deposits of greisen and vein type (Chapter 13). The tin and tungsten deposits of the Hercynides of Europe (Cornwall, Erzegebirge, Portugal, etc.) are good examples. An excellent study of this activity is to be found in Beckinsale (1979) who discusses the generation of tin deposits in south-east Asia in the context of collision tectonics.

Uranium deposits, particularly of vein type, may be associated with S-type granites as in the Hercynides (Fig. 5.5) and Damarides (Rössing deposit, Chapter 10). Unfortunately, not all S-type granites have associated mineralization, for example those found in a belt stretching from Idaho to Baja California (Miller & Bradfish 1980). Sawkins (1984) has suggested that the reason that some have associated tin mineralization, some uranium and others neither, presumably reflects the geochemistry of the protolith from which these granites were derived, and their subsequent magmatic history.

Of course, the various collision possibilities described above can result in deposits developed in arcs of various types being incorporated into collision belts. The other features, such as intermontane basins and foreland basins, which are present in arc orogens (described in previous sections), will have similar mineralization potential.

In some collision belts where there has been only restricted development of oceanic crust, such as the European Alps (Hsü 1972, Dercourt & Paquet 1985), there was a plate margin setting where the plate motion was dominantly lateral. Subduction of oceanic crust was not, therefore, developed on any substantial scale, and this had important results as far as the evolution of the Alps is concerned. It accounts for several problems of Alpine geology: the distinctive tectonic style, the subordinate development of tectonic mélanges, the longevity of the Alpine flysch trenches and the very small amount of calc-alkaline igneous activity of either plutonic or volcanic type. This has resulted in a virtual absence of porphyry copper deposits and Besshi- and Kuroko-type massive sulphide deposits. Indeed, by comparison with the Andean belt or mature island arc terranes, the paucity of post-Hercynian mineral deposits in the Alps is most marked (Evans 1975). Despite the fairly common occurrence of ophiolites in parts of the Alps, this remark also applies to chromite deposits which are insignificant in size. The most important metalliferous mineralizations of the Alps are uranium-sulphide deposits in the Permian, which may have developed in a graben environment, and lead-zinc deposits in Triassic carbonate sequences of the eastern Alps. The latter appear to be syngenetic. The occurrence of baryte, celestine and anhydrite in the ores, the presence of evaporite beds in the host limestones and the general geological setting invite a comparison with the lead-zinc ores of the Red Sea region. Thus, as Evans (1975) has shown, by considering the possible nature of the evolving Alpine geosyncline from a plate tectonic point of view it is possible to account for the ore deposits which are present and the absence of others. These, and other deposits likely to be developed in Mediterranean-type geosynclines, are the ones upon which the search for ore deposits should be concentrated in the post-Hercynian rocks of the Alpine region.

Space does not allow an extension of this chapter to deal with Precambrian deposits, but it is clear from the literature that plate tectonic interpretations are applicable, at least in part, to the problems of the geology of Precambrian shields; the genesis of many Precambrian orebodies in the light of this theory has already been considered by a number of workers. Some useful summaries are given in Hutchison (1983) and Windley (1984), with a shorter review in the next chapter.

23

Ore Mineralization through Geological Time

It is now well known to geologists that the earth, and its crust in particular, have passed through an evolutionary sequence of changes throughout geological time (Windley 1984). These changes have been so considerable that we must expect them to have had some influence on the nature and extent of mineralization. Reference has already been made in this book (Chapter 6) to the association of most of the world's tin mineralization with Mesozoic and late Palaeozoic granites, to the virtual restriction of banded iron formation to the Precambrian and the bulk of it to the interval 2600–1800 Ma ago, and to the importance of the Precambrian for nickel and orthomagmatic ilmenite deposits. In Chapter 12, it was noted that the lack of numerous Phanerozoic nickel sulphide deposits may be due to depletion of the mantle in sulphur during the Archaean, and in Chapter 16 attention was drawn to the fact that volcanic massive sulphide deposits show important changes with time; these are discussed in detail in Hutchinson (1983). We will now examine such changes in the type and style of mineralization in a little more detail. These changes can be conveniently discussed in terms of the Archaean, Proterozoic and Phanerozoic intervals and the environments which prevailed during them.

The Archaean

This interval, 3800–2500 Ma ago, is notable both for the abundance of certain metals and the absence of others. Metals and metal associations developed in significant amounts include Au, Ag, Sb, Fe, Mn, Cr, Ni-Cu and Cu-Zn-Fe. Notable absentees are Pb, U, Th, Hg, Nb, Zr, REE and diamonds.

Two principal tectonic environments are found in the Archaean; the high grade regions and the greenstone belts. The former are not important for their mineral deposits which include Ni-Cu in amphibolites, e.g. Pikwe, Botswana, and chromite in layered anorthositic complexes, e.g. Fiskenæsset (west Greenland) as well as chromitite seams in dunite lenses known to be at least 3800 Ma old (Chadwick & Crewe 1986). The greenstone belts on the other hand are very rich in mineral deposits whose diversity has been described by Watson (1976). The principal mineral deposits are related to the major rock groups of the greenstone belts and their adjoining granitic terranes as follows:

(1) ultramafic flows and intrusions: Cr, Ni-Cu;
(2) mafic to felsic volcanics: Au, Ag, Cu-Zn;

(3) sediments: Fe, Mn and baryte;
(4) granites and pegmatites: Li, Ta, Be, Sn, Mo, Bi.

CHROMITE

This is not common in greenstone belts but a very notable exception is present at Selukwe in Zimbabwe. This is a very important occurrence of high grade chromite in serpentinites and talc-carbonate rocks intruded into schists which lie close to the Great Dyke. It resembles the podiform class of deposit, though it is in a very different tectonic environment from that in which this class is normally found.

NICKEL-COPPER

These deposits are mainly composed of massive and disseminated ores developed in or near the base of komatiitic and tholeiitic lava flows and sills as described in Chapter 12. Only four important fields occur, those of south-western Australia (Kalgoorlie belt), southern Canada (Abitibi belt), Zimbabwe and the Baltic Shield of northern USSR. As these metals and their host rocks are probably mantle-derived, this suggests the existence of metallogenic provinces controlled by inhomogeneities in the mantle, though the anomaly may not consist of an excess of nickel but rather of a concentration of sulphur which led to the extraction of nickel from silicate minerals.

GOLD

Gold has been won in smaller or larger amounts from every greenstone belt of any size and its occurrence is the principal reason for the early prospecting and mapping of these belts. The gold is principally in vein deposits cutting basic or intermediate igneous rocks — both intrusions and lava flows — but the more competent intrusives such as the Golden Mile Dolerite of Kalgoorlie are more important. Some gold deposits show an association with banded iron formation and these appear to have been deposited from subaqueous brines to form exhalites. The greatest concentration of gold mineralization occurs in the marginal zones of the greenstone belts near the bordering granite plutons and it decreases towards the centre of the belts. This suggests that it has been concentrated from the ultrabasic-basic volcanics by the action of thermal gradients set up by the intrusive plutons.

Silver is usually present with gold in the greenstone belts. In the Abitibi Belt, Canada it is found in a Au-Ag-Cu-Zn association at granite-mafic volcanic contacts. These granites also have porphyry-style Cu-Mo mineralization (refs. in Windley 1984).

COPPER-ZINC

Volcanogenic massive sulphide deposits are very common in the Archaean, especially in the Abitibi Orogen of southern Canada (Sangster & Scott 1976). These deposits are principally sources of copper and zinc but they are of Primitive type and their lead content is normally very low. The virtual absence of lead mineralization from these greenstone belts may to some extent reflect the fact that during the Archaean there had been insufficient time for much lead to be generated by the decay of uranium and thorium in the mantle.

IRON

Banded iron formation is common throughout Archaean time but not in the quantities in which it appears in the Proterozoic. It is the Algoma type (Chapter 19) which is present. There is some production from Archaean iron ore in Western Australia and from the Michipicoten Greenstone Belt, Canada where, at the Helen Mine, there occurs one of the world's largest siderite deposits. Morton & Nebel (1984) have recently published evidence favouring an exhalative origin for this.

The early to mid Proterozoic

The beginning of the Proterozoic about 2500 Ma ago was marked by a great change in tectonic conditions. The first stable lithospheric plates developed, although these seem to have been of small size. Their appearance permitted the formation of sedimentary basins, the deposition of platform sediments and the development of continental margin geosynclines.

GOLD-URANIUM CONGLOMERATES

The establishment of sedimentary basins allowed the formation of these deposits. The best known example is that of the Witwatersrand Basin with its widespread gold-uranium conglomerates (Chapter 19, Fig. 19.7), but other examples are known along the north shore of Lake Huron in Canada (Blind River area), at Serra de Jacobina in Brazil and at localities in Australia and Ghana. These deposits represent a unique metallogenic event which many feel has not been repeated because a reducing atmosphere was a *sine qua non* for the preservation of the detrital uranium minerals and pyrite, but this view is disputed (Chapter 19).

SEDIMENTARY MANGANESE DEPOSITS

There appears to have been an appreciable concentration of manganese in carbonate sediments over the period 2300–2000 Ma ago. These concentrations are known in the Republic of South Africa, Brazil and India.

SEDIMENT-HOSTED STRATIFORM LEAD-ZINC DEPOSITS

By about 1700 Ma ago, the CO_2 content in the hydrosphere had reached a level that permitted the deposition of thick dolomite sequences. In a number of localities these host syngenetic base metal sulphide orebodies such as those of the Balmat-Edwards and Franklin Furnace districts, USA. In other sedimentary hosts varying from dolomitic shales to siltstones there are deposits such as McArthur River (Pb-Zn-Ag) and Mount Isa (Pb-Zn-Ag and separate Cu orebodies) in Australia and Sullivan, Canada. Various exhalative and biogenic origins have been suggested for these ores. They should not be confused with the carbonate-hosted lead-zinc deposits described in Chapter 18.

THE CHROMIUM-NICKEL-PLATINUM-COPPER ASSOCIATION

The presence of small crustal plates permitted the development of large-scale fracture systems and the intrusion at this time of giant dyke-like layered bodies, such as the Great Dyke of Zimbabwe, and enormous layered stratiform igneous complexes like that of the Bushveld in South Africa. These are the repositories of enormous quantities of chromium and platinum with other important by-products (Chapter 11). Though similar intrusions occur in other parts of the world, the great concentration of chromium is in southern Africa, and this has led some workers to postulate the presence of chromium-rich mantle beneath this region.

TITANIUM-IRON ASSOCIATION

About the middle of the Proterozoic many anorthosite plutons were emplaced in two linear belts which now lie in the northern and southern hemispheres when plotted on a pre-Permian continental drift reconstruction. A number of these carry ilmenite orebodies which are exploited in Norway and Canada. This was a unique magmatic event that has not been repeated. It suggests the gathering of reservoirs of magma (in the top of the mantle) which were able to penetrate upwards along deep fractures in the crust. The strength of the crust and the thermal conditions seem to have reached the point where magma could accumulate at the base of the crust in this extensive manner, rather like the accumulation of magma beneath present day rift valleys.

DIAMONDS

Diamantiferous kimberlites appear for the first time in the Proterozoic. This suggests that the geothermal gradients had decreased considerably permitting the development of thick lithospheric plates, because diamonds, requiring extreme pressure for their formation, cannot crystallize unless the lithosphere is at least 120 km thick.

BANDED IRON FORMATION (BIF)

The greatest development of BIF occurred during the interval 2600–1800 Ma ago (Goldich 1973). Although this rock-type is important in the Archaean, it could not be developed on the large scale seen in the early Proterozoic because stable continental plates were not present. Following the development of stable lithospheric plates, BIF could be laid down synchronously over very large areas; this took place possibly in intra-plate basins and certainly on continental shelves. The weathering of basic volcanics in the greenstone belts would have yielded ample iron and silica. If the atmosphere was essentially CO_2-rich, the iron could have travelled largely in ionic solution. It is now suspected that iron-precipitating bacteria may have played an important part in depositing the iron and oxidizing it to the ferric state as modern iron bacteria are able to oxidize ferrous iron at very low levels of oxygen concentration. Although BIF appears at later times in the Proterozoic, its development is very restricted compared with that in the early Proterozoic and this fall off in importance has been correlated by some workers with the evolution of an oxidizing atmosphere. In the Phanerozoic the place of BIF is taken by the Clinton and Minette ironstones.

Mid-late Proterozoic

HIGH GRADE LINEAR BELTS

It has been suggested (Piper 1974, 1975, 1976) that a supercontinent existed through much of Proterozoic time, and Davies & Windley (1976) have plotted the trends of major high grade linear belts on this supercontinent showing that they lie on small circles having a common point of rotation. These linear belts affect middle to late Proterozoic as well as older rocks and include shear belts, mobile belts and linear zones of transcurrent displacements of magnetic and gravity anomaly patterns. They contain some deep dislocations that penetrate right down to the mantle and form channelways for uprising magma. Nickel mineralization occurs in some of these belts, e.g. the Nelson River Gneissic Belt of Manitoba.

SEDIMENTARY COPPER

Watson (1973) has drawn attention to the anomalously high concentrations of copper in some late Proterozoic sediments in many parts of the world. These represent the oldest large sedimentary copper accumulations. Examples include the Katanga System of Zambia and Shaba (Chapter 16) and the Belt Series of the north-western USA.

SEDIMENTARY MANGANESE DEPOSITS

A second important period of manganese deposition occurred during the late Proterozoic, and manganese-rich sediments were laid down on or along the margins of cratonic blocks. The most important deposits are in central India and Namibia. These deposits have been metamorphosed but the Indian examples, now spessartite-quartz rocks, appear to have originally been manganiferous argillaceous and arenaceous sediments.

TIN

Watson (1973) observed that tin mineralization does not appear in major quantities in the crust until the late Proterozoic, where it is associated with high level alkaline and peralkaline anorogenic granite and pegmatities. This is particularly the case in Africa where these deposits lie in three north-south belts (Fig. 6.1). Another belt passes through the Rhondônia district of western Brazil.

The Phanerozoic

Towards the end of the Proterozoic a new tectonic pattern developed which gave rise to Phanerozoic fold belts formed by continental drift. There was large scale recycling of oceanic crust, which greatly increased the number and variety of ore-forming environments by producing long chains of island and continental margin arcs, back-arc basins, rift-bordered basins and other features described in Chapter 22. Consequently some Archaean volcanic types and Proterozoic sedimentalogical types reappear together with a few types dependent perhaps on more evolved plate mechanisms or extreme geochemical evolution of siliceous magmas. The latter process, plus recycling of crustal accumulations, may account for the important development of molybdenum, tin and tungsten ores in the Phanerozoic (Meyer 1981).

Cyprus-type copper-pyrite deposits and sandstone uranium deposits first appear in the Phanerozoic, while podiform chromite, found in the Archaean but not in the Proterozoic (Meyer 1985), becomes much more common. Lead becomes increasingly important in the volcanic-associated massive sulphide deposits. Some of the largest epigenetic metal concentrations of the Phanerozoic are those of the porphyry copper and molybdenum deposits of the continental margin and island arcs. Much of this mineralization activity has been covered in the previous chapter. I hope that a study of these two brief chapters will reveal the importance of taking an account of plate tectonic settings, and the effect of continental drift and geological time in designing mineral exploration programmes.

> *'For as birds are born to fly freely through the air, so are fishes born to swim through the waters, while to other creatures Nature has given the earth that they might live in it, and particularly to man that he might cultivate it and draw out of its caverns metals and other mineral products.'*
>
> Georgius Agricola in
> *De Re Metallica*, 1556

Appendix 1

Formulae of some Minerals mentioned in the Text

Almandine	$Fe_3Al_2(SiO_4)_3$
Alunite	$KAl_3(SO_4)_2(OH)_6$
Amblygonite	$(Li,Na)Al(PO_4)(F,OH)$
Anhydrite	$CaSO_4$
Apatite	$Ca_5(F,Cl)(PO_4)_3$
Arsenopyrite	$FeAsS$
Baddeleyite	ZrO_2
Baryte	$BaSO_4$
Bastnäsite	$(Ce,La)(CO_3)F$
Bertrandite	$Be_4Si_2O_7(OH)_2$
Beryl	$Be_3Al_2Si_6O_{18}$
Boehmite	$AlO(OH)$
Bornite	Cu_5FeS_4
Carnotite	$K_2(UO_2)_2(VO_4)_2.3H_2O$
Cassiterite	SnO_2
Chalcocite	Cu_2S
Chalcopyrite	$CuFeS_2$
Chromite	$FeCr_2O_4$
Cinnabar	HgS
Coffinite	$(USiO_4)_{1-x}(OH)_{4x}$
Columbite-tantalite	$(Fe,Mn)(Nb,Ta)_2O_6$
Copper	Cu
Covellite	CuS
Diamond	C
Diaspore	$AlO(OH)$
Digenite	Cu_9S_5
Enargite	Cu_3AsS_4
Eucryptite	$LiSiAlO_4$
Fluorite	CaF_2
Francolite	$Ca_5(PO_4,CO_3)_3(OH,F)$
Galena	PbS
Garnierite	$(Ni,Mg)_3Si_2O_5(OH)_4$
Gibbsite	$Al(OH)_3$
Gold	Au
Graphite	C
Gypsum	$CaSO_4.2H_2O$

Hematite	Fe_2O_3
Ilmenite	$FeTiO_3$
Jarosite	$KFe_3(SO_4)_2(OH)_6$
Kaolinite	$Al_2Si_2O_5(OH)_4$
Lepidolite	$KLi_2Al(Si_4O_{10})(OH)_2$
Limonite	'Sack' term for brown, hydrous iron oxides such as goethite and lepidocrocite
Löllingite	$FeAs_2$
Magnesite	$MgCO_3$
Magnetite	Fe_3O_4
Marcasite	FeS_2
Microlite	$(Na,Ca)_2Ta_2O_6(O,OH,F)$
Molybdenite	MoS_2
Monazite	$(Ce,La,Y,Th)PO_4$
Montroseite	$VO(OH)$
Muscovite	$KAl_2Si_3O_{10}(OH)_2$
Nepheline	$NaAlSiO_4$
Parisite	$(Ce,La)_2Ca(CO_3)_3F_2$
Pentlandite	$(Fe,Ni)_9S_8$
Petalite	$LiAlSi_4O_{10}$
Pitchblende	UO_2
Pollucite	$CsSi_2AlO_6$
Pyrite	FeS_2
Pyrochlore	$NaCaNb_2O_6F$
Pyrolusite	MnO_2
Pyrrhotite	$Fe_{1-x}S$
Quartz	SiO_2
Rhodochrosite	$MnCO_3$
Rutile	TiO_2
Scheelite	$CaWO_4$
Sericite	Fine-grained muscovite
Serpentine	$Mg_3Si_2O_5(OH)_4$
Siderite	$FeCO_3$
Silver	Ag
Sphalerite	$(Zn,Fe)S$
Spodumene	$LiAlS_2O_6$
Stannite	Cu_2FeSnS_4
Strontianite	$SrCO_3$
Talc	$Mg_3Si_4O_{10}(OH)_2$
Tantalite	$(Fe,Mn)(Ta,Nb)_2O_6$
Tetrahedrite-tennanite	$(Cu,Fe,Ag)_{12}(Sb,As)_4S_{13}$
Uraninite	UO_2
Uranothorite	$(Th,U)SiO_4$

Wad	'Sack' term for black manganese oxides and hydroxides
Wolframite	$(Fe,Mn)WO_4$
Xenotime	YPO_4
Zircon	$ZrSiO_4$

Appendix 2

Postscript 1

Richardson (1986) has determined the ages of inclusions in diamonds from the Premier kimberlite, southern Africa and the Argyle lamproite to be 1150 and 1580 Ma respectively, compared with the hot diatreme emplacement of 1100–1200 Ma. He suggested that these data indicate a second genetically distinct origin of diamonds from that described above, apparently related in time and space to kimberlitic or lamproitic magmatism.

Postscript 2

The Miocene age quoted on page 104 is still correct for the West Kimberley pipes but Skinner *et al.* (1985) recorded 1126±9 Ma for the Argyle Pipe. This is consistent with the ages of a large number of other kimberlites, lamproites and related alkalic rocks from world-wide localities, suggesting an important period of alkalic intrusive activity about this time.

Skinner E.M.W., Smith C.B., Bristow J.W., Scott Smith B.H. & Dawson J.B. (1985) Proterozoic Kimberlites and Lamproites and a Preliminary Age for the Argyle Lamproite Pipe, Western Australia. *Trans. geol. Soc. S. Afr.*, **88**, 335–40.

References

Adamek P.M. & Wilson M.R. (1979) The Evolution of a Uranium Province in Northern Sweden. *Phil. Trans. R. Soc. London*, **A291**, 355–68.

Alderton D.H.M. (1978) Fluid Inclusion Data for Lead-zinc Ores from South-west England. *Trans. Instn Min. Metall. (Sect. B: Appl. earth Sci.)*, **87**, B132–5.

Amade E. (1983) Caractéristiques Comparées des Quatre Principaux 'Porphyry Copper' de Niugini. *Chron. rech. min.*, **472**, 3–22.

Anderson C.A. (1948) Structural Control of Copper Mineralization, Bagdad, Arizona. *Amer. Inst. Min. Metall. Engng Trans.*, **178**, 170–80.

Anderson G.M. (1975) Precipitation of Mississippi Valley-type Ores, *Econ. Geol.*, **70**, 937–42.

Anderson G.M. (1977) Thermodynamics and Sulfide Solubilities. In Greenwood H.J. (ed.), *Application of Thermodynamics to Petrology and Ore Deposits*. Mineralogical Association of Canada, Toronto.

Anderson G.M. (1983) Some Geochemical Aspects of Sulphide Precipitation in Carbonate Rocks. In Kisvarsanyi G., Grant S.K., Pratt W.P. & Koenig J.W. (eds), *International Conference on Mississippi Valley Type Lead-Zinc Deposits*, 77–85, Univ. of Missouri-Rolla, Rolla.

Anderson P. (1980) Regional Time-Space Distribution of Porphyry Deposits — a Decisive Test for the Origin of Metals in Magma-related Ore Deposits. In Ridge J.D. (ed.), *Proceedings of the Fifth Quadrennial IAGOD Symposium*, 35–48, E. Schweitzerbart'sche Verlagsbuchhandlung, Stuttgart.

Anhaeusser C.R. (1976) The Nature and Distribution of Archaean Gold Mineralization in Southern Africa. *Miner. Sci. Engng*, **8**, 46–84.

Annels A.E. (1979) Mufulira Greywackes and their Associated Sulphides. *Trans. Instn Min. Metall. (Sect. B: Appl. earth sci.)*, **88**, B15–B23.

Annels A.E. (1984) The Geotectonic Environment of Zambian Copper-cobalt Mineralization. *J. geol. Soc. London*, **141**, 279–89.

Anon. (1979) Nchanga Consolidated Copper Mines. *Eng. Ming J.*, **180** (Nov.), 150.

Anon. (1982a) Argyle Mine, Australia will be one of the World's Major Diamond Producers. *Ming Mag. London*, **147**, 261.

Anon. (1982b) The Rio Tinto-Zinc Corporation PLC Annual Report and Accounts.

Anon. (1983a) Diamonds. *Eng. Ming J.*, **184** (Nov.), 119–23.

Anon. (1983b) Ashton Mining Limited Annual Report and Accounts.

Anon. (1983c) New Zealand: Ironsand to Steel. *Ming Mag. London*, **148**, 346–53.

Anon. (1984) *Ming Mag. London*, **151**, 292.

Anon. (1985a) *Ming J. London*, **7802**, 146.

Anon. (1985b) Synthetic Diamond Breakthrough. *Ind. Minerals*, **10**, (June).

Anon. (1985c) Hard Times for Tin Mines. *Ming. Mag. London*, **305**, 125–6.

Anon. (1985d) Boddington Possible Go Ahead. *Ming J.*, **304**, 99.

Appel P.W.U. (1979) Stratabound Copper Sulphides in a Banded Iron-Formation and in Basaltic Tuffs in the Early Precambrian Isua Supracrustal Belt, West Greenland. *Econ. Geol.*, **74**, 45–52.

Arndt N.T. & Nisbet E.G. (1982a) What is a Komatiite? In Arndt N.T. & Nisbet E.G. (eds.), *Komatiites*, 19–27, George Allen & Unwin, London.

Ardnt N.T. & Nisbet E.G. (eds.) (1982b) *Komatiites*, George Allen & Unwin, London.

Arribas A. & Gumiel P. (1984) First Occurrence of a Strata-bound Sb-W-Hg Deposit in the Spanish Hercynian Massif. In Wauschkuhn A., Kluth C. & Zimmerman R.A. (eds.) *Syngenesis and Epigenesis in the Formation of Mineral Deposits*, 468–481, Springer-Verlag, Berlin.

Atkinson W.J., Hughes F.E. & Smith C.B. (1984) A Review of the Kimberlitic Rocks of Western Australia. In Kornprobst J. (ed.), *Kimberlites I: Kimberlites and Related Rocks*, 195–224, Elsevier, Amsterdam.

Awramik S.M. & Barghoorn E.S. (1977) The Gunflint Microbiota. *Precambrian Res.*, **5**, 121–42.

Aye F. (1982) Contrôles Géologiques des Gîtes Stratiformes de Pb, Zn, Cu, Ag de la Bordure du Bassin de Châteaulin. *Mémoire du BRGM*, **120**, Orléans.
Ayres L.D. & Černý P. (1982) Metallogeny of Granitoid Rocks in the Canadian Shield. *Can. Mineral.*, **20**, 439–536.
Badham J.P.N. (1978) Slumped Sulphide Deposits at Avoca, Ireland, and Their Significance. *Trans. Instn Min. Metall. (Sect. B: Appl. earth sci.)*, **87**, B21–B26.
Bagby W.C. & Berger B.R. (1986) Geologic Characteristics of Sediment-hosted, Disseminated Precious-metal Deposits in the Western United States. In Berger B.R. & Bethke P.M. (eds.) *Geology and Geochemistry of Epithermal Systems*, 169–202, Society of Economic Geologists, El Paso.
Baker P.E. (1968) Comparative Volcanology and Petrology of the Atlantic Island Arcs. *Bull. Volcanol.*, **32**, 189–206.
Baldwin J.A. & Pearce J.A. (1982) Discrimination of Productive and Nonproductive Porphyritic Intrusions in Chilean Andes. *Econ. Geol.* **77**, 664–74.
Ball T.K. & Bland D.J. (1985) The Cae Coch Volcanogenic Massive Sulphide Deposit, Trefriw, North Wales. *J. geol. Soc. London*, **142**, 889–98.
Banks D.A. (1985) A Fossil Hydrothermal Worm Assemblage from the Tynagh Lead-zinc Deposit in Ireland. *Nature*, **313**, 128–31.
Barbier J. (1982) Géochimie en Roche sur le Gîte de Scheelite de Salau. *Bull. BRGM* (2), section II, No. 1, 25–44.
Bardossy G. (1982) *Karst Bauxite: Bauxite Deposits on Carbonate Rocks*. Elsevier, Amsterdam.
Barker D.S. (1969) North American Felspathoidal Rocks in Space and Time. *Bull. geol. Soc. Am.*, **80**, 2369–72.
Barnes H.L. (1975) Zoning of Ore Deposits: Types and Causes. *Trans. R. Soc. Edinburgh*, **69**, 295–311.
Barnes H.L. (ed.) (1979a) *Geochemistry of Hydrothermal Ore Deposits*, Second Edition, Wiley, New York.
Barnes H.L. (1979b) Solubilities of Ore Minerals. In Barnes H.L. (ed.) *Geochemistry of Hydrothermal Ore Deposits, Second Edition*, 404–60, Wiley, New York.
Barnes H.L. (1983) Ore-depositing Reactions in Mississippi Valley-type Deposits. In Kisvarsanyi G., Grant S.K., Pratt W.P. & Koenig J.W. (eds.), *International Conference on Mississippi Valley Type Lead-Zinc Deposits*, 77–85, Univ. of Missouri-Rolla, Rolla.
Barnes R.G. (1983) Stratiform and Stratabound Tungsten Mineralization in the Broken Hill Block, N.S.W. *J. geol. Soc. Australia*, **30**, 225–39.
Barton P.B. & Skinner B.J. (1979) Sulphide Mineral Stabilities. In Barnes H.L. (ed.), *Geochemistry of Hydrothermal Ore Deposits*, Second Edition, 278–403, Wiley, New York.
Barton P.B. & Toulmin P. (1963) Sphalerite Phase Equilibria in the System Fe-Zn-S between 580°C and 850°C. *Econ. Geol.*, **58**, 1191–2.
Batchelor B.C. (1979) Geological Characteristics of Certain Coastal and Off-shore Placers as Essential Guides for Tin Exploration in Sundaland, S.E. Asia. In Yeap C.H. (ed.), *Geology of Tin Deposits*, Bull. Geol. Soc. Malaysia, **II**, 283–314.
Bateman A.M. (1950) *Economic Mineral Deposits*. Wiley, New York.
Baumann L. (1965) Zur Erzführung und regionalen Verbreitung des 'Felsithorizontes' von Halbrücke. *Freib. Forsch.*, **C186**, 63–81.
Baumann L. (1970) Tin Deposits of the Erzgebirge. *Trans. Instn Min. Metall. (Sect. B: Appl. earth sci.)* **79**, B68–B75.
Baumann L. & Krs M. (1967) Paläomagnetische Altersbestimmungen an einigen Mineralparagenesen des Freiberger Lagerstättenbezirkes. *Geologie*, **16**, 765–80.
Bayley R.W. & James H.L. (1973) Precambrian Iron-Formations of the United States. *Econ. Geol.*, **68**, 934–59.
Beales F.W. & Jackson S.A. (1966) Precipitation of Lead-zinc Ores in Carbonate Reservoirs as Illustrated by Pine Point Ore Field, Canada. *Trans. Instn Min. Metall. (Sect. B. Appl. earth Sci.)*, **75**, B278–85.
Beaufort D. & Meunier A. (1983) Petrographic Characterization of an Argillic Hydrothermal Alteration Containing Illite, K-rectorite, K-beidellite, Kaolinite and Carbonates in a Cupromolybdic porphyry at Sibert (Rhone, France). *Bull. mineral.*, **106**, 535–51.
Beckinsale R.D. (1979) Granite Magmatism in the Tin Belt of South-east Asia. In Atherton M.P. & Tarney J. (eds.), *Origin of Granite Batholiths*, 34–44, Shiva, Orpington.
Berning J. Cooke R., Hiemstra S.A. & Hoffman U. (1976) The Rössing Uranium Deposit, South-west Africa. *Econ. Geol.*, **71**, 351–68.
Bernstein L.R. (1986) Geology and Mineralogy of the Apex Germanium-Gallium Mine, Washington County, Utah. *Bull. 1577 U.S. Geol. Surv.* Washington.

Besson M., Boyd R., Czamanske G., Foose M., Groves D., van Gruenewaldt G., Naldrett A., Nilsson G., Page N., Papunen H. & Peredery W. (1979) IGCP Project No. 161 and a Proposed Classification of Ni-Cu-PGE Sulphide Deposits. *Can. Mineral.*, **17**, 143–4.
Best J.L. & Brayshaw A.C. (1985) Flow Separation — a Physical Process for the Concentration of Heavy Mineral within Alluvial Channels. *J. geol. Soc. London*, **142**, 747–55.
Best M.C. (1982) *Igneous and Metamorphic Petrology*. Freeman, San Francisco.
Bichan R. (1969) Origin of Chromite Seams in the Hartley Complex of the Great Dyke, Rhodesia. In Wilson H.D.B. (ed.), *Magmatic Ore Deposits, Econ. Geol.* Monograph **4**, 95–113.
Bischoff J.L., Radtke A.S. & Rosenbauer R.J. (1981) Hydrothermal Alteration of Greywacke by Brine and Seawater: Roles of Alteration and Chloride Complexing on Metal Solubilization at 200° and 350°C. *Econ. Geol.*, **76**, 659–76.
Binda P.L. (1975) Detrital Bornite Grains in the Late Precambrian B Greywacke of Mufulira, Zambia. *Mineral. Deposita*, **10**, 101–7.
Bjrlykke A. & Sangster D.F. (1981) An Overview of Sandstone Lead Deposits and Their Relation to Red-bed Copper and Carbonate-hosted Lead-zinc Deposits. *Econ Geol.*, **75th. Anniv. Vol.**, 179–213.
Blissenbach E.B. & Fellerer R. (1973) Continental Drift and the Origin of Certain Mineral Deposits. *Geol. Rundsch.*, **62**, 812–39.
Boast A.M., Coleman M.L. & Halls C. (1981) Textural and Stable Isotopic Evidence for the Genesis of the Tynagh Base Metal Deposit, Ireland. *Econ. Geol.*, **76**, 27–55.
Bonatti E., Guernstein-Honnorez B.-M. & Honnorez J. (1976) Copper-Iron Sulphide Mineralizations from the Equatorial Mid-Atlantic Ridge. *Econ. Geol.*, **71**, 1515–25.
Bowden P. (1982) Magmatic Evolution and Mineralization in the Nigerian Younger Granite Province. In Evans A.M. (ed.), *Metallization Associated with Acid Magmatism*, 51–61, Wiley, Chichester.
Bowles J.F.W. (1985) A Consideration of the Development of Platinum-group Minerals in Laterite. *Can. Mineral.*, **23**, 296.
Boyle R.W. (1959) The Geochemistry, Origin and Role of Carbon Dioxide, Sulfur, and Boron in the Yellowknife Gold Deposits Northwest Territories, Canada. *Econ. Geol.*, **54**, 1506–24.
Boyle R.W. (1970) Regularities in Wall-rock Alteration Phenomena Associated with Epigenetic Deposits. In Pouba Z. & Štemprok M. (eds), *Problems of Hydrothermal Ore Deposition*, 233–260. E. Schweizerbart'sche Verlagsbuchandlung, Stuttgart.
Boyle R.W. (1979) The Geochemistry of Gold and Its Deposits. *Geol Surv. Canada*, Bull. **280**.
Brady L.L. & Jobson H.E. (1973) An Experimental Study of Heavy-mineral Segregation under Alluvial-flow Conditions. Prof. Pap. 562-K, *U.S. Geol. Surv.*, Washington.
Bray C.J., Spooner E.T.C., Golightly J.P. & Saracoglu N. (1982) Carbon and Sulphur Isotope Geochemistry of Unconformity-related Uranium Mineralization, McClean Lake Deposits, Northern Saskatchewan, Canada. *Geol. Soc. Amer. Abstr. Vol., New Orleans Meeting*, **451**.
Brigo L. & Omenetto P. (1983) Scheelite-bearing Occurrences in the Italian Alps: Geotectonic and Lithostratigraphic Setting. In Schneider H-J. (ed.), *Mineral Deposits of the Alps and of the Alpine Epoch in Europe*, 41–50. Springer-Verlag, Berlin.
Brimhall G.H. (1979) Lithologic Determination of Mass Transfer Mechanisms of Multiple-Stage Porphyry Copper Mineralization at Butte, Montana: Vein formation by Hypogene Leaching and Enrichment of Potassium-Silicate Protore. *Econ. Geol.* **74**, 556–89.
Brimhall G.H. & Ghiorso M.S. (1983) Origin and Ore-forming Consequences of the Advanced Argillic Alteration Process in Hypogene Environments by Magmatic Gas Contamination of Meteoric Fluids. *Econ. Geol.*, **78**, 73–90.
Brocoum S.J. & Dalziel I.W.D. (1974) The Sudbury Basin, the Southern Province, the Grenville Front and the Penokean Orogeny. *Bull. geol. Soc. Am.*, **85**, 1571–80.
Brown A.C. (1978) Stratiform Copper Deposits — Evidence for their Post-sedimentary Origin. *Miner. Sci. Engng*, **10**, 172–81.
Brown C. & Williams B. (1985) A Gravity and Magnetic Interpretation of the Structure of the Irish Midlands and Its Relation to Ore Genesis. *J. geol. Soc. London*, **142**, 1059–75.
Brownlow A.H. (1979) *Geochemistry*. Prentice-Hall, Englewood Cliffs, N.J.
Brunt D.A. (1978) Uranium in Tertiary Stream Channels, Lake Frome Area, South Australia. *Proc. Australas. Inst. Min. Metall.*, No. **266**, 79–90.
Bryce J.D., Thompson J.M. & Staff (1958) The Bicroft Operation. *Western Miner and Oil Review*, **31**, **4**, 79–92.
Buchanan D.L. & Rouse J.E. (1984) Role of Contamination in the Precipitation of Sulphides in the Platreef of the Bushveld Complex. In Buchanan D.L. & Jones M.J. (eds.), *Sulphide Deposits in Mafic and Ultramafic Rocks*, 141–6, Instn Min. Metall., London.
Burger J.R. (1985) More Desert Gold: Newmount Finds Yet Another Five Million oz. at Carlin,

Nevada. *Eng. Ming J.* (October) **17**.
Burke K. & Dewey J.F. (1973) Plume-generated Triple Junctions: Key Indicators in Applying Plate Tectonics to Old Rocks. *J. Geol. London*, **81**, 406–33.
Burn R.G. (1971) Localized Deformation and Recrystallization of Sulphides in an Epigenetic Mineral Deposit. *Trans. Instn Min. Metall. (Sect. B: Appl. earth sci.)*, **80**, B116–B119.
Burnham C.W. (1959) Metallogenic Provinces of the Southwestern United States and Northern Mexico. *Bull. State Bur. Mines Min. Resources, New Mexico*, **65**, 76 pp.
Burnham C.W. (1979) Magmas and Hydrothermal Fluids. In Barnes H.L. (ed.), *Geochemistry of Hydrothermal Ore Deposits, Second Edition*, 71–136, Wiley, New York.
Burnham C.W. & Ohmoto H. (1980) Late Stage Processes of Felsic Magmatism. In Ishihari S. & Takenouchi S (eds.), *Granitic Magmatism and Related Mineralization*, 1–11. *Mining Geol Spec.* **8**, *Soc. Ming. Geologists Japan*.
Burnie S.W., Schwarcz H.P. & Crocket J.H. (1972) A sulfur isotopic study of the White Pine Mine. Michigan, *Econ. Geol.*, **67**, 895–914.
Burrows D.R., Wood P.C. & Spooner E.T.C. (1986) Carbon Isotope Evidence for a Magmatic Origin for Archaean Gold-quartz Vein Ore Deposits. *Nature*, **321**, 851–54.
Burt D.M. (1981) Acidity-Salinity Diagrams — Application to Greisen and Porphyry Deposits. *Econ. Geol.*, **76**, 832–43.
Button A. (1976) Transvaal and Hamersley Basins — Review of Basin Development and Mineral Deposits. *Mineral. Sci. Engng*, **8**, 262–93.
Callahan W.H. (1967) Some Spatial and Temporal Aspects of the Localization of Mississippi Valley-Appalachian Type Ore Deposits. In Brown J.S. (ed.), *Genesis of Stratiform Lead-zinc-barite-fluorite Deposits*, 14–19. Economic Geology Publishing Co., Lancaster, Pennsylvania.
Cameron E.N. & Emerson M.E. (1959) The Origin of Certain Chromite Deposits in the Eastern Part of the Bushveld Complex. *Econ. Geol.* **54**, 1151–1213.
Cameron E.N., Jahns R.H., McNair A.H. & Page L.R. (1949) Internal Structure of Granitic Pegmatites. *Econ. Geol.* Mon. **2**, 115.
Campbell J.D. (1953) The Triton Gold Mine, Reedy, Western Australia. In Edwards A.B. (ed.), *Geology of Australian Ore Deposits*, 195–207, Australasian Inst. Min. Metall, Melbourne.
Cann J.R. (1970) New Model for the Structure of the Ocean Crust. *Nature*, **266**, 928–30.
Cann J.R. & Strens M.R. (1982) Black Smokers Fuelled by Freezing Magma. *Nature*, **298**, 147–9.
Capuano R.M. & Cole D.R. (1982) Fluid-mineral Equilibria in a Hydrothermal System, Roosevelt Hot Springs, Utah. *Geochim. cosmochim. Acta.*, **46**, 1353–64.
Card K.D., Gupta V.K., McGrath P.H. & Grant F.S. (1984) The Sudbury Structure: Its Regional Geological and Geophysical Setting. In Pye E.G., Naldrett A.J. & Giblin P.E. (eds.), *The Geology and Ore Deposits of the Sudbury Structure*, 25–43, **Spec. Vol. 1**, Ontario Geol. Surv., Toronto.
Carmichael I.S.E., Turner F.J. & Verhoogen J. (1974) *Igneous Petrology*. McGraw-Hill, New York.
Cassard D., Nicolas A., Rabinovitch M., Moutte J., Leblanc M. & Prinzhofer A. (1981) Structural Classification of Chromite Pods in Southern New Caledonia. *Econ. Geol.*, **76**, 805–31.
Cathelineau M. (1982) Signification de la fluorine dans les gisements d'uranium de la chaine hercynienne. *Bull. BRGM* (2), sect. II, (**4**) 407–13.
Cathles L.M. & Smith A.T. (1983) Thermal Constraints on the Formation of Mississippi Valley-Type Lead-Zinc Deposits and Their Implications for Episodic Basin Dewatering and Deposit Genesis. *Econ. Geol.*, **78**, 983–1002.
Cavarretta G., Gianelli G. & Puxeddu M. (1982) Formation of Authigenic Minerals and Their Use as Indicators of the Physicochemical Parameters of the Fluid in the Largerello-Travale Geothermal Field. *Econ. Geol.* **77**, 1071–84.
Cawthorn R.G., Barton J.W. Jr & Viljoen M.J. (1985) Interaction of Floor Rocks with the Platreef on Overysel, Potgietersrus, Northern Transvaal. *Econ. Geol.*, **80**, 988–1006.
Cayeux L. (1906) Structure et origine probable du minerai de fer magnétique de Dielette (Manche). *C.R. Acad. Sci. Paris*, **142**, 716–8.
Černý P. (1982a) Anatomy and Classification of Granitic Pegmatites. In Černý P. (ed.), *Short course in Granitic Pegmatites in Science and Industry*, Mineral. Assoc. Canada, Winnipeg, 1–39.
Černý P. (1982b) Petrogenesis of Granitic Pegmatites. In Černý P. (ed.), *Short Course in Granitic Pegmatites in Science and Industry*, Mineral. Assoc. Canada, Winnipeg, 405–461.
Černý P. (1982c) The Tanco Pegmatite at Bernic Lake, Southeastern Manitoba. In Černý P. (ed.), *Short Course in Granitic Pegmatites in Science and Industry*, 527–43 Mineral. Assoc. Canada, Winnipeg.

Černý P. (ed.) (1982d) *Short Course in Granitic Pegmatites in Science and Industry*, Mineral. Assoc. Canada, Winipeg.

Chadwick B. & Crewe M.A. (1986) Chromite in the Early Archaean Akilia Association (ca. 3000 Ma), Ivisârtoq Region, Inner Godthåbsfjord, West Greenland. *Econ. Geol.*, **81**, 184–91.

Champigny N. & Sinclair A.J. (1982) Cinola Gold Deposit, Queen Charlotte Islands, B.C. — A Geochemical Case History. In Levinson A.A. (ed.), *Precious Metals in the Northern Cordillera*, 121–37.

Chartrand F.M. & Brown A.C. (1985) The Diagenetic Origin of Stratiform Copper Mineralization, Coates Lake, Redstone Copper Belt, N.W.T., Canada. *Econ. Geol.*, **80**, 325–43.

Chatterjee A.K. & Strong D.F. (1984) Rare-earth and Other Element Variations in Greisens and Granites Associated with East Kemptville Tin Deposit, Nova Scotia, Canada. *Trans. Instn Min. Metall. (Sect. B: Appl. earth sci.)*, **93**, B59–70.

Chételat E. de (1947) La Genèse et l'évolution des gisements de nickel de la Nouvelle-Calédonie. *Bull. Soc. geol. Fr.*, ser. 5, **17**, 105–60.

Chivas R. & Wilkins W.T. (1977) Fluid Inclusion Studies in Relation to Hydrothermal Alteration and Mineralization at the Koloula Porphyry Copper Prospect, Guadalcanal. *Econ. Geol.*, **72**, 153–69.

Clark G.S. (1982) Rubidium-Strontium Isotope Systematics of Complex Granitic Pegmatites. In Černý P. (ed.), *Short Course in Granitic Pegmatites in Science and Industry*, Mineral. Assoc. Canada, Winnipeg, 347–71.

Clark K.F. (1972) Stockwork Molybdenum Deposits in the Western Cordillera of North America. *Econ. Geol.*, **67**, 731–58.

Clark R.J. McH., Homeniuk L.A. & Bonnar R. (1982) Uranium Geology in the Athabaska and a Comparison with Other Canadian Proterozoic Basins. *CIM Bull.*, **75**, April, 91–8.

Clarke M.C.G. (1983) Current Chinese thinking on the South China Tungsten Province. *Trans Instn Min. Metall. (Sect. B. Appl, earth sci.)*, **92**, B10–15.

Clement C.R., Skinner E.M.W. & Scott Smith B.H. (1984) Kimberlite Redefined. *J. Geol.*, **92**, 223–8.

Coleman R.J. (1975) Savage River Magnetite Deposits. In Knight C.L. (ed.), *Economic Geology of Australia and Papua New Guinea, 1, Metals*, 598–604, Australasian Inst. Min. Metall., Parkville.

Colvine A.C. (1983) (ed.) *The Geology of Gold in Ontario*. Ont. Geol. Surv. Misc. Pap. 110.

Comrate (Committee on Mineral Resources and the Environment) (1975) *Mineral Resources and the Environment*. National Academy of Sciences, Washington.

Conn H.K. (1979) The Johns-Manville Platinum-palladium Prospect, Stillwater Complex, Montana, U.S.A. *Can. Mineral*, **17**, 463–8.

Cook P.J. (1984) Spatial and Temporal Controls on the Formation of Phosphate Deposits. In Nriagu J.O. & Moore P.B. (eds.), *Phosphate Minerals*, 242–74, Springer-Verlag, Berlin.

Coomer P.G. & Robinson B.W. (1976) Sulphur and Sulphate-oxygen Isotopes and the Origin of the Silvermines Deposits, Ireland. *Mineral. Deposita*, **11**, 155–69.

Craig J.R. & Scott S.D. (1974) Sulphide Phase Equilibria. In Ribbe P.H. (ed.), *Sulphide Mineralogy*. Min. Soc. Am., Short Course Notes, Vol. 1, Ch. 5.

Craig J.R. & Vaughan D.J. (1981) *Ore Microscopy and Ore Petrography*, Wiley, New York.

Criss R.E., Ekren E.B. & Hardyman R.F. (1984) Casto Ring Zone: A 4500 km^2 Fossil Hydrothermal System in the Challis Volcanic Field, Central Idaho, *Geology*, **12**, 331–4.

Cuney M. (1978) Geologic Environment, Mineralogy and Fluid Inclusions of the Bois Noirs-Limouzat Uranium Vein, Forez, France. *Econ. Geol.*, **73**, 1567–610.

Czamanske G.K., Haffty J. & Nabbs S.W. (1981) Pt, Pd and Rh Analyses and Beneficiation of Mineralized Mafic Rocks from the La Perouse Layered Gabbro, Alaska. *Econ. Geol.*, **76**, 2001–11.

Dahlkamp F.J. (1978) Classification of Uranium Deposits. *Mineralium Deposita*, **13**, 83–104.

Dahlkamp F.J. (1984) Characteristics and Problematics of the Metallogenesis of Proterozoic Vein-like Type Uranium Deposits. In Wauschkuhn A., Kluth C. & Zimmermann R.A. (eds.), *Syngenesis and Epigenesis in the Formation of Mineral Deposits*, 183–92, Springer-Verlag, Berlin.

Damon P.E., Shafiqullah M. & Clark K.F. (1983) Geochemistry of the Porphyry Copper Deposits and Related Mineralization of Mexico. *Can. J. Earth Sci.*, **20**, 1052–71.

Danielson M.J. (1975) King Island Scheelite Deposits. In Knight C.L. (ed.), *Economic Geology of Australia and Papua New Guinea*. 592–98. Australas. Inst. Min. Metall., Parkville.

Davies F.B. & Windley B.F. (1976) Significance of Major Proterozoic High Grade Linear Belts in Continental Evolution. *Nature*, **263**, 383–5.

Davis W.J. & Williams-Jones A.E. (1985) A Fluid Inclusion Study of the Porphyry-greisen, Tungsten-Molybdenum Deposit at Mount Pleasant, New Brunswick, Canada. *Mineralium Deposita*, **20**, 94–101.
Dawson J.B. (1971) Advances in Kimberlite Geology, *Earth Sci. Rev.*, **7**, 187–214.
Dawson J.B. (1980) *Kimberlites and Their Xenoliths*. Springer-Verlag, Berlin.
Degens E.T. & Ross D.A. (1969) (eds.) *Hot Brines and Recent Heavy Metal Deposits in the Red Sea*. Springer, New York.
Degens E.T. & Ross D.A. (1976) Strata-bound Metalliferous Deposits Found in or near Active Rifts. In Wolf K.H. (ed.), *Handbook of Strata-Bound and Stratiform Ore Deposits*, Vol. 4, 165–202. Elsevier, Amsterdam.
Dejonghe L. & de Walque L. (1981) Pétrologie et Géochimie du Filon Sulfuré de Heure (Belgique) du Chapeau de Fer Associé et de l'encaissant Carbonaté. *Bull. BRGM*, (3), Sect. II, 1980–81 165–91.
Dercourt J. & Paquet J. (1985) *Geology: Principles and Methods*, Graham & Trotman, London.
De Villiers J.E. (1983) The Manganese Deposits of Griqualand West, South Africa: some Mineralogic Aspects. *Econ. Geol.*, **78**, 1108–18.
De Vore G.W. (1955) The Role of Adsorption in the Fractionation and Distribution of Elements. *J. Geol.*, **63**, 159–90.
De Wit, M.J. (1985) *Minerals and Mining in Antarctica*. Clarendon Press, Oxford.
Dietz R.S. (1964) Sudbury Structure as an Astrobleme. *J. Geol.*, **72**, 412–34.
Dimroth E. (1977) Facies Models 5 — Models of Physical Sedimentation of Iron Formations. *Geosci. Can.*, **4**, 23–30.
Dines H.G. (1956) The Metalliferous Mining Region of South-west England. *Mem. Geol. Surv. G.B.* vol. 1.
Dixon C.J. (1979) *Atlas of Economic Mineral Deposits*. Chapman & Hall, London.
Dmitriev L., Barsukov V. & Udintsev G. (1971) Rift Zones of the Ocean and the Problem of Ore-Formation. *Proc. IMA-IAGOD Meetings '70*, Spec. Issue 3 (IAGOD Vol.), Soc. Mining Geol. Japan, 65–9.
Doe B.R. & Delavaux M.H. (1972) Source of Lead in Southeast Missouri Galena Ores. *Econ. Geol.*, **67**, 409–25.
Duhovnik J. (1967) Facts For and Against a Syngenetic Origin of the Stratiform Ore Deposits of Lead and Zinc. In Brown J.S. (ed.), *Genesis of Stratiform Lead-zinc-barite-fluorite Deposits, 108–25, Econ. Geol. Monogr. 3.*
Duke J.M. (1983) Ore Deposit Models 7, Magmatic Segregation Deposits of Chromite. *Geoscience Can.* **10**, 15–24.
Dulski P., Möller P., Villalpando A. & Schneider H.J. (1982) Correlation of Trace Element Fractionation in Cassiterites with the Genesis of the Bolivian Metallotect. In Evans A.M. (ed.), *Metallization Associated with Acid Magmatism*, 71–83, Wiley, Chichester.
Dunham K.C. (1959) Non-ferrous Mining Potentialities of the Northern Pennines. In *Future of Non-ferrous Mining in Great Britain and Ireland*, 115–147. Instn. Min. Metall., London.
Dunnet D. (1971) Some Aspects of the Panantarctic Cratonic Margin in Australia. *Philos. Trans. R. Soc. London*, **A280**, 641–54.
Dunsmore H.E. (1973) Diagenetic Processes of Lead-zinc Emplacement in Carbonates. *Trans. Instn Min. Metall. (Sect. B: Appl. earth sci.)*, **82**, B168–B173.
Dunsmore H.E. & Shearman D.J. (1977) Mississippi Valley-type Lead-zinc Orebodies: a Sedimentary and Diagenetic Origin. In *Proceedings of the Forum on Oil and Ore in Sediments*, 189–205. Geology Dept., Imperial College, London.
Dupré B., Chauvel C. & Arndt N.T. (1984) Pb and Nd Isotopic Study of Two Archaean Komatiitic Flows from Alexo, Ontario. *Geochim. Cosmochim. Acta.*, **48**, 1965–72.
Durocher M.E. (1983) The Nature of Hydrothermal Alteration Associated with the Madsen and Starratt-Olsen Gold Deposits, Red Lake Area. In Colvine A.C. (ed.), *The Geology of Gold in Ontario*, 123–40, Geol. Surv. Ontario, Misc. Pap. **110**.
Du Toit A.L. (1954) *The Geology of South Africa*. Oliver and Boyd, Edinburgh.
Dybdahl I. (1960) Ilmenite Deposits of the Egersund Anorthosite Complex. In Vokes F. (ed.), *Mines in South and Central Norway*. Guide to Excursion No. C10. *Int. geol. Congr.* 21st, Norden.
Eadington P.J. (1983) A Fluid Inclusion Investigation of Ore Formation in a Tin-mineralized Granite, New England, New South Wales. *Econ. Geol.* **78**, 1204–21.
Eales H.V. & Reynolds I.M. (1985) Cryptic Variations within Chromites of the Upper Critical Zone, Northwestern Bushveld Complex. *Can. Mineral*, **23**, 302.
Easthoe C.J. (1978) A Fluid Inclusion Study of the Panguna Porphyry Copper Deposit, Bougainville, Papua New Guinea. *Econ. Geol.*, **73**, 721–42.

Eckstrand O.R. (1984), (ed.) *Canadian Mineral Deposit Types: A Geological Synopsis*, 39–42, Geol. Surv. Canada Economic Report **36**, Ottawa.

Edwards A.B. (1952) The Ore Minerals and Their Textures. *J. Proc. R. Soc. New South Wales*, **85**, 26–46.

Edwards A.B. (1960) Textures of the Ore Minerals and their Significance, *Australas. Inst. Min. Metall., Melbourne*.

Edwards A.B., Baker G. & Callow K.J. (1956) Metamorphism and Metasomatism at King Island Scheelite Mine. *J. geol. Soc. Aust.* **3**, 55–98.

Edwards A.B. & Lyon R.J.P. (1957) Mineralization at Aberfoyle Tin Mine, Rossarden, Tasmania. *Proc. Australas. Inst. Min. Metall.*, **181**, 93–145.

Edwards R. & Atkinson K. (1986) *Ore Deposit Geology*. Chapman & Hall, London.

Eilenberg S. & Carr M.J. (1981) Copper Contents of Lavas from Active Volcanoes in El Salvador and Adjacent Regions in Central America. *Econ. Geol.*, **76**, 2246–8.

Einaudi M.T. & Burt D.M. (1982) Introduction — Terminology, Classification and Composition of Skarn Deposits. *Econ. Geol.*, **77**, 745–54.

Einaudi M.T., Meinert L.D. & Newberry R.J. (1981) Skarn Deposits. *Econ Geol.* **75th Anniv. Vol.**, 317–91.

Ellis A.J. (1979) Explored Geothermal Systems. In Barnes H.L. (ed.), *Geochemistry of Hydrothermal Ore Deposits, Second Edition*, 632–83, Wiley, New York.

Elmore R.D. (1984) The Copper Harbor Conglomerate: A Late Precambrian Fining-upward Alluvial Fan Sequence in Northern Michigan. *Bull. geol. Soc. Am.*, **95**, 610–7.

El Shazly E.M., Webb J.S. & Williams D. (1957) Trace Elements in Sphalerite, Galena and Associated Minerals from the British Isles. *Trans. Instn Min. Metall.*, **66**, 241–71.

Eriksson K.A. & Truswell J.F. (1978) Geological Processes and Atmospheric Evolution in the Precambrian. In Tarling D.H. (ed.), *Evolution of the Earth's Crust*, 219–38, Academic Press, London.

Evans A.M. (1962) Geology of the Bicroft Uranium Mine, Ontario. *Ph.D thesis Queen's University*, Kingston, Ontario.

Evans A.M. (1966) The Development of *Lit-par-lit* Gneiss at the Bicroft Uranium Mine, Ontario. *Can. Mineral*, **8**, 593–609.

Evans A.M. (1975) Mineralization in Geosynclines — the Alpine Enigma. *Mineralium Deposita*, **10**, 254–60.

Evans A.M. (1976a) Genesis of Irish Base-metal Deposits. In Wolf K.H. (ed.), *Handbook of Strata-Bound and Stratiform Deposits*, Vol. 5, 231–55, Elsevier, Amsterdam.

Evans A.M. (1976b) Mineralization in Geosynclines. In Wolf K.H. (ed.), *Handbook of Strata-Bound and Stratiform Deposits*, Vol. 4, 1–29, Elsevier, Amsterdam.

Evans A.M. (1980) *An Introduction to Ore Geology*, First Edition. Blackwell Scientific Publications, Oxford.

Evans A.M. (1982) (ed.) *Metallization Associated with Acid Magmatism*, Vol. 6, Wiley, Chichester.

Evans A.M. & El-Nikhely A. (1982) Some Palaeomagnetic Dates from the West Cumbrian Hematite Deposits, England. *Trans. Instn Min. Metall. (Sect. B. Appl. earth Sci.)*, **91**, B41–3.

Evans A.M. & Evans N.D.M. (1977) Some Preliminary Palaeomagnetic Studies of Mineralization in the Mendip Orefield. *Trans. Instn. Min. Metall. (Sect. B. Appl. earth sci.)*, **86**, B149–B151.

Evans A.M. & Maroof S.I. (1976) Basement Controls on Mineralization in the British Isles. *Ming Mag. London*, **134**, 401–11.

Ewers G.R., Ferguson J. & Donnelly T.H. (1983) The Narbarlek Uranium Deposit, Northern Territory, Australia: Some Petrologic and Geochemical Constraints on Genesis. *Econ. Geol.*, **78**, 823–37.

Ewers G.R. & Keays R.R. (1977) Volatile and Precious Metal Zoning in the Broadlands Geothermal Field, New Zealand. *Econ. Geol.*, **72**, 1337–54.

Farr P. (1984) Beryllium, *Mining Annual Review*, *Ming J.* London, **88**.

Feiss P.G. (1978) Magmatic Sources of Copper in Porphyry Copper Deposits. *Econ. Geol.*, **73** 397–404.

Ferguson J. (1980a) Tectonic Setting and Palaeogeotherms of Kimberlites with Particular Emphasis on Southeastern Australia. In Glover J.E. & Groves D.I., (eds.), *Kimberlites and Diamonds*, 1–14, Extension Service, University of Western Australia, Nedlands.

Ferguson J. (1980b) Kimberlite and Kimberlitic Intrusives of Southeastern Australia. *Mineralog. Mag.* London, **43**, 727–31.

Fleischer V.D., Garlick W.G. & Haldane R. (1976) Geology of the Zambian Copperbelt. In Wolf K.H. (ed.), *Handbook of Strata-Bound and Stratiform Ore Deposits*, Vol. 6, 223–352. Elsevier, Amsterdam.

Fletcher C.J.N. (1984) Strata-bound, Vein and Breccia-pipe Tungsten Deposits of South Korea. *Trans. Instn Min. Metall. (Sect. B: Appl. earth sci.)*, **93**, B176–84.
Fletcher K. & Couper J. (1975) Greenvale Nickel Laterite, North Queensland. In Knight C.L. (ed.), *Economic Geology of Australia and Papua New Guinea, I. Metals, Australas.* Inst. Min. Metall., Parkville, 995–1001.
Förster H. & Jafarzadeh A. (1984) The Chador Malu Iron Ore Deposit (Bafq District, Central Iran) — Magnetite Filled Pipes. *Neues Jahrb. Geol. Palaeontol. Abhandlungen.* **168**, 524–34.
Forsythe D.L. (1971) Vertical Zoning of Gold-silver Tellurides in the Emperor Gold Mine, Fiji. *Proc. Australas. Inst. Min. Metall.*, No. **240**, 25–31.
Foster R.P. (1984) (ed.) *GOLD '82: The Geology, Geochemistry and Genesis of Gold Deposits.* Balkema, Rotterdam.
Fournier R.O. (1967) The Porphyry Copper Deposit Exposed in the Liberty Open-pit Mine near Ely, Nevada. Part II. The Formation of Hydrothermal Alteration Zones. *Econ. Geol.*, **62**, 207–27.
Fowler A.D. & Doig R. (1983) The Age and Origin of Grenville Province Uraniferous Granites and Pegmatites. *Can. J. Earth Sci.*, **20**, 92–104.
Fox J. S. (1984) Besshi-type Volcanogenic Sulphide Deposits — A Review. *CIM Bull.*, **77** (April), 57–68.
Francheteau J. and 14 co-authors (1979) Massive Deep-sea Sulphide Ore Deposits Discovered on the East Pacific Rise. *Nature*, **277**, 523–8.
Francis P.W., Halls C. & Baker M.C.W. (1983) Relationships between Mineralization and Silicic Volcanism in the Central Andes. *Geotherm. J. Volc. Res.*, **18**, 165–90.
Franklin J.M. Lydon J.W. & Sangster D.F. (1981) Volcanic-associated Massive Sulphide Deposits. *Econ. Geol., 75th Anniv. Vol.*, 485–627.
Frater K.M. (1985) Mineralization at the Golden Grove Cu-Zn Deposit, Western Australia. *Can. J. Earth Sci.*, **22**, 1–26.
Frietsch R. (1978) On the Magmatic Origin of Iron Ores of the Kiruna Type. *Econ. Geol.*, **73**, 478–85.
Fripp R.E.P. (1976) Stratabound Gold Deposits in Archaean Banded Iron-formation, Rhodesia. *Econ Geol.*, **71**, 58–75.
Frutos J. (1982) Andean Metallogeny Related to the Tectonic and Petrologic Evolution of the Cordillera. Some Remarkable Points. In Amstutz G.C., El Goresy A., Frenzel G., Khuth C., Moh G., Wauschkuhn A. & Zimmerman R.A. (eds.), *Ore Genesis: the State of the Art*, 493–507. Springer-Verlag, Berlin.
Fulp M.S. & Renshaw J.L. (1985) Volcanogenic-exhalative Tungsten Mineralization of Proterozoic Age near Santa Fe, New Mexico, and implications for exploration. *Geology*, **13**, 66–9.
Fyfe W.S. & Henley R.W. (1973) Some Thoughts on Chemical Transport Processes, with particular reference to Gold. *Mineral Sci. Engng*, **5**, 295–303.
Gain S.B. (1985) The Geologic Setting of the Platiniferous UG-2 Chromitite Layer on the Farm Maandagshoek, Eastern Bushveld Complex. *Econ. Geol.*, **80**, 925–43.
Gandhi S.S. (1978) Geological Setting and Genetic Aspects of Uranium occurrences in the Kaipokok Bay-Big River Area, Labrador. *Econ. Geol.*, **73**, 1492–522.
Garven G. (1985) The Role of Regional Fluid Flow in the Genesis of the Pine Point Deposit, Western Canada Sedimentary Basin. *Econ. Geol.*, **80**, 307–24.
Gee R.D. (1975) Regional Geology of the Archaean Nucleii of the Western Australian Shield. In Knight C.L. (ed.), *Economic Geology of Australia and Papua New Guinea — 1, Metals*, Mon. 5, 43–55. Australas. Inst. Min. Metall., Parkville.
George E., Pagel M. & Dusausoy Y. (1983) α-U_3O_7 à Brousse Broquiès (Avreyon, France): une Forme Quadratique Exceptionnelle parmi les Oxydes d'Uranium Naturels. *Terra Cognita*, **3**, 173.
Gilluly J. (1932) Geology and Ore Deposits of the Stockton and Fairfield Quadrangles, Utah. Prof. Pap. **173**, *U.S. Geol. Surv.*, Washington.
Gilluly J. (1976) Lineaments — Ineffective Guides to Ore Deposits. *Econ. Geol.*, **71**, 1507–14.
Gilluly J., Waters A.C. & Woodford A.O. (1959) *Principles of Geology*. Freeman, San Francisco.
Gilmour P. (1982) Grades and Tonnages of Porphyry Copper Deposits. In Titley S.R. (ed.), *Advances in Geology of the Porphyry Copper Deposits of Southwestern North America*, Univ. of Arizona Press, Tucson, 7–36.
Ginsburg A.I., Timofeyev I.N. & Feldman L.G. (1979) *Principles of the Geology of Granitic Pegmatites*, Nedra, Moscow, in Russian. (Some aspects of this book are covered in Černý 1982a & 1982c).
Giordano T.H. (1985) A Preliminary Evaluation of Organic Ligands and Metal-Organic Complexing in Mississippi Valley-Type Ore Solutions. *Econ. Geol.*, **80**, 96–106.

Giordano T.H. & Barnes H.L. (1981) Lead Transport in Mississippi Valley-Type Ore Solutions. *Econ. Geol.*, **76**, 2200–11.

Glazkovsky A.A., Gorbunov G.I. & Sysoev F.A. (1977) Deposits of Nickel. In Smirnov V.I. (ed.), *Ore Deposits of the USSR Vol. II*, 3–79, Pitman, London.

Gold D.P. (1984) A Diamond Exploration Philosophy for the 1980s. *Earth miner. Sci. Pennsylvania State Univ.*, **53**, 37–42.

Gold D.P., Vallee M. & Charette J.P. (1967) Economic Geology and Geophysics of the Oka Alkaline Complex, Quebec. *Canadian Mining and Metallurgical Bull.*, **60**, 1131–1144.

Goldich S.S. (1973) Ages of Precambrian Banded Iron Formations. *Econ. Geol.*, **68**, 1126–34.

Goldsmith J.R. & Newton R.C. (1969) P-T-X Relations in the System $CaCO_3$-$MgCO_3$ at High Temperature and Pressures. *Am. J. Sci.*, **267A**, 160–90.

Gole M.J. (1981) Archaean Banded Iron formations, Yilgarn Block, Western Australia. *Econ. Geol.*, **76**, 1954–74.

Golightly J.P. (1981) Nickeliferous Laterite Deposits. *Econ Geol., 75th. Anniv. Vol.* 710–35.

Goodwin A.M. (1982) Distribution and Origin of Precambrian Banded Iron Formations. *Rev. Bras. Geosci.*, **12**, 457–62.

Goodwin A.M. (1973) Archaean Iron-formations and Tectonic Basins of the Canadian Shield. *Econ. Geol.*, **68**, 915–33.

Gouanvic Y. & Gagny C. (1983) Étude d'une aplo-pegmatite litée à cassiterite et wolframite, magma différencié de l'endogranite de la mine de Santa Comba, (Galice, Espagne). *Bull. Soc. geol. Fr.*, Ser. 7, *25*, 335–48.

Grant J.N., Halls C., Avila W. & Avila G. (1977) Igneous Geology and the Evolution of Hydrothermal Systems in some sub-volcanic Tin Deposits of Bolivia. In *Volcanic Processes in Ore Genesis*. Spec. Pub. No. 7, Geol. Soc. London, 117–126.

Grant J.N., Halls C., Sheppard S.M.F. & Avila W. (1980) Evolution of the Porphyry Tin Deposits of Bolivia. *Mining Geology* Special Issue **8**, 151–73.

Gregory P.W. & Robinson B.W. (1984) Sulphur Isotope Studies of the Mt. Molloy, Dianne and O.K. Stratiform Sulphide Deposits, Hodgkinson Province, North Queensland, Australia. *Mineralium Deposita*. **19**, 36–43.

Greig J.A., Baadsgaard H., Cumming G.L., Folinsbee R.E., Krouse H.R., Ohmoto H., Sasaki A. & Smejkal V. (1971) Lead and Sulphur Isotopes of the Irish Base Metal Mines in Carbonate Host Rocks. *Society Mining Geologists Japan Spec. Issue 2 Proc.* (IMA-IAGOD Mtgs, 1970, Joint Symp. Vol.), 84–92.

Griffiths J.R. & Godwin C.I. (1983) Metallogeny and Tectonics of Porphyry Copper-molybdenum Deposits in British Columbia. *Can. J. Earth Sci.*, **20**, 1000–18.

Gross G.A. (1965) Geology of Iron Deposits in Canada, I. General Geology and Evaluation of Iron Deposits. *Geol. Surv., Can. Econ. Geol. Rep.*, **22**.

Gross G.A. (1970) Nature and Occurrence of Iron Ore Deposits. In *Survey of World Iron Ore Resources*, United Nations, New York, 13–31.

Gross G.A. (1980) A Classification of Iron Formations Based on Depositional Environments. *Can. Mineral.*, **18**, 215–22.

Groves D.I. & Hudson D.R. (1981) The Nature and Origin of Archaean Strata-bound Volcanic-associated Nickel-iron-copper Sulphide Deposits. In Wolf K.H. (ed.), *Handbook of Strata-bound and Stratiform Ore Deposits*, 305–410, Elsevier, Amsterdam.

Groves D.I. & Lesher C.M. (1982) (eds.) *Regional Geology and Nickel Deposits of the Norseman-Wiluna Belt, Western Australia*, Geology Department and Extension Service, Univ. of Western Australia, East Perth.

Groves D.I. & McCarthy T.S. (1978) Fractional Crystallization and the Origin of Tin Deposits in Granitoids. *Mineralium Deposita*, **13**, 11–26.

Groves D.I. & Solomon M. (1969) Fluid Inclusion Studies at Mount Bischoff, Tasmania. *Trans. Instn Min. Metall. (Sect. B: Appl. earth sci.)*, **78**, B1–B11.

Groves D.I., Solomon M. & Rafter T.A. (1970) Sulphur Isotope Fractionation and Fluid Inclusion Studies at the Rex Hill Mine, Tasmania. *Econ. Geol.*, **65**, 459–69.

Grubb P.L.C. (1973) High-level and Low-level Bauxitization: a Criterion for Classification. *Mineral. Sci. Engng*, **5**, 219–31.

Guild P.W. (1978) Metallogenesis in the Western United States. *J. geol. Soc. London*, **135**, 355–76.

Gurney J.J. (1985) A Correlation between Garnets and Diamonds in Kimberlites. In Glover J.E. & Harris P.G. (eds.), *Kimberlite Occurrence and Origin: a Basis for Conceptual Models in Exploration*, 143–66, Geology Department and University Extension, University of Western Australia, Nedlands.

Gustafson L.B. & Curtis L.W. (1983) Post-Kombolgie Metasomatism at Jabiluka, Northern

Territory, Australia, and Its Significance in the Formation of High Grade Uranium Mineralization in Lower Proterozoic Rocks. *Econ. Geol.*, **78**, 26–56.
Gustafson L.B. & Hunt J.P. (1975) The Porphyry Copper Deposit at El Salvador, Chile. *Econ. Geol.*, **70**, 857–912.
Gustafson L.B. & Williams N. (1981) Sediment-hosted Stratiform Deposits of Copper, Lead and Zinc. *Econ. Geol., 75th. Anniv. Vol.*, 139–78.
Hass J.L. Jr (1971) The Effect of Salinity on the Maximum Thermal Gradient of a Hydrothermal System at Hydrostatic Pressure. *Econ. Geol.*, **66**, 940–6.
Haase C.S. (1982) Phase Equilibria in Metamorphosed Iron Formations: Qualitative T-X(CO_2) Petrogenetic Grids. *Am. J. Sci.*, **282**, 1623–54.
Hagner A.F. & Collins L.G. (1967) Magnetite Ore Formed during Regional Metamorphism. *Econ. Geol.*, **62**, 1034–71.
Hall A.L. (1932) The Bushveld Igneous Complex of the Central Transvaal. *Geol. Surv. S. Africa*, Mem. **28**, 560.
Hall W.E., Friedman I. & Nash J.T. (1974) Fluid Inclusion and Light Stable Isotope Study of the Climax Molybdenum Deposits, Colorado. *Econ. Geol.*, **69**, 884–901.
Hamilton J.M., Delaney G.D., Hauser R.L. & Ransom P.W. (1983) Geology of the Sullivan Deposit, Kimberley, B.C. Canada. In Sangster D.F. (ed.) *Sediment-hosted Stratiform Lead-zinc Deposits*, Short Course Handbook, Vol. 8, 31–83. Mineralogical Association of Canada. Victoria.
Hanor J.S. (1979) The Sedimentary Genesis of Hydrothermal Fluids. In Barnes H.L. (ed.), *Geochemistry of Hydrothermal Ore Deposits*, 137–72, Wiley, New York.
Harben P.W. & Bates R.L. (1984) *Geology of the Nonmetallics.* Metal Bulletin, New York.
Harder E.C. & Creig E.W. (1960) Bauxite. In Gillson J.L. *et al.* (eds.), *Industrial Minerals and Rocks*, Amer. Inst. Ming Eng., New York, 65–85.
Harris P.G. (1985) Kimberlite volcanism. In Glover J.E. & Harris P.G. (eds) *Kimberlite Occurrence and Origin: a Basis for Conceptual Models in Exploration*, 125–42, Geology Department and University Extension, University of Western Australia, Nedlands.
Hatcher M.I. & Bolitho B.C. (1982) The Greenbushes Pegmatite, South-west Australia. In Černý P. (ed.), *Short Course in Granitic Pegmatites in Science and Industry*, Mineral. Assoc. Canada, Winnipeg, 513–525.
Hattori K. & Muehlenbachs K. (1980) Marine Hydrothermal Alteration at a Kuroko Ore Deposit, Kosaka, Japan. *Contrib. Mineral. Petrol.*, **74**, 285–92.
Hawley J.E. (1962) The Sudbury Ores: Their Mineralogy and Origin. *Can. Mineral*, 7, i-xiv & 1–207.
Hayba D.O., Bethke P.M., Heald P. & Foley N.K. (1986) Geologic, Mineralogic and Geochemical Characteristics of Volcanic-hosted Epithermal Precious Metal Deposits. In Berger B.R. & Bethke P.M. (eds.), *Geology and Geochemistry of Epithermal Systems*, 129–67, Society of Economic Geologists, El Paso.
Haymon R.M., Koski R.A. & Sinclair C. (1984) Fossils of Hydrothermal Vent Worms from Cretaceous Sulphide Ores of the Samail Ophiolite, Oman. *Science*, **223**, 1407–9.
Heaton T.H.E. & Sheppard S.M.F. (1977) Hydrogen and Oxygen Isotope Evidence for Sea Water-Hydrothermal Alteration and Ore Deposition, Troodos Complex, Cyprus. In *Volcanic Processes in Ore Genesis*, Spec. Publ. No. 7, Geol. Soc. London.
Hekinian R. & Fouquet Y. (1985) Volcanism and Metallogenesis of Axial and Off-Axial Structures on the East Pacific Rise near 13°N. *Econ. Geol.*, **80**, 221–49.
Hendry D.A.F., Chivas A.R., Long J.V.P. & Reed S.J.B. (1985) Chemical Differences between Minerals from Mineralizing and Barren Intrusions from Some North American Porphyry Copper Deposits. *Contrib. Mineral. Petrol.*, **89**, 317–29.
Henley R.W. (1986) The Geological Framework of Epithermal Deposits. In Berger B.R. & Bethke P.M. (eds.), *Geology and Geochemistry of Epithermal Systems*, 1–24, Society of Economic Geologists, El Paso.
Henley R.W. & Ellis A.J. (1983) Geothermal Systems Ancient and Modern: a Geochemical Review. *Earth Sci. Rev.*, **19**, 1–50.
Henley R.W., Truesdell A.H., Barton P.B. & Whitney J.A. (1984) *Fluid-Mineral Equilibria in Hydrothermal Systems*, Reviews in *Econ. Geol.* **1**, Soc. of Econ. Geologists.
Hewitt D.F. (1967) *Pegmatite Mineral Resources of Ontario,* Ont. Dept Mines Industrial Mineral report 21, Toronto.
Heyl A.V. (1969) Some Aspects of Genesis of Zinc-lead-barite-fluorite deposits in the Mississippi Valley, U.S.A. *Trans. Instn Min. Metall. (Sect. B. Appl. earth sci.)*, **78**, B148–B160.
Heyl A.V. (1972) The 38th Parallel Lineament and Its Relationship to Ore Deposits. *Econ. Geol.*, **67**, 879–94.

Heyl A.V., Delevaux M.H., Zartman R.E. & Brock M.R. (1966) Isotopic Study of Galenas from the Upper Mississippi Valley, the Illinois-Kentucky and some Appalachian Valley Mineral Districts. *Econ. Geol.*, **61**, 933–61.

Heyl A.V., Landis G.P. & Zartman R.E. (1974) Isotopic Evidence for the Origin of Mississippi Valley-type Mineral Deposits: A Review. *Econ. Geol.*, **69**, 992–1006.

Hill R.E.T. & Gole M.J. (1985) Characteristics of Centres of Archaean Komatiitic Volcanism, Exemplified by Lithologies in the Agnew Area, Yilgarn Block, Western Australia. *Can. Mineral.*, **23**, 327.

Hills E.S. (1953) Tectonic Setting of Australian Ore Deposits. In Edwards A.B. (ed.), *Geology of Australian Ore Deposits*, Australas. Inst. Min. Metall., Melbourne, 41–61.

Hildebrand R.S. (1986) Kiruna-type Deposits: Their Origin and Relationship to Intermediate Subvolcanic Plutons in the Great Bear Magmatic Zone, Northwest Canada. *Econ. Geol.*, **81**, 640–59.

Hodder R.W. & Petruk W. (1982) *Geology of Canadian Gold Deposits*. Spec. vol. 24, Canad. Inst. Min. Metall., Montreal.

Hoeve J. (1984) Host Rock Alteration and Its Application as an Ore Guide at the Midwest Lake Uranium Deposit, Northern Saskatchewan. *Canadian Mining and Metallurgical Bull*, 77, (August) 63–72.

Höll R. (1985) Geothermal systems and Active Ore formation in the Taupo Volcanic Zone, New Zealand. In Germann K. (ed.), *Geochemical Aspects of Ore Formation in Recent and Fossil Sedimentary Environments*, 55–71, Monogr. Ser. on Mineral Deposits 25, Gebrüder Borntraeger, Berlin.

Höll R. & Maucher A. (1976) The Strata-Bound Ore Deposits in the Eastern Alps. In Wolf K.H. (ed.), *Handbook of Strata-Bound and Stratiform Ore Deposits*, Vol. 5. 1–36. Elsevier, Amsterdam.

Hollister V.F. (1975) An Appraisal of the Nature of some Porphyry Copper Deposits. *Mineral. Sci. Engng*, **7**, 225–33.

Hollister V.F., Potter R.R. & Barker A.L. (1974) Porphyry type deposits of the Appalachian Orogen. *Econ. Geol.*, **69**, 618–30.

Hollister V.F. (1978) *Geology of the Porphyry Copper Deposits of the Western Hemisphere*. Amer. Inst. Min. Metall and Pet. Engrs, New York.

Hopwood T. (1981) The Significance of Pyritic Black Shales in the Genesis of Archaean Nickel Sulphide Deposits. In Wolf K.H. (ed.), *Handbook of Strata-bound and Stratiform Ore Deposits*, 411–67, Elsevier, Amsterdam.

Horikoshi E. & Sato T. (1970) Volcanic Activity and Ore Deposition in the Kosaka Mine. In Tatsumi T. (ed.), *Volcanism and Ore Genesis*. 181–95. University of Tokyo Press, Tokyo.

Hose H.R. (1960) The Genesis of Bauxites, the Ores of Aluminium. *Int. geol. Congr.* 21st. Pt. **16**, 237–47.

Hosking K.F.G. (1951) Primary Ore Deposition in Cornwall. *Trans. R. geol. Soc. Cornwall*, **18**, 309–56.

Howd F.H. & Barnes H.L. (1975) Ore Solution Chemistry IV. Replacement of Marble by Sulphides at 450°C. *Econ. Geol.*, **70**, 968–81.

Hsü K.J. (1972) The Concept of the Geosyncline, Yesterday and Today. *Trans. Leicester Lit. Philos. Soc.*, **66**, 26–48.

Hsü K.J. (1984) A Nonsteady State Model for Dolomite, Evaporite and Ore Genesis. In Wauschkuhn A., Kluth C. & Zimmermann R.A. (eds.), *Syngenesis and Epigenesis in the Formation of Mineral Deposits*, Springer-Verlag, Berlin.

Huppert H.E. & Sparks R.S. (1985) Komatiites I: Eruption and Flow. *J. Petrology*, **26**, 694–725.

Hutchinson R.W. (1980) Massive Base Metal Sulphide Deposits as Guides to Tectonic Evolution. In Strangeway D.W. (ed.), *The Continental Crust and Its Mineral Deposits*, 659–684, Geol. Assoc., Canada Spec. Pap. 20.

Hutchinson R.W. (1983) Mineral Deposits, Time and Evolution. *Proc. Denver Region Exploration Geologists Society Symposium — The Genesis of Rock Mountain Ore Deposits: Changes with Time and Tectonics*, 1–9, Denver.

Hutchison C.S. (1983) *Economic Deposits and Their Tectonic Setting*. Macmillan, London.

Hutchison M.N. & Scott S.D. (1981) Sphalerite Geobarometry in the Cu-Fe-Zn-S System. *Econ. Geol.*, **76**, 143–53.

Ihlen P.M., Trønnes R. & Vokes F.M. (1982) Mineralization, Wall Rock Alteration and Zonation of Ore Deposits Associated with the Drammen Granite in the Oslo Region, Norway. In Evans A.M. (ed.), *Metallization Associated with Acid Magmatism*, 111–36, Wiley, Chichester.

Irvine T.N. & Smith C.H. (1969) Primary Oxide Minerals in the Layered Series of the Muskox Intrusion. *Econ. Geol. Monograph 4*, 76–94.

Ivosevic S.W. (1984) *Gold and Silver Handbook*. Ivosevic, Denver.
Ixer R.A. & Townley R. (1979) The Sulphide Mineralogy and Paragenesis of the South Pennine Orefield, England. *Mercian Geol.*, **7**, 51–64.
Jackson E.D. (1961) Primary Textures and Mineral Associations in the Ultramafic Zone of the Stillwater Complex, Montana. *U.S. Geol. Surv.* Prof. Pap. 358.
Jackson E.D. & Thayer T.P. (1972) Some Criteria for Distinguishing between Stratiform, Concentric and Alpine Peridotite-Gabbro Complexes. *Int. geol. Congr.*, 24th. session, section, 2, 289–96.
Jackson N.J., Halliday A.N., Sheppard S.M.F. & Mitchell J.G. (1982) Hydrothermal Activity in the St Just Mining District, Cornwall, England. In Evans A.M. (ed.), *Metallization Associated with Acid Magmatism*, 137–79, Wiley, Chichester.
Jackson R. (1982) *Ok Tedi: the Pot of Gold*. University of Papua New Guinea.
Jackson S.A. & Beales F.W. (1967) An Aspect of Sedimentary Basin Evolution; the Concentration of Mississippian Valley-type Ores during Late Stages of Diagenesis. *Bull. Can. Pet. Geol.*, **15**, 383–433.
Jacobsen J.B.E. (1975) Copper Deposits in Time and Space. *Miner. Sci. Engng*, **7**, 337–71.
Jacobsen J.B.E. & McCarthy T.S. (1976) The Copper-bearing Breccia Pipes of the Messina District, South Africa. *Mineralium Deposita*, **11**, 33–45.
Jahns R.H. (1955) The Study of Pegmatites. *Econ. Geol.* (**50th Anniv. Vol.**), 1025–130.
Jahns R.H. (1982) Internal Evolution of Pegmatite Bodies. In Černý P. (ed.), *Short Course in Granitic Pegmatites in Science and Industry*, Mineral. Assoc. Canada, Winnipeg, 293–327.
James H.L. (1954) Sedimentary Facies of Iron formation. *Econ. Geol.*, **49**, 235–93.
James H.L. (1955) Zones of Regional Metamorphism in the Precambrian of Northern Michigan. *Bull. geol. Soc. Amer.*, **66**, 1455–88.
James H.L. (1983) Distribution of Banded Iron Formations in Space and Time. In Trendall A.F. & Morris R.C. (eds.), *Iron Formation: Facts and Problems*, 471–86, Elsevier, Amsterdam.
James H.L. & Sims P.K. (1973) Precambrian Iron-formations of the World. *Econ. Geol.*, **68**, 913–4.
Jaques A.L. & Ferguson J. (1983) Diamondiferous Kimberlitic Rocks, West Kimberley. *BMR 83*, 44 (Yearbook of the Bureau of Mineral Resources, Canberra).
Jaques A.L., Lewis J.D., Smith C.B., Gregory G.P., Ferguson J., Chappell B.W. & McCulloch M.T. (1984) The diamond-bearing Ultrapotassic (Lamproitic) Rocks of the West Kimberley Region, Western Australia. In Kornprobst J. (ed.), *Kimberlites I: Kimberlites and Related Rocks*, 225–54, Elsevier, Amsterdam.
Jensen M.L. & Bateman A.M. (1979) *Economic Mineral Deposits*, Wiley, New York.
Kay A. & Strong D.F. (1983) Geologic and Fluid controls on As-Sb-Au Mineralization in the Moretons Harbour Area, Newfoundland. *Econ. Geol.*, **78**, 1590–1604.
Keays R.R. (1982) Palladium and Iridium in Komatiites and Associated Rocks: Application to Petrogenetic Problems. In Arndt N.T. & Nisbet E.G. (eds.), *Komatiites*, 435–57, George Allen & Unwin, London.
Keays R.R. (1984) Archaean Gold Deposits and Their Source Rocks: the Upper Mantle Connection. In Foster R.P. (ed.), *Gold '82: The Geology, Geochemistry and Genesis of Gold Deposits*, 17–51, Balkema, Rotterdam.
Kerrich R. & Fryer B.J. (1979) Archaean Precious Metal Hydrothermal Systems, Dome Mine, Abitibi Greenstone H Belt. II. REE and Oxygen Isotope Relations. *Can. J. Earth Sci.*, **16**, 440–58.
Kesler S.E. (1968) Contact-localized Ore Formation at the Memé Mine, Haiti. *Econ. Geol.*, **63**, 541–52.
Kesler S.E. (1973) Copper, Molybdenum and Gold Abundances in Porphyry Copper Deposits. *Econ. Geol.*, **68**, 106–112.
Khitarov N.I., Malinin S.P., Lebedev Ye.B. & Shibayeva N.P. (1982) The Distribution of Zn, Cu, Pb and Mo between a Fluid Phase and a Silicate Melt of Granitic composition at High Temperatures and Pressures. *Geochemistry International*, **19** (*4*) 123–36.
Knittel U. & Burton C.K. (1985) Polillo Island (Philippines): Molybdenum Mineralization in an Island Arc. *Econ. Geol.*, **80**, 2013–8.
Kontak D.J. & Clark A.K. (1985) Exploration Criteria for Sn and W Mineralization in the Cordillera Oriental of SE Peru. In Taylor R.P. & Strong D.F. (eds.), *Granite-related Mineral Deposits*, 173–8, CIM Geology Division, Halifax, Canada.
Kretschmar U. & Scott S.D. (1976) Phase Relations Involving Arsenopyrite in the System Fe-As-S and Their Application. *Can. Mineral.*, **14**, 364–86.
Krs M. & Štovicková N. (1966) Palaeomagnetic Investigation of Hydrothermal Deposits in the Jáchymov (Joachimsthal) Region, Western Bohemia. *Trans. Instn Min. Metall.* (*Sect. B, Appl. earth sci.*), **75**, B51-B57.

Kruger F.J. & Marsh J.S. (1985) The Mineralogy, Petrology and Origin of the Merensky Cyclic Unit in the Western Bushveld Complex. *Econ. Geol.*, **80**, 958–74.
Kullerud G. (1953) The FeS-ZnS System: a Geological Thermometer. *Nor. geol. Tiddskr.*, **32**, 61–147.
Kwak T.A.P. (1978a) Mass Balance Relationships and Skarn-forming Processes at the King Island Scheelite Deposits, King Island, Tasmania, Australia. *Am. J. Sci.*, **278**, 943–68.
Kwak T.A.P. (1978b) The Conditions of Formation of the King Island Scheelite Contact Skarn, King Island, Tasmania, Australia. *Am. J. Sci.*, **278**, 969–99.
Lamey C.A. (1966) *Metallic and Industrial Mineral Deposits*, McGraw-Hill, New York.
Lang A.H. (1970) Prospecting in Canada. *Geol. Surv. Canada, Econ. Geol. Rep.*, **7**.
Lapham D.M. (1968) Triassic Magnetite and Diabase at Cornwall, Pennsylvania. In Ridge J.D. (ed.), *Ore Deposits of the United States 1933–1967*, Vol. 1, 72–94. Am. Inst. Min. Metall. Pet. Engrs, New York.
Large D.E. (1983) Sediment-hosted Massive Sulphide Lead-zinc Deposits: An Empirical Model. In Sangster D.F. (ed.), *Short Course in Sediment-hosted Stratiform Lead-zinc Deposits*, 1–29, Mineral. Assoc. Canada, Victoria.
Lawrence L.J. (1972) The Thermal Metamorphism of a Pyritic Sulphide Ore. *Econ. Geol.*, **67**, 487–96.
Laznicka P. (1976) Porphyry Copper and Molybdenum Deposits of the USSR and their Plate Tectonic Settings. *Trans. Instn Min. Metall. (Sect. B: Appl. earth sci.)*, **85**, B14–B32.
Laznicka P. (1985) *Empirical Metallogeny*, Elsevier, Amsterdam.
Le Bas M.J. (1977) *Carbonatite-nephelinite Volcanism*. Wiley, London.
Lehman B. (1985) Formation of the Strata-bound Kellhuani Tin Deposits, Bolivia. *Mineralium Deposita*. **20**, 169–76.
Leroy J. (1978) The Margnac and Fanay Uranium Deposits of the La Crouzille District (Western Massif Central, France): Geologic and Fluid Inclusion Studies. *Econ. Geol.*, **73**, 1611–34.
Leroy J. (1984) Episyénitisation dans le Gisement d'Uranium du Bernardan (Marche): Comparaison avec des Gisements similaires du Nord-Ouest du Massif Central Français. *Mineralium Deposita.*, **19**, 26–35.
Levin E.M., Robbins C.R. & McMurdie H.F. (1969) *Phase Diagrams for Ceramicists*. Am. Ceram. Soc., Columbus, Ohio.
Lincoln T.N. (1981) The Redistribution of Copper during Low-Grade Metamorphism of the Karmutsen Volcanics, Vancouver Island, British Columbia. *Econ. Geol.*, **76**, 2147–61.
Lindberg P.A. (1985) A Volcanogenic Interpretation for Massive Sulphide Origin, West Shasta District, California. *Econ. Geol.*, **80**, 2240–54.
Lindgren W. (1913) (second edition, 1933). *Mineral Deposits*. McGraw-Hill, New York.
Lindgren W. (1922) A Suggestion for the Terminology of Certain Mineral Deposits. *Econ. Geol.*, **17**, 292–4.
Lindgren W. (1924) Contact Metamorphism at Bingham, Utah. *Geol. Soc. Amer. Bull.*, **35**, 507–34.
London D. (1984) Experimental Phase Equilibria in the System $LiAlSiO_4$-SiO_2-H_2O: a Petrogenetic Grid for Lithium-rich Pegmatites. *Am. Mineral.*, **69**, 995–1004.
Lonsdale P. & Becker K. (1985) Hydrothermal Plumes, Hot Springs and Conductive Heat Flow in the Southern Trough of Guaymas Basin. *Earth planet. Sci. Lett.* **73**, 211–25.
Lowell J.D. (1974) Regional Characteristics of Porphyry Copper Deposits of the Southwest. *Econ. Geol.*, **69**, 601–17.
Lowell J.D. & Guilbert J.M. (1970) Lateral and Vertical Alteration Mineralization Zoning in Porphyry Ore Deposits. *Econ. Geol.*, **65**, 373–408.
Lumbers S.B. (1979) The Grenville Province of Ontario. 5th. Ann. Meeting Int. Union Geol. Sci., Subcomm. Precamb. Stratigraphy. *Geol. Surv. Minnesota* and *Univ. Minnesota Guidebook* **Ser. 13**, 1–35.
Lusk J., Campbell F.A. & Krouse H.R. (1975) Application of Sphalerite Geobarometry and Sulphur Isotope Geothermometry to Ores of the Quemont Mine, Noranda, Quebec. *Econ. Geol.*, **70**, 1070–83.
Macdonald E.H. (1983) *Alluvial Mining*. Chapman & Hall. London.
Mainwaring P.R. & Naldrett A.J. (1977) Country Rock Assimilation and the Genesis of Cu-Ni Sulphides in the Water Hen Intrusion, Duluth Complex, Minnesota. *Econ. Geol.*, **72**, 1269–84.
Malyutin R.S. & Sitkovskiy I.N. (1968) Structural Features of the Gyumushlug Lead-zinc Deposit. *Geologiya Rudnykh Mestorozhdeniy*, **10**, 96–99. (In Russian.)
Manning D.A.C. (1984) Volatile Control of Tungsten Partitioning in Granitic Melt-Vapour Systems. *Trans. Instn Min, Metall. (Sect. B Appl. earth Sci.)*, **93**, B185–94.
Manning D.A.C. (1986) Contrasting types of Sn-W Mineralization in Peninsular Thailand and

SW England. *Mineralium Deposita.*, **21**, 44–52.
Marcoux E. (1982) Étude géologique et métallogénique du district plombozincifère de Pontivy (Massif armoricain, France). *Bull. BRGM (2), sect. II*, (**1**) 1–24.
Marmont S. (1983) The Role of Felsic Intrusions in Gold Mineralization In Colvine A.C. (ed.), *The Geology of Gold in Ontario*, 38–47, Geol. Surv. Ont. Misc. Pap. 110.
Martin J.E. & Allchurch P.D. (1975) Perservance Nickel Deposit, Agnew. In Knight C.L. (ed.), *Economic Geology of Australia and Papua New Guinea — 1, Metals.* Mon. 5, 149–155. Australas. Inst. Min. Metall., Parkville.
Martin P.L. (1966) Structural Analysis of the Chisel Lake Orebody. *CIM Bull.*, **59**, 630–36.
Mason A.A.C. (1953) The Vulcan Tin Mine. In Edwards A.B. (ed.), *Geology of Australian Ore Deposits*, 718–721. Australas. Inst. Min. Metall., Melbourne.
Mason D.R. & Feiss P.G. (1979) On the Relationship between Whole Rock Chemistry and Porphyry Copper Mineralization. *Econ. Geol.*, *74*, 1506–10.
Maynard J.B. (1983) *Geochemistry of Sedimentary Ore Deposits*, Springer-Verlag, New York.
McConchie D. (1984) A Depositional Environment for the Hamersley Group: Palaeogeography and Geochemistry. In Muhling J.R., Groves D.I. & Blake T.S. (eds.), *Archaean and Proterozoic Basins of the Pilbara, Western Australia: Evolution and Mineralization Potential*, 144–77, **Pubn. No. 9**, Geol. Dept & Univ. Extension, Univ. of Western Australia, Nedlands.
McKinstry H.E. (1948) *Mining Geology*, Prentice-Hall, New York.
McMillan W.J. & Panteleyev A. (1980) Ore Deposit Models: 1. Porphyry Copper Deposits. *Geosci. Canada* **7** 52–63.
Melcher G.C. (1966) The Carbonatites of Jacupiranga, Sao Paulo, Brazil. In Tuttle O.F. & Gittins J. (eds.), *Carbonatites*, 169–81, Wiley, New York.
Mertie J.B. (1969) *Economic Geology of the Platinum Metals.* Prof. Pap. 630, *U.S. Geol. Surv.*, Washington.
Meyer C. (1981) Ore-forming Processes in Geologic History. *Econ. Geol., 75th. Anniv. Vol.*, 6–41.
Meyer C. (1985) Ore Metals through Geologic History. *Science*, **227**, 1421–8.
Meyer C. & Hemley J.J. (1967) Wall Rock Alteration. In Barnes H.L. (ed.), *Geochemistry of Hydrothermal Ore Deposits*, 166–235. Holt, Rinehart and Winston, New Yrok.
Meyer C., Shea E.P., Goddard Jr. C.C. & Staff (1968) Ore Deposits at Rutte, Montana. In Ridge J.R. (ed.), *Ore Deposits of the United States, 1933–1967*, Vol. II. 1373–416. Am. Inst. Min. Metall. Pet. Engns, New York.
Meyer H.O.A. (1985) Genesis of Diamond: a Mantle Saga. *Am. Mineral.*, **70**, 344–55.
Milledge H.J., Mendelssohn M.J., Seal M., Rouse J.E., Swart P.K. & Pillinger C.T. (1983) Carbon Isotopic Variation in Spectral Type II Diamonds. *Nature*, **303**, 79102.
Miller C.F. & Bradfish L.J. (1980) An Inner Cordilleran Belt of Muscovite-bearing Plutons. *Geology*, **8**, 412–6.
Miller R.G. & O'Nions R.K. (1985) Sources of Precambrian Chemical and Clastic Sediments. *Nature*, **314**, 325–30.
Milovskiy G.A., Zlenko B.F. & Gubanov A.M. (1978) Conditions of Formation of Scheelite Ores in the Chorukh-Dayron Mineralized Area (As Revealed by a Study of Gas-liquid Inclusions). *Geochem. Int.*, **15**, 45–52.
Mitcham T.W. (1974) Origin of Breccia Pipes. *Econ. Geol.*, **69**, 412–13.
Mitchell A.H.G. (1973) Metallogenic Belts and Angle of Dip of Benioff Zones. *Nat. Phys. Sci.*, **245**, 49–52.
Mitchell A.H.G. & Garson M.S. (1972) Relationship of Porphyry Copper and Circum-Pacific Tin Deposits to Palaeo-Benioff Zones. *Trans. Instn Min. Metall.*, (*Sect. B: Appl. earth sci.*), **81**, B10–B25.
Mitchell A.H.G. & Garson M.S. (1976) Mineralization at Plate Boundaries. *Miner. Sci. Engng*, **8**, 129–69.
Mitchell A.H.G. & Garson M.S. (1981) *Mineral Deposits and Global Tectonic Settings.* Academic Press, London.
Mitchell A.H.G. & Reading H.G. (1986) Sedimentation and Tectonics. In Reading H.G. (ed.), *Sedimentary Environments and Facies*, 471–519, Blackwell Scientific Publications, Oxford.
Möller P. (1985) Development and Application of the Ga/Ge-Geo Thermometer for Sphalerite from Sediment-hosted Deposits. In Germann K. (ed.), *Geochemical Aspects of Ore Formation in Recent and Fossil Sedimentary Environments*, 15–30, Gebrüder Borntraeger, Berlin.
Moorbath S., O'Nions R.L. & Pankhurst R.J. (1973) Early Archaean Age for the Isua Iron Formation, West Greenland, *Nature*, **245**, 138–9.
Moore A.C. (1973) Carbonatites and Kimberlies in Australia: a Review of the Evidence. *Miner. Sci. Engng*, **5**, 81–91.

Moore J.M. (1982) Mineral Zonation near the Granitic Batholiths of South-west and Northern England and Some Geothermal Analogues. In Evans A.M. (ed.), *Metallization Associated with Acid Magmatism*, 229–41, Wiley, Chichester.
Morton R.L. & Nebel M.L. (1984) Hydrothermal Alteration of Felsic Volcanic Rocks at the Helen Siderite Deposit, Wawa, Ontario. *Econ. Geol.*, **79**, 1319–33.
Mosig R.W. (1980) Morphology of Indicator Minerals as a Guide to Proximity of Source. In Glover J.E. and Groves D.I., (eds.), *Kimberlites and Diamonds*, 81–88, Extension Service, Univ. of Western Australia, Nedlands.
Naldrett A.J. (1973) Nickel Sulphide Deposits — Their Classification and Genesis, with Special Emphasis on Deposits of Volcanic Association. *Can. Inst. Min. Met. Trans.*, **76**, 183–201.
Naldrett A.J. (1981) Nickel Sulphide Deposits: Classification, Composition and Genesis. In Skinner B.J. (ed.), *Econ. Geol.*, **75th Anniv. Vol.**, 628–85, Economic Geology Publishing Co., El Paso.
Naldrett A.J. & Cabri L.J. (1976) Ultramafic and Related Mafic Rocks: Their Classification and Genesis with Special Reference to the Concentration of Nickel Sulphides and Platinum Group Elements. *Econ. Geol.*, **71**, 1131–58.
Naldrett A.J. & Campbell I.A.H. (1982) Physical and Chemical Constraints on Genetic Models for Komatiite-related Ni-sulphide Deposits. In Arndt N.T. & Nisbet E.G. (eds.), *Komatiites*, 423–34, George Allen & Unwin, London.
Naldrett A.J., Duke J.M., Lightfoot P.C. & Thompson J.F.H. (1984) Quantitative Modelling of the Segregation of Magmatic Sulphides: an Exploration Guide. *Canadian Mining and Metallurgical Bull.*, **77**, (April), 46–56.
Naldrett A.J., Gasparrini C., Barnes S.J., Sharpe M.R. & von Gruenewaldt G. (1985) The Origin of the Merensky Reef. *Can. Mineral.*, **23**, 308–9.
Naldrett A.J. & Hewins R.H. (1984) The Main Mass of the Sudbury Igneous Complex. In Pye E.G., Naldrett A.J. & Giblin P.E. (eds.), *The Geology and Ore Deposits of the Sudbury Structure*, 235–51, Spec. Vol. 1, Geol. Surv. Ontario, Toronto.
Nash J.T. (1976) Fluid Inclusion Petrology – Data from Porphyry Copper Deposits and Applications to Exploration. *Prof. Pap. 907–D, U.S. Geol. Surv.*, Washington.
Nash J.T., Granger H.C. & Adams S.S. (1981) Geology and Concepts of Genesis of Important Types of Uranium Deposits. *Econ. Geol.*, **75th Anniv. Vol.**, 63–116.
Nickel K.B. & Green D.H. (1985) Empirical Geothermobarometry for Garnet Peridotites and Implications for the Nature of the Lithosphere, Kimberlites and Diamonds. *Earth. Planet. Sci. Lett.* **73**, 158–170.
Nisbet B.W., Devlin S.P. & Joyce P.J. (1983) Geology and Suggested Genesis of Cobalt-Tungsten Mineralization at Mt. Cobalt, North Western Queensland. *Proc. Australas. Inst. Min. Metall.*, **No. 287**, 9–17.
Nixon P.H. (1973) (ed.), *Lesotho Kimberlites*. Lesotho National Development Corporation, Maseru.
Nixon P.H. (1980a) The Morphology and Mineralogy of Diamond Pipes. In Glover J.E. & Groves D.I. (eds.), *Kimberlites and Diamonds*, 32–47. Extension Service, Univ. of Western Australia. Nedlands.
Nixon P.H. (1980b) Regional Diamond Exploration — Theory and Practice. In Glover J.E. & Groves D.I. (eds.), *Kimberlites and Diamonds*, 64–80. Extension Service, Univ. of Western Australia, Nedlands.
Nixon P.H., Boyd F.R. & Boctor N.Z. (1983) East Griqualand Kimberlites. *Trans. geol. Soc. S. Afr.*, **86**, 221–36.
Noble J.A., (1980) Two Metallogenic Maps for North America. *Geol. Rundsch.*, **69**, 594–609.
Noble S.R., Spooner E.T.C. & Harris F.R. (1984) The Logtung Large Tonnage Low-grade W (Scheelite)-Mo Porphyry Deposit, South-central Yukon Territory. *Econ. Geol.*, **79**, 848–68.
Norman D.I. & Trangcotchasan Y. (1982) Mineralization and Fluid Inclusion Study of the Yod Nam Tin Mine, Southern Thailand. In Evans A.M. (ed.), *Metallization Associated with Acid Magnetism*, Wiley, Chichester, 261–72.
Norton J.J. (1983) Sequence of Mineral assemblages in Differentiated Granitic Pegmatites. *Econ. Geol.*, **78**, 854–74.
Notholt A.J.G. (1979) The Economic Geology and Development of Igneous Phosphate Deposits in Europe and the USSR. *Econ. Geol.*, **74**, 339–50.
Ohmoto H. & Rye R.O. (1974) Hydrogen and Oxygen Isotopic Compositions of Fluid Inclusions in the Kuroko Deposits, Japan. *Econ. Geol.*, **69**, 947–53.
Ohmoto H. & Skinner B.J. (1983) *The Kuroko and Related Volcanogenic Massive Sulphide Deposits*. *Econ. Geol.*, Mon. 5.
Olade M.A. (1980) Plate Tectonics and Metallogeny of Intracontinental Rifts and Aulacogens in Africa — A Review. In Ridge J.D. (ed.), *Proc. 5th Quadriennial IAGOD Symp.*, **Vol. 1**,

81–9. E. Schweizer-bart'sche Verlagbuchhandlung, Stuttgart.
Olson J.C. & Pray L.C. (1954) The Mountain Pass Rare-earth Deposits. In *Geology of Southern California*, Chap. VIII, *Mineral Deposits and Mineral Industry*, Bull. 170, Division of Mines, State of California, 23–29.
Olson J.C., Shawe D.R., Pray L.C. & Sharp W.N. (1954) Rare-earth Mineral Deposits of the Mountain Pass District, San Bernardino County, California. *U.S. Geol. Surv.*, Prof. Pap. 261.
Open University S333 Course Team (1976) *Porphyry Copper Case Study*. Open University, Milton Keynes.
Ostwald J. (1981) Evidence for a Biogeochemical Origin of the Groote Eylandt Manganese Ores. *Econ. Geol.*, **76**, 556–67.
Oudin E. & Constantinou G. (1984) Black Smoker Chimney Fragments in Cyprus Sulphide Deposits. *Nature*, **308**, 349–53.
Owen H.B. & Whitehead S. (1965) Iron Ore Deposits of Iron Knob and the Middleback Ranges. In McAndrew J. (ed.), *Geology of Australian Ore Deposits*, 301–8. Aust. Inst. Min. Metall., Melbourne.
Page R.W. & McDougall I. (1972) Ages of Mineralization in Gold and Prophyry Copper Deposits in the New Guinea Highlands. *Econ. Geol.*, **67**, 1034–48.
Pagel M. (1975) Détermination des Conditions Physicochimiques de la Silicification Diagénétique des Grès Athabaska (Canada) au Moyen des Inclusions Fluides. *C.R. Acad. Sci. Paris*, **280**, ser. D., 2301–4.
Pagel M. & Jafferezic H. (1977) Analyses Chimiques des Saumures des Inclusions du Quartz et de la Dolomite du Gisement d'Uranium de Rabbit Lake (Canada). *C.R. Acad. Sci. Paris*, **284**, ser D. 113–6.
Pagel M., Poty B. & Sheppard S.M.F. (1980) Contributions to Some Saskatchewan Uranium Deposits Mainly from Fluid Inclusions and Isotopic Data. In Ferguson J. & Goleby A.B. (eds.), *Uranium in the Pine Creek Geosyncline*, 639–54, IAEA, Vienna.
Palabora Mining Company Limited Mine Geological and Mineralogical Staff (1976) The Geology and the Economic Deposits of Copper, Iron and Vermiculite in the Palabora Igneous Complex: a Brief Review. *Econ. Geol.*, **71**, 177–92.
Parák T. (1975) Kiruna Iron Ores Are Not 'Intrusive Magmatic Ores of the Kiruna Type.' *Econ. Geol.*, **70**, 1242–58.
Parák T. (1985) Phosphorus in Different Types of Ore, Sulfides in the Iron Deposits, and the Type and Origin of Ores at Kiruna. *Econ. Geol.*, **80**, 646–65.
Park C.F. (1961) A Magnetite 'Flow' in Northern Chile. *Econ. Geol.*, **56**, 431–41.
Park C.F. Jr. & MacDiarmid R.A. (1975) *Ore Deposits*. Freeman, San Francisco.
Pavlov N.V. & Grigor'eva I.I. (1977) Deposits of Chromium. In Smirnov V.I. (ed.), *Ore Deposits of the USSR* **1** 179–236, Pitman, London.
Pearce J.A., Lippard S.J. & Roberts S. (1984) Characteristics and Tectonic Significance of Supra-subduction Zone Ophiolites. In Kokelaar, B.P. & Howells M.F. (eds), Marginal Basin Geology, 77–94 Blackwell Scientific Publications, Oxford.
Peterson U. (1970) Metallogenic Provinces in South America. *Geol. Rundsch.*, **59**, 834–97.
Phillips G.N. & Groves D.I. (1983) The Nature of Archaean Gold-bearing Fluids as Deduced from Gold Deposits of Western Australia. *J. geol. Soc. Australia*, **30**, 25–39.
Phillips G.N., Groves D.I. & Clark M.E. (1983) The Importance of Host-rock Mineralogy in the Location of Archaean Epigenetic Gold Deposits. In De Villiers J.P.R. & Cawthorn P.A. (eds.), *ICAM 81*, Spec. Publ. geol. Soc. S. Afr., **7**, 79–86.
Phillips G.N., Groves D.I. & Martyn J.E. (1984) An Epigenetic Origin for Archaean Banded Iron Formation Hosted Gold Deposits. *Econ. Geol.*, **79**, 162–71.
Phillips W.J. (1973) Mechanical Effects of Retrograde Boiling and its Probable Importance in the Formation of Some Porphyry Ore Deposits. *Trans. Instn Min. Metall. (Sec. B: Appl. earth sci.)*, **82**, B90–B98.
Piper J.D.A. (1974) Proterozoic Crustal Distribution, Mobile Belts and Apparent Polar Movement, *Nature*, **251**, 381–4.
Piper J.D.A. (1974) Proterozoic Supercontinent: Time Duration and the Grenville Problem. *Nature*, **256**, 519–20.
Piper J.D.A. (1976) Palaeomagnetic Evidence for a Proterozoic Supercontinent. *Phil. Trans. R. Soc. Lond.*, **A280**, 469–90.
Pitcher W.S. (1983) Granite: Typology, Geological Environments and Melting Relationships. In Atherton M.P. & Gribbe C.D. (eds.), *Migmatites Melting and Metamorphism*, 277–85, Shiva, Nantwich.
Plimer I.R. (1985) Broken Hill Pb-Zn-Ag Deposit — A Product of Mantle Metasomatism. *Mineralium Deposita*, **20**, 147–53.

Premoli C. (1985) The Future of Large, Low Grade, Hard-rock Tin Deposits. *Natural Resources Forum*, **9**, 107–119.

Preto V.A. (1978) Setting and Genesis of Uranium Mineralization at Rexspar. *Canad. Inst. Min. Bull.*, **71**, 82–8.

Pretorius D.A. (1975) The Depositional Environment of the Witwatersrand Goldfields: a Chronological Review of Speculations and Observations. *Miner. Sci. Engng*, **7**, 18–47.

Pretorius D.A. (1981) Gold and Uranium in Quartz-pebble Conglomerates. *Econ. Geol. 75th Anniv Vol.*, 117–38.

Prouhet J.-P. (1983) Les Minéralisations de Skarns des Pyrénées Françaises. (Field guide to tungsten mineralization in the Pyrenees). In French. *BRGM*

Pye E.G., Naldrett A.J. & Giblin P.E. (eds.), (1984) *The Geology and Ore Deposits of the Sudbury Structure*. Spec. Vol 1, Geol. Surv. Ontario, Toronto.

Rackley R.I. (1976) Origin of Western States-Type Uranium Mineralization. In Wolf K.H. (ed.), *Handbook of Strata-Bound and Stratiform Deposits*, Vol. 7, 89–156. Elsevier, Amsterdam.

Radkevich E.A. (1972) The Metallogenic Zoning in the Pacific Ore Belt. *Int. geol. Congr.* 24th Session, Sect. 4, 52–59.

Radtke A.S. (1985) *Geology of the Carlin Gold Deposit, Nevada*. USGS. Prof. Pap. 1267.

Raedeke L.D. & Vian R.W. (1985) A Three-dimensional View of Mineralization in the Stillwater J-M Reef. *Can. Mineral.*, **23**, 312.

Ramdohr P. (1969) *The Ore Minerals and Their Intergrowths*. Pergamon Press, Oxford.

Ransome F.L. (1919) The Copper Deposits of Ray and Miami, Arizona. *U.S. Geol. Surv.* Prof. Pap. **115**.

Ranta D.E., Ward A.D. & Ganster M.W. (1984) Ore Zoning Applied to Geologic Reserve Estimation of Molybdenum Deposits. In Erickson A.J. (ed.), *Applied Mining Geology*, 83–114. Amer. Inst. Min. Metall. Pet. Engrs, New York.

Raybould J.G. (1978) Tectonic Controls on Proterozoic Stratiform Copper Mineralization. *Trans. Instn Min. Metall. (Sect. B: Appl. earth sci.)*, **87**, B79–B86.

Reading H.G. (1986) *Sedimentary Environments and Facies*. Blackwell Scientific Publications, Oxford.

Reedman A.J., Colman T.B., Campbell S.D.G. & Howells M.F. (1985) Volcanogenic Mineralization Related to the Snowdon Volcanic Group (Ordovician), Gwynedd, North Wales. *J. geol. Soc. London*, **142**, 875–88.

Reedman J.H. (1984) Resources of Phosphate, Niobium, Iron and Other Elements in Residual Soils over the Sukulu Carbonatite Complex, Southeastern Uganda. *Econ. Geol.*, **79**, 716–24.

Reid I. & Frostick L.E. (1985) Role of Settling, Entrainment and Dispersive Equivalence and of Interstice Trapping in Placer Formation. *J. geol. Soc. London*, **142**, 739–46.

Reynolds I.M. (1985) The Nature and Origin of Titaniferous-rich Layers in the Upper Zone of the Bushveld Complex: A Review and Synthesis. *Econ. Geol.*, **80**, 1089–1108.

Reynolds R.L. & Goldhaber M.B. (1978) Origin of a South Texas Roll-type Uranium Deposit: I. Alteration of Iron-titanium Oxide Minerals. *Econ. Geol.*, **73**, 1677–89.

Richards J.R. & Pidgeon R.T. (1983) Some Age Measurements on Micas from Broken Hill, Australia. *J. geol. Soc. Aust.*, **10**, 664–78.

Richardson J.M.G., Spooner E.T.C. & McAuslan D.A. (1982) The East Kemptville Tin Deposit, Nova Scotia: An Example of a Large Tonnage, Low Grade, Greisen-hosted Deposit in the Endocontact of a Granite Batholith. In *Current Research, Part B*. Geo. Surv. Canada. Pap. 82–1Bm 27–32.

Richardson S.H. (1986) Latter-day Origin of Diamonds of Eclogitic Paragenesis. *Nature*, **322**, 623–7.

Richardson S.H., Gurney J.J., Erlank A.J. & Harris J.W. (1984) Origin of Diamonds in Old Enriched Mantle. *Nature*, **310**, 198–202.

Richard D.T., Willden M.Y., Marde Y. & Ryhage R. (1975) Hydrocarbons Associated with Lead-zinc Ores at Laisvall, Sweden. *Nature*, **255**, 131–3.

Rimskaya-Korsakov O.M. (1964) Genesis of the Kovdor Iron Ore Deposit (Kola Peninsula). *Internat. Geol. Rev.*, **6**, 1735–46.

Ringwood A.E. (1974) The Petrological Evolution of Island Arc Systems. *J. Geol. Soc. London*, **130**, 183–204.

Rittenhouse G. (1943) Transportation and Deposition of Heavy Minerals. *Geol. Soc. Am. Bull.*, **54**, 1725–80.

Robertson D.S. (1962) Thorium and Uranium Variations in the Blind River Ores. *Econ. Geol.*, **57**, 1175–84.

Robinson A. & Spooner E.T.C. (1984) Can the Elliot Lake Uraninite-bearing Quartz Pebble Conglomerates Be Used to Place Limits on the Oxygen Content of the Early Proterozoic Atmosphere? *J. geol. Soc. London*, **141**, 221–8.

Robinson B.W. & Ohmoto H. (1973) Mineralogy, Fluid Inclusions and Stable Isotopes of the Echo Bay U-Ni-Ag-Cu Deposits, Northwest Territories, Canada. *Econ. Geol.*, **68**, 635–56.
Robinson D.N. (1978) The Characteristics of Natural Diamonds and Their Interpretation. *Miner. Sci. Engng.*, **10**, 55–72.
Roedder E. (1972) Composition of Fluid Inclusions. Prof. Pap. 440-JJ. *U.S. Geol. Surv.*
Roedder E. (1984) *Fluid Inclusions.* Reviews in *Mineralogy* **Vol. 12**, Miner. Soc. Am., Resten, Virginia.
Roedder E. & Bodnar R.J. (1980) Geologic Pressure Determinations from Fluid Inclusion Studies. *Ann. Rev. Earth Planet. Sci.*, **8**, 263–301.
Rogers N. & Hawkesworth C. (1984) New Date for Diamonds. *Nature*, **310**, 187–8.
Rona P.A., Klinkhammer G., Nelson T.A., Trefry J.H. & Elderfield H. (1986) Black Smokers, Massive Sulphides and Vent Biota at the Mid-Atlantic Ridge. *Nature*, **321**, 33–7.
Ronoy A.B. (1964) Common Tendencies in the Chemical Evolution of the Earth's Crust, Ocean and Atmosphere. *Geochem. Int.*, **1**, 713–37.
Roscoe S.M. (1968) Huronian Rocks and Uraniferous Conglomerates of the Canadian Shield. *Geol. Surv. Can. Pap. 68–40.*
Rose A.W. & Burt D.M. (1979) Hydrothermal Alteration. In Barnes H.L. (ed.), *Geochemistry of Hydrothermal Ore Deposits, Second Edition*, 173–235. Wiley, New York.
Ross J.R. & Hopkins G.M.F. (1975) Kambalda Nickel Sulphide Deposits. In Knight C.L. (ed.), *Economic Geology of Australia and Papua New Guinea–1, Metals.* Mon, 5, 100–121. Australas. Inst. Min. Metall., Parkville.
Routhier P. (1963) *Les Gisements Métallifères.* Vol. 1. Masson et Cie, Paris.
Roy S. (1976) Ancient Manganese Deposits. In Wolf K.H. (ed.), *Handbook of Strata-Bound and Stratiform Deposits*, Vol. 7, 395–474. Elsevier, Amsterdam.
Roy S. (1981) *Manganese Deposits.* Academic Press, London.
Rubey W.W. (1933) The Size Distribution of Heavy Minerals within a Waterlain Sandstone. *J. Sediment. Petrol.*, **3**, 3–29.
Ruckmick J.C. (1963) The Iron Ores of Cerro Bolivar, Venezuela. *Econ. Geol.*, **58**, 218–36.
Ruiz J., Kelly W.C. & Kaiser C.J. (1985) Strontium Isotopic Evidence for the Origin of Barites and Sulfides from the Mississippi Valley-type Ore Deposits in Southeast Missouri — A Discussion. *Econ. Geol.*, **80**, 773–5.
Russell M.J. (1983) Major Sediment-hosted Exhalative Zinc and Lead Deposits: Formation from Hydrothermal Convection Cells That Deepen During Crustal Extension. In Sangster D.F. (ed.), *Short Course in Sediment-hosted Stratiform Lead-zinc Deposits.* 251–82, Mineral. Assoc. Canada, Victoria.
Russell M.J., Solomon M. & Walshe J.L. (1981) The Genesis of Sediment-hosted, Exhalative Zinc and Lead Deposits. *Mineralium Deposita*, **16**, 113–27.
Russell N., Seaward M., Rivera J.A., McCurdy K., Kesler S.E. & Cloke P.L. (1981) Geology and Geochemistry of the Pueblo Viejo Gold-silver Oxide Ore Deposit, Dominican Republic. *Trans. Instn Min. Metall. (Sect. B: Appl. earth sci.)*, **90**, B153–62.
Rye R.O. (1985) A Model for the Formation of Carbonate-hosted Disseminated Gold Deposits Based on Geologic, Fluid-inclusion, Geochemical and Stable-isotope Studies of the Carlin and Cortez Deposits, Nevada. In Tooker E.W. (ed.), *Geologic Characteristics of Sediment- and Volcanic-hosted Disseminated Gold Deposits — Search for an Occurrence Model.* USGS Bull. 1646.
Rye R.O. & Ohmoto H. (1974) Sulfur and Carbon Isotopes and Ore Genesis: A Review. *Econ. Geol.*, **69**, 826–42.
Saager R., Meyer M. & Muff R. (1982) Gold Distribution in Supracrustal Rocks from Archaean Greenstone Belts of Southern Africa and from Paleozoic Ultramafic Complexes of the European Alps: Metallogenic and Geochemical Implications. *Econ. Geol.*, **77**, 1–24.
Samoilov V.S. & Plyusnin G.S. (1982) The Source of Material for Rare-earth Carbonatites. *Geochem. Int.*, **19**, No. 5, 13–25.
Sangster D.F. (1976) Carbonate-hosted Lead-zinc Deposits. In Wolf K.H. (ed.), *Handbook of Strata-Bound and Stratiform Deposits*, Vol. 6, 447–56, Elsevier, Amsterdam.
Sangster D.F. (1983a) *Short Course in Sediment-hosted Stratiform Lead-zinc Deposits.* Mineral. Assoc. Canada, Victoria.
Sangster D.F. (1983b) Mississippi Valley-type Deposits: a Geological Mélange. In Kisvarsanyi G., Grant S.K., Pratt W.P. & Koenig J.W. (eds) *International Conference on Mississippi Valley Type Lead-zinc Depoists*, 7–19, Univ. of Missouri-Rolla, Rolla.
Sangster D.F. & Scott S.D. (1976) Precambrian, Strata-bound, Massive Cu-Zn-Pb Sulphide Ores in North America. In Wolf K.H. (ed.), *Handbook of Strata-Bound and Stratiform Ore Deposits*, Vol. 6, 129–222. Elsevier, Amsterdam.
Sato T. (1977) Kuroko Deposits: Their Geology, Geochemistry and Origin. In *Volcanic Processes*

in Ore Genesis. Spec. Pub. No. 7, Geol. Soc. London.
Sawkins F.J. (1972) Sulphide Ore Deposits in Relation to Plate Tectonics. *J. Geol.*, **80**, 377–97.
Saupé F. (1973) La Géologie du Gîsement de Mercure d'Almaden. *Sciences de la Terre, Mem.* 29.
Sawkins F.J. (1984) *Metal Deposits in Relation to Plate Tectonics*. Springer-Verlag, Berlin.
Sawkins F.J. & Rye D.M. (1974) Relationship of Homestake-type Gold Deposits to Iron-rich Precambrian Sedimentary Rocks. *Trans. Instn Min. Metall. (Sect. B: Appl. earth sci.)*, **83**, B56–9.
Sawkins F.J. & Scherkenbach D.A. (1981) High Copper Content of Fluid Inclusions in Quartz from Northern Sonora: Implications for Ore-genesis Theory. *Geology*, **9**, 37–40.
Schneiderhöhn H. (1955) *Erzlagerstätten*. Gustav Fischer-Verlag, Stuttgart.
Schuiling R.D. (1967) Tin Belts on Continents around the Atlantic Ocean. *Econ. Geol.*, **62**, 540–50.
Scott S.D. (1973) Experimental Calibration of the Sphalerite Geobarometer. *Econ. Geol.*, **68**, 466–74.
Scott S.D. (1974) Experimental Methods in Sulphide Synthesis. In Ribbe P.H. (ed.), *Sulphide Mineralogy*, Min. Soc. Am. Short Course Notes, Vol. 1, Ch. 4.
Scott S.D. & Barnes H.L. (1971) Sphalerite Geothermometry and Geobarometry. *Econ. Geol.*, **66**, 653–69.
Scott Smith B.H. & Skinner E.M.W. (1984) Diamondiferous Lamproites. *J. Geol.*, **92**, 433–438.
Scratch R.B., Watson G.P., Kerrich R. & Hutchinson R.W. (1984) Fracture-controlled Antimony-quartz Mineralization, Lake George Deposit, New Brunswick: Mineralogy, Geochemistry, Alteration and Hydrothermal Regimes. *Econ. Geol.*, **79**, 1159–86.
Selkman S.O. (1984) A Deformation Study at the Saxerget Sulphide Deposit, Sweden. *Forh. geol. Foren. Stockholm*, **106**, 235–44.
Selley R.C. (1976) *An Introduction to Sedimentology*. Academic Press, London.
Sharp Z.D., Essene E.J. & Kelly W.C. (1985) A Re-examination of the Arsenopyrite Geothermometer: Pressure Considerations and Applications to Natural Assemblages. *Can. Mineral.*, **23**, 517–34.
Shaw D.M. (1954) Trace Elements in Pelitic Rocks. *Geol. Soc. Amer. Bull.*, **65**, 1151–82.
Shaw S.E. (1968) Rb-Sr Isotopic Studies of the Mine Sequence Rocks at Broken Hill. In Radmanovich M. & Woodcock J.T. (eds.), *Broken Hill Mines — 1968, Australas. Inst. Min. Metall. Mongr.* Ser. **3**, 185–98.
Shepherd T.J. & Allen P.M. (1985) Metallogenesis in the Harlech Dome, North Wales: a Fluid Inclusion Interpretation. *Mineralium Deposita*, **20**, 159–68.
Sheppard S.M.F. (1977) Identification of the Origin of Ore-forming Solutions by the Use of Stable Isotopes. In *Volcanic Processes in Ore Genesis*. Spec. Publ. No. 7, Geol. Soc. London.
Shilo N.A., Milov A.P. & Sobolev A.P. (1983) *Mesozoic Granitoids of North-east Asia*. **Mem. 159**, Geol. Soc. Amer., 149–57.
Shimazaki H. & MacLean W.H. (1976) An Experimental Study on the Partition of Zinc and Lead between the Silicate and Sulphide Liquids. *Mineralium Deposita*, **11**, 125–32.
Shmakin B.M. (1983) Geochemistry and Origin of Granitic Pegmatites. *Geochem. Int., 20*, **6**, 1–8.
Sibson R.H., Moore & Rankin A.H. (1975) Seismic Pumping — a Hydrothermal Fluid Transport Mechanism. *J. geol. Soc. London*, **131**, 653–9.
Silberman M.L. (1985) Geochemistry of Hydrothermal Mineralization and Alteration: Tertiary Epithermal Precious-metal Deposits in the Great Basin. In Tooker E.W. (ed.), *Geologic Characteristics of Sediment- and Volcanic-hosted Disseminated Gold Deposits — Search for an Occurrence Model* USGS Bull. 1646, 55–70.
Sillitoe R.H. (1972a) Formation of Certain Massive Sulphide Deposits at Sites of Sea-floor Spreading. *Trans. Instn Min. Metall. (Sect. B: Appl. earth sci.)*, **81**, B141–B148.
Sillitoe R.H. (1972b) A Plate Tectonic Model for the Origin of Porphyry Copper Deposits. *Econ. Geol.*, **67**, 184–97.
Sillitoe R.H. (1973) The Tops and Bottoms of Porphyry Copper Deposits. *Econ. Geol.*, **68**, 799–815.
Sillitoe R.H. (1976) Andean Mineralization: a Model for the Metallogeny of Convergent Plate Margins. In Strong D.F. (ed.), *Metallogeny and Plate Tectonics*, 59–100, Geol. Assoc. Can. Spec. Pap. 14.
Sillitoe R.H. (1980a) Types of Porphyry Molybdenum Deposits. *Ming Mag.*, **142**, 550–553.
Sillitoe R.H. (1980b) Strata-bound Ore Deposits Related to Infracambrian Rifting along Northern Gondwanaland. In Ridge J.D. (ed.), *Proc. 5th IAGOD Symp. Vol. 1*, 163–72, E. Schweizerbart'sche Verlagsbuchhandlung, Stuttgart.
Sillitoe R.H. (1981a) Regional Aspects of the Andean Porphyry Copper Belt in Chile and Argentina. *Trans. Instn Min. Metall. (Sect. B Appl. earth Sci.)* **90**, B15–36.
Sillitoe R.H. (1981b) Ore Deposits in Cordilleran and Island Arc Settings. *Ariz. Geol. Soc. Dig.*

14, 49–70, cited by Sawkins (1984).
Sillitoe R.H., Halls C. & Grant J.N. (1975) Porphyry Tin Deposits in Bolivia. *Econ. Geol.*, **70**, 913–27.
Sillitoe R.H. & Hart S.R. (1984) Lead-isotope Signatures of Porphyry Copper Deposits in Oceanic and Continental setting, Colombian Andes. *Geochim. Cosmochim. Acta*, **48**, 2135–42.
Simpson P.R. & Bowles J.F. (1977) Uranium Mineralization of the Witwatersrand and Dominion Reef systems. *Phil. Trans. R. Soc. London, A. 286*, 527–48.
Sinclair A.J., Drummond A.D., Carter N.C. & Dawson K.M. (1982) A Preliminary Analysis of Gold and Silver Grades of Porphyry-type Deposits in Western Canada. In Levinson A.A. (ed.), *Precious Metals in the Northern Cordillera*, 157–72, Assoc. Exploration Geochemists, Rexdale.
Skinner B.J. (1979) The Many Origins of Hydrothermal Mineral Deposits. In Barnes H.L. (ed.), *Geochemistry of Hydrothermal Ore Deposits Second Edition*, 1–21, Wiley, New York.
Skirrow R. & Coleman M.L. (1982) Origin of Sulphur and Geothermometry of Hydrothermal Sulphides from the Galapagos Rift 86°W. *Nature*, **299**, 142–4.
Smith C.S. (1964) Some Elementary Principles of Polycrystalline Microstructure. *Metall. Rev.*, **9**, 1–48.
Smith T.E., Miller P.M. & Huang C.H. (1982) Solidification and Crystallization of a Stanniferous Granitoid Pluton, Nova Scotia, Canada. In Evans A.M. (ed.), *Metallization Associated with Acid Magmatism*, Vol. 6, Wiley, Chichester, 301–20.
Sokolov G.A. & Grigor'ev V.M. (1977) Deposits of Iron. In Smirnov V.I. (ed.), *Ore Deposits of the USSR*, Pitman, London, 7–113.
Solomon M. (1976) 'Volcanic' Massive Sulphide Deposits and their Host Rocks — a Review and an Explanation. In Wolf K.H. (ed.), *Handbook of Strata-Bound and Stratiform Ore Deposits*, Vol. 6, 21–54. Elsevier, Amsterdam.
Solomon M. & Walshe J.L. (1979) The formation of Massive Sulfide Deposits on the Sea Floor. *Econ. Geol.*, **74**, 797–813.
Souch B.E., Podolsky T. & Geological Staff (1969) The Sulphide Ores of Sudbury: Their Particular Relationship to a Distinctive Inclusion-bearing Facies of the Nickel Irruptive. In Wilson H.D.B. (ed.), *Magmatic Ore Deposits — A Symposium*, Econ. Geol. Mon. **4**, 252–61.
South B.C. & Taylor B.E. (1985) Stable Isotope Geochemistry and Metal Zonation at the Iron Mountain Mine, West Shasta District, California. *Econ. Geol.*, **80**, 2177–95.
Spooner E.T.C. (1977) Hydrodynamic Model for the Origin of Ophiolitic Cupriferous Pyrite Ore Deposits of Cyprus. In *Volcanic Process in Ore Genesis*. Spec. Publ. No. 7, Geol. Soc. London.
Spooner E.T.C., Bray C.J. & Chapman H.J. (1977) A Sea Water Source for the Hydrothermal Fluid which formed the Ophiolitic Cupriferous Pyrite Ore Deposits of the Troodos Massif, Cyprus. *J. geol. Soc. Lond.*, **134**, 395.
Springer J. (1983) Invisible Gold. In Colvine A.C. (ed.), *The Geology of Gold in Ontario*, 240–50, Geol. Surv. Ont., Misc. Pap. 110.
Stacey J.S., Zartman R.E. & Nkomo I.T. (1968) A Lead Isotope Study of Galenas and Selected Feldspars from Mining Districts in Utah. *Econ. Geol.*, **63**, 796–814.
Stanton R.L. (1972) *Ore Petrology*. McGraw-Hill, New York.
Stanton R.L. (1978) Mineralization in Island Arcs with Particular Reference to the South-west Pacific Region. *Proc. Australas. Inst. Min. Metall.*, No. **268**, 9–19.
Stanworth C.W. & Badham J.P.N. (1984) Lower Proterozoic Red Beds, Evaporites and Secondary Sedimentary Uranium Deposits from the East Arm, Great Slave Lake, Canada. *J. Geol. Soc. London,* **141**, 235–42.
Steele I.M., Bishop F.C., Smith J.V. & Windley B.F. (1977) The Fiskenæsset Complex, West Greenland, Part III, *Grønlands geol. Unders. Bull.* **124**.
Stein H.J. (1985) Genetic Traits of Climax-type Granites and Molybdenum Mineralization, Colorado Mineral Belt. In Taylor R.P. & Strong D.F. (eds.), *Granite-related Mineral Deposits*, 242–7, CIM Geology Division, Halifax, Canada.
Štemprok M. (1985) Vertical Extent of Greisen Mineralization in the Krusnehory/Erzgebirge Granite Pluton of Central Europe. In Halls C. (ed.), *High Heat Production (HHP) Granites, Hydrothermal Circulation and Ore Genesis*, Instn. Min. Metall., London, 383–91.
Streckeisen A.L. (1976) To Each Plutonic Rock Its Proper Name. *Earth Sci. Rev.*, **12**, 1–33.
Strong D.F. (1981) Ore Deposit Models: 5. A Model for Granophile Mineral Deposits. *Geosci. Can.* **8**, 155–60.
Susak N.J. & Crerar D.A. (1982) Factors Controlling Mineral Zoning in Hydrothermal Ore Deposits. *Econ. Geol.*, **77**, 476–82.
Sutherland D.G. (1985) Geomorphological Controls on the Distribution of Placer Deposits. *J. geol. Soc. London*, **142**, 727–31.

Sverjensky D.A. (1984) Oilfield Brines as Ore-forming Solution. *Econ. Geol.*, **79**, 23–37.
Symons R. (1961) Operation at Bikita Minerals (Private) Ltd., Southern Rhodesia. *Trans. Instn Min. Metall.*, **71**, 129–72.
Tanelli G. & Lattanzi P. (1985) The Cassiterite-polymetallic Sulfide Deposits of Dachang (Guangxi, People's republic of China). *Mineralium Deposita*, **20**, 102–6.
Tarney J., Dalziel I. & De Wit M. (1976) Marginal Basin 'Rocas Verdes' Complex from S. Chile: a Model for Archaean Greenstone Belt Formation In Windley B.F. (ed.), *The Early History of the Earth*, 131–46, Wiley, London.
Taylor R.G. (1979) *Geology of Tin Deposits*, Elsevier, Amsterdam.
Taylor R.P. (1981) Isotope Geology of the Bakircay Porphyry Copper Prospect, Northern Turkey. *Mineralium Deposita*, **16**, 375–90.
Taylor R.P. & Frayer B.J. (1982) Rare Earth Element Geochemistry as an Aid to Interpretating Hydrothermal Ore Deposits. In Evans A.M. (ed.), *Metallization Associated with Acid Magmatism*, 357–65, Wiley, Chichester.
Taylor R.P. & Fryer B.J. (1983) Strontium Isotope Geochemistry of the Santa Rita Porphyry Copper Deposit, New Mexico, *Econ. Geol.*, **78**, 170–4.
Taylor R.P. & Strong D.F. (1985) *Granite-related Mineral Deposits*, CIM Geology Division, Halifax, Canada.
Taylor S. (1984) Structural and Paleotopographic Controls of Lead-zinc Mineralization in the Silvermines Orebodies, Republic of Ireland. *Econ. Geol.*, **79**, 529–48.
Taylor S. & Andrew C.J. (1978) Silvermines Orebodies, County Tipperary, Ireland. *Trans. Instn Min. Metall. (Sect. B: Appl. earth sci.)*, **87**, B111–B124.
Taylor S.R. (1955) The Origin of Some New Zealand Metamorphic Rocks as shown by their Major and Trace Element Compositions. *Geochim. cosmochim. Acta.*, **8**, 182–97.
Thayer T.P. (1964) Principal Features and Origin of Podiform Chromite Deposits, and some Observations on the Guleman-Soridag District, Turkey. *Econ. Geol.*, **59**, 1497–1524.
Thayer T.P. (1967) Chemical and Structural Relations of Ultramafic and Feldspathic Rocks in Alpine Intrusive Complexes. In Wyllie P.J. (ed.), *Ultramafic and Related Rocks*. 222–39. Wiley, New York.
Thayer T.P. (1969a) Gravity Differentiation and Magmatic Re-emplacement of Podiform Chromite Deposits. *Econ. Geol.*, Mon., **4**, 132–46.
Thayer T.P. (1969b) Peridotite-gabbro Complexes as Keys to Petrology of Mid-ocean Ridges. *Geol. Soc. Am. Bull.*, **80**, 1515–22.
Thayer T.P. (1971) Authigenic, Polygenic and Allogenic Ultramafic and Gabbroic Rocks as Hosts for Magmatic Ore Deposits. *Geol. Soc. Aust.*, Spec. Publ. 3, 239–51.
Thayer T.P. (1973) *Chromium*. In Probst D.A. & Pratt W.P. (eds.), United States Mineral Resources, *Prof. Pap. 820*, 111–121. *U.S. Geol. Surv.*, Washington.
Theis N.J. (1979) Uranium-bearing and Associated Minerals in their Geochemical and Sedimentological Context, Elliot Lake, Ontario. *Geol. Surv. Can. Bull.* 304.
Theodore T.G. (1977) Selected Copper-bearing Skarns and Epizonal Granitic Intrusions in the South-western United States. *Geol. Soc. Malaysia*, Bull. **9**, 31–50.
Thomas L.J. (1978) *An Introduction to Mining*. Methuen of Australia, Sydney.
Thurlow J.G. (1977) Occurrences, Origin and Significance of Mechanically Transported Sulphide Ores at Buchans, Newfoundland. In *Volcanic Processes in Ore Genesis*. Spec. Publ. No. 7, Geol. Soc. London.
Tilsley J.E. (1981) Ore Deposit Models: 3. Genetic Considerations Relating to Some Uranium Deposits. *Geosci. Canada*, **8**, 3–7.
Titley S.R. (1978) Copper, Molybdenum and Gold Content of Some Porphyry Copper Systems of the Southwestern and Western Pacific. *Econ. Geol.*, **73**, 977–81.
Titley S.R. (1982) (ed.), *Advances in Geology of the Porphyry Copper Deposits*, Univ. Arizona Press, Tucson.
Titley S.R. & Beane R.E. (1981) Porphyry Copper Deposits. *Econ. Geol.*, 75th Anniv. Vol., 214–69.
Tooker E.W. (1985) (ed.), *Geologic Characteristics of Sediment- and Volcanic-hosted Disseminated Gold Deposits — Search for an Occurrence Model*. USGS Bull. 1646.
Touray J.-C. & Guilhaumou (1984) Characterization of H_2S Bearing Fluid Inclusions. *Bull. Mineral.*, **107**, 181–8.
Trendall A.F. (1968) Three Great Basins of Precambrian Iron Formation: a Systematic Comparison. *Bull. geol. Soc. Amer.*, **79**, 1527–44.
Trendall A.F. (1973) Iron Formations of the Hamersley Group of Westeren Australia: Type Examples of Varved Precambrian Evaporites. In *Genesis of Precambrian Iron and Manganese Deposits*, 257–70. Proc. Kiev Symp. 1970, Unesco, Paris.
Trendall A.F. & Blockley J.G. (1970) The Iron Formation of the Precambrian Hamersley Group,

Western Australia. *Bull. geol. Surv. W. Australia*, **119**.

Trocki L.K., Curtis D.B., Gancarz A.J. & Banar J.C. (1984) Ages of Major Uranium Mineralization and Lead Loss in the Key Lake Uranium Deposit, Northern Saskatchewan, Canada. *Econ. Geol.*, **79**, 1378–86.

Trueman D.L. & Černý P. (1982) Exploration for Rare-element Granitic Pegmatites. In Černý P. (ed.), *Short course in Granitic Pegmatites in Science and Industry*, Mineral. Assoc. Canada, Winnipeg, 463–93.

Tucker M.E. (1981) *Sedimentary Petrology*. Blackwell Scientific Publications, Oxford.

Turneaure F.S. (1960) A Comparative Study of Major Ore Deposits of Central Bolivia. Parts I and II. *Econ. Geol.*, **55**, 217–254 and 574–606.

Ullmer E. (1985) The Results of Exploration for Unconformity-type Uranium Deposits in East Central Minnesota. *Econ. Geol.*, **80**, 1425–35.

Van Gruenewaldt G. (1977) The Mineral Resources of the Bushveld Complex. *Miner. Sci. Engng*, **9**, 83–95.

Van Gruenewaldt G., Sharpe M.R. & Hatton C.J. (1985) The Bushveld Complex: Introduction and Review. *Econ. Geol.*, **80**, 803–12.

Van Staal C.R. & Williams P.F. (1984) Structure, Origin and Concentration of the Brunswick 12 and 6 Orebodies. *Econ. Geol.*, **79**, 1669–92.

Varentsov I.M. (1964) *Sedimentary Manganese Ores*. Elsevier, Amsterdam.

Varentsov I.M. & Rakhmanov V.P. (1977) Deposits of Manganese. In Smirnov V.I. (ed.), *Ore Deposits of the USSR*, Vol. 1, 114–78, Pitman, London.

Vartiainen H. & Parma H. (1979) Geological Characteristics of the Sokli Carbonatite Complex, Finland. *Econ. Geol.*, *74*, 1296–1306.

Vermaak C.F. (1976) The Merensky Reef — Thoughts on Its Environment and Genesis. *Econ. Geol.*, **71**, 1270–98.

Verwoerd W.J. (1964) South African Carbonatites and Their Probable Mode of Origin. *Ann. Univ. Stellenbosch.*, Ser. A, **41**, 115–233.

Vokes F.M. (1968) Regional Metamorphism of the Palaeozoic Geosynclinal Sulphide Ore Deposits of Norway. *Trans. Instn Min. Metall.* (*Sect. B: Appl. earth sci.*) 77, B53–B59.

Wallace S.R., MacKenzie W.B., Blair R.G. & Muncaster N.K. (1978) Geology of the Urad and Henderson Molybdenite Deposits, Clear Creek County, Colorado, with a Section on a Comparison of These Deposits with Those at Climax, Colorado. *Econ. Geol.*, **73**, 325–68.

Wallis R.H., Saracoglu N., Brummer J.J. & Golightly J.P. (1984) The Geology of the McClean Uranium Deposits, Northern Saskatchewan. *Canadian Mining and Metallurgical Bull.* **77**, (April) 69–96.

Walthier T.N., Sirvas E. & Araneda R. (1985) The El Indio Gold, Silver, Copper Deposit. *Eng. & Min. J.*, **186**, No. 10, 38–42.

Watson J.V. (1973) Influence of Crustal Evolution on Ore Deposition. *Trans. Instn Min. Metall.* (*Sect. B: Appl. earth sci.*), **82**, B107–B114.

Watson J.V. (1976) Mineralization in Archaean Provinces. In Windley B.F. (ed.), 443–53. *The Early History of the Earth*, Wiley, London.

Weissberg B.G., Browne P.R.L. & Seward T.M. (1979) Ore Metals in Active Geothermal Systems. In Barnes H.L. (ed.), *Geochemistry of Hydrothermal Ore Deposits, Second Edition*, 738–80, Wiley, New York.

Westra G. & Keith S.B. (1981) Classification and Genesis of Stockwork Molybdenum Deposits. *Econ. Geol.*, **76**, 844–73.

White D.E. (1974) Diverse Origins of Hydrothermal Ore Fluids. *Econ. Geol.*, **69**, 954–73.

White D.E. (1981) Active Geothermal systems and Hydrothermal Ore Deposits. In Skinner B.J. (ed.), *Econ. Geol., 75th Anniv. Vol.*, 392–423, Econ. Geol. Publishing Co., Lancaster, Penn.

White W.H., Bookstrom A.A., Kamilli R.J., Ganster M.W., Smith R.P., Ranta D.E. & Steininger R.C. (1981) Character and Origin of Climax-Type Molybdenum Deposits. *Econ. Geol., 75th Anniv Vol.*, 270–316.

White W.S. (1971) A Paleohydrologic Model for Mineralization of the White Pine Copper Deposit, Northern Michigan. *Econ. Geol.*, **66**, 1–13.

Whitney J.A., Hemley J.J. & Simon F.O. (1985) The Concentration of Iron in Chloride Solutions Equilibrated with Synthetic Granitic Compositions: the Sulphur-Free System. *Econ. Geol.*, **80**, 444–60.

Williams D. (1969) Ore Deposits of Volcanic Affiliation. In James C.H. (ed.), *Sedimentary Ores Ancient and Modern* (*Revised*), 197–206, Spec. Pub. No. 1, Geol. Dept, Leicester.

Williams H.R. & Williams R.A. (1977) Kimberlites and Plate Tectonics in West Africa. *Nature*, **270**, 507–8.

Wilson A.F. (1983) The Economic Significance of Non-hydrothermal Transport of Gold, and of the accretion of Large Gold Nuggets in Laterite and Other Westhering Profiles in Australia. In De Villiers J.P.R. & Cawthorn P.A. (eds.), *ICAM 81*, 229–34, Spec. Pubn. No. 7, geol. Soc. S. Afr.

Wilton D.H.C. (1985) REE and Background Au/Ag Evidence Concerning the Origin of Hydrothermal Fluids in the Cape Ray Electrum Deposits, Southwestern Newfoundland. *Canadian Mining and Metallurgical Bull.* **78**, (February) 48–59.

Windley B.F. (1984) *The Evolving Continents*. Wiley, London.

Winkler H.G.F. (1979) *Petrogenesis of Metamorphic Rocks*. Springer-Verlag, New York.

Winward K. (1975) Quaternary Coastal Sediments. In Marham N.L. & Basden H. (eds.), *The Mineral Deposits of New South Wales*, 595–621. Geol. Surv., New South Wales, Dept. of Mines, Sydney.

Wolf K.H. (1981) Terminologies, Structuring and Classifications in Ore and Host-rock Petrology. In Wolf K.H. (ed.), *Handbook of Strata-bound and Stratiform Ore Deposits*, 1–337, Elsevier, Amsterdam.

Wolfe J.A. (1984) *Mineral Resources — A World Review*. Chapman & Hall, New York.

Woodall R. (1979) Gold — Australia and the World. In Glover J.E. & Groves D.I. (eds.), *Gold Mineralization*, 1–34, Geology Department and Extension Service, Univ. Western Australia, Nedlands.

Wright J.B. & McCurry P. (1973) Magmas, Mineralization and Seafloor Spreading. *Geol. Rundsch.*, **62**, 116–25.

Wu Y. & Beales F. (1981) A Reconnaissance Study by Palaeomagnetic Methods of the Age of Mineralization along the Viburnum Trend, Southeast Missouri. *Econ. Geol.*, **76**, 1879–94.

Zachrisson E. (1984) Lateral Metal Zonation and Stringer Zone Development, Reflecting Fissure-controlled Exhalations at the Stekenjokk-Levi Strata-bound Sulphide Deposit, Central Scandinavian Caledonides. *Econ. Geol.*, **79**, 1643–59.

Zantop H. (1981) Trace Elements in Volcanogenic Manganese Oxides and Iron Oxides: The San Francisco Manganese Deposit, Jalisco, Mexico. *Econ. Geol.*, *76*, 545–55.

Index

Page numbers in **bold** refer to tables and in *italic* refer to figures.